# Unfolding Irish landscapes

# Unfolding Irish landscapes

*Tim Robinson, culture and environment*

Edited by Derek Gladwin and Christine Cusick

Manchester University Press

Copyright © Manchester University Press 2016

While copyright in the volume as a whole is vested in Manchester University Press, copyright in individual chapters belongs to their respective authors, and no chapter may be reproduced wholly or in part without the express permission in writing of both author and publisher.

Published by Manchester University Press
Altrincham Street, Manchester M1 7JA
www.manchesteruniversitypress.co.uk

British Library Cataloguing-in-Publication Data
A catalogue record for this book is available from the British Library

Library of Congress Cataloging-in-Publication Data applied for

ISBN 978 17849 9278 1 hardback
ISBN 978 07190 9947 2 paperback

First published 2016

The publisher has no responsibility for the persistence or accuracy of URLs for any external or third-party internet websites referred to in this book, and does not guarantee that any content on such websites is, or will remain, accurate or appropriate.

Typeset by Out of House Publishing
Printed in Great Britain
by CPI Group (UK) Ltd, Croydon, CR0 4YY

# Contents

| | |
|---|---|
| List of figures | vii |
| List of contributors | x |
| Acknowledgements | xiv |
| Foreword: Tim Robinson<br>  *Robert Macfarlane* | xvi |
| Map of Aran Islands and neighbouring coasts | xxii |
| Introduction: Ireland's 'ABC of earth wonders'<br>*Derek Gladwin and Christine Cusick* | 1 |

## I Explorations in cartography and geography — 19

1. *Genius loci*: the geographical imagination of Tim Robinson — 21
   *Patrick Duffy*

2. Catchments — 41
   *John Elder*

3. 'The fineness of things': the deep mapping projects of Tim Robinson's art and writings, 1969–72 — 53
   *Nessa Cronin*

4. Documentary map-making and film-making in Pat Collins's *Tim Robinson: Connemara* — 73
   *Derek Gladwin*

## II Topographic writing and narrative — 87

5. 'And now intellect, discovering its own effects': Tim Robinson as narrative scholar — 89
   *Christine Cusick*

6  Not knowing as aesthetic imperative in Tim Robinson's *Stones of Aran*   103
   Kelly Sullivan

7  Thirteen ways of looking at a landscape: the poetic in the work
   of Tim Robinson                                                          119
   Moya Cannon

8  Tim Robinson and Chris Arthur: in defence of the Irish essay             127
   Karen Babine

## III  Place and the Irish cultural imagination                            145

9  'But his study is out of doors': Tim Robinson's place in Irish Studies   147
   Eamonn Wall

10 Maps, movements and migrants: reading Tim Robinson through
   Gluaiseacht Chearta Sibhialta na Gaeltachta                               158
   Jerry White

11 'About nothing, about everything': listening in/to Tim Robinson          173
   Gerry Smyth

12 'Another half-humanized boulder lying on unprofitable ground'?:
   the visual art of Tim Robinson/Timothy Drever                            190
   Catherine Marshall

13 'An ear to the earth': matrixial gazing in Tim Robinson's
   walk-art-text practice                                                   202
   Moynagh Sullivan

14 Essayist of place: postcolonialism and ecology in the work of
   Tim Robinson                                                             218
   Eóin Flannery

   Epilogue: On the rocks road                                              237
   Andrew McNeillie

   Bibliography                                                             241
   Index                                                                    251

# Figures

1 Map of the Burren by Robinson (cropped left corner overview). Reproduced with the permission of the Tim Robinson Archive, The James Hardiman Library Archive, NUI Galway.    6

2 Robinson's mapping process of Aill na nGlasóg, Árainn (April 1982). Reproduced with the permission of the Tim Robinson Archive, The James Hardiman Library Archive, NUI Galway.    7

3 Going to Inis Meáin (April 1982). Reproduced with the permission of the Tim Robinson Archive, The James Hardiman Library Archive, NUI Galway.    9

4a Ordnance Survey, first edition six-inch map of Aran section (1840s). Reproduced with the permission of the Board of Trinity College Dublin.    23

4b Map of Árainn by Robinson. Reproduced with the permission of Tim Robinson.    23

5a Map of south Connemara by Robinson. Reproduced with the permission of Tim Robinson.    25

5b Ordnance Survey, first edition map of south Connemara. Reproduced with the permission of the Board of Trinity College Dublin.    26

6 Aran: 'a wilderness of stones and stone walls'. Photo by Cotton Coulson – Keen Press/National Geographic Creative. Reproduced with the permission of *National Geographic*.    28

7 Connemara topography (from Pat Collins's film *Tim Robinson: Connemara*, photo by Colm Hogan). Reproduced with the permission of Collins and Hogan.    42

8 Map of Little Otter Creek in Bristol (map from USGS with markings by Elder). Reproduced with open access permission from the United States Geological Survey.    43

| 9 | *Moonfield* (Timothy Drever, Camden Arts Centre, London, 1969). Reproduced with the permission of Tim Robinson. | 61 |
| --- | --- | --- |
| 10 | Garden at Nimmo House, Roundstone, Co. Galway (photo by Nessa Cronin) | 63 |
| 11 | *To the Centre* (IMMA Installation, 1997, from *The View from the Horizon*, Robinson/Drever, 1997). Reproduced with the permission of Simon Cutts at Coracle Press. | 64 |
| 12 | Connemara topography (from Pat Collins's film *Tim Robinson: Connemara*, photo by Colm Hogan). Reproduced with the permission of Collins and Hogan. | 74 |
| 13 | Bark of tree (close up resembling a map) (from Pat Collins's film *Tim Robinson: Connemara*, photo by Colm Hogan). Reproduced with the permission of Collins and Hogan. | 84 |
| 14 | Robinson's library (photo by Nessa Cronin). Reproduced with the permission of Cronin. | 96 |
| 15 | *Dry stone wall* (drawing by Angie Shanahan). Reproduced with the permission of Shanahan. | 98 |
| 16 | Map of Árainn by Robinson. Reproduced with the permission of Tim Robinson. | 106 |
| 17 | Cliffs of Árainn (photo by Kelly Sullivan). Reproduced with the permission of Sullivan. | 110 |
| 18 | Stone walls of Árainn (photo by Kelly Sullivan). Reproduced with the permission of Sullivan. | 120 |
| 19 | *Sand dune graveyard, Aran Islands* (drawing by Angie Shanahan). Reproduced with the permission of Shanahan. | 121 |
| 20 | Robinson writing at his desk in Roundstone, Co. Galway (from Pat Collins's film *Tim Robinson: Connemara*, photo by Colm Hogan). Reproduced with the permission of Collins and Hogan. | 131 |
| 21 | The Robinsons' Nimmo House in Roundstone (photo by Nessa Cronin). Reproduced with the permission of Cronin. | 148 |
| 22 | Robinson in a boat going to Deer Island (Árainn) (March 1984). Reproduced with the permission of the Tim Robinson Archive, The James Hardiman Library Archive, NUI Galway. | 160 |
| 23 | Robinson 'listening to the wind' in the woods of Connemara (from Pat Collins's film *Tim Robinson: Connemara*, photo by Colm Hogan). Reproduced with the permission of Collins and Hogan. | 174 |
| 24 | *To the Sun* (oil on canvas, 1969, Timothy Drever). Reproduced with the permission of Tim Robinson. | 192 |
| 25 | Map of Cleggan Head, Connemara, by Robinson (cropped from original sized map). Reproduced with the permission of the Tim Robinson Archive, The James Hardiman Library Archive, NUI Galway. | 207 |

26  Essayist of place – Robinson at his desk juxtaposed against the Connemara landscape (from Pat Collins's film *Tim Robinson: Connemara*, photo by Colm Hogan). Reproduced with the permission of Collins and Hogan. 219
27  Bothar na gCreag (photo by Andrew McNeillie). Reproduced with the permission of McNeillie. 237

# Contributors

**Karen Babine** is the author of *Water and What We Know: Following the Roots of a Northern Life* (University of Minnesota Press, 2015) and editor of *Assay: A Journal of Nonfiction Studies*. Her work has appeared in *River Teeth*, *Ascent*, *North Dakota Quarterly*, *Fugue* and other literary journals. She is Assistant Professor of English at Concordia College (USA).

**Moya Cannon** was born in Co. Donegal, Ireland. She has published four collections of poetry. Her most recent collection, *Hands* (Carcanet Press, Manchester), was shortlisted for the 2012 Irish Times/Poetry Now award. Carcanet Press will publish her forthcoming collection, *Keats Lives on the Amtrak*, in 2015. A winner of the Brendan Behan Award and the Lawrence O'Shaughnessy Award, she has been editor of *Poetry Ireland Review* and was 2011 Heimbold Professor of Irish Studies at Villanova University.

**Nessa Cronin** is Lecturer in Irish Studies and Director of the MA in Irish Studies, Centre for Irish Studies, NUI Galway. She has published widely in the fields of Irish cultural cartography and Irish literary geographies. She is Co-director of the Ómós Áite: Space, Place Research Network, and the Irish Co-convenor of the Mapping Spectral Traces International Collaborative, and is currently co-editing an interdisciplinary essay collection with Tim Collins entitled *Lifeworlds: Space, Place and Irish Culture*.

**Christine Cusick** is Associate Professor of English and Director of the Honours Program at Seton Hill University (USA). Her research focuses on the intersections of ecology, natural history and cultural memory. She has published ecocritical studies of contemporary Irish poetry, fiction, bogland photography and nature writing and has been nationally recognised for her creative non-fiction. She edited *Out of the Earth: Ecocritical Readings of Irish Texts* (Cork University Press, 2010), which includes her interview with Tim Robinson. Her essay 'Mapping

Placelore: Tim Robinson's Ambulation and Articulation of Connemara as Bioregion' appears in *The Bioregional Imagination: New Perspectives on Literature, Ecology and Place* (University of Georgia Press, 2012).

**Patrick Duffy** is Emeritus Professor of Geography in Maynooth University. He is author of *Landscapes of South Ulster: A Parish Atlas of the Diocese of Clogher* (Institute of Irish Studies, 1993); *Exploring the History and Heritage of Irish Landscapes* (Four Courts, 2007); editor of *To and from Ireland: Planned Migration Schemes c.1600–2000* (Geography Publications, 2004); *Gaelic Ireland: Land, Lordship and Settlement c.1250–1650* (Four Courts, 2004); *At the Anvil: Essays in Historical Geography in Honour of W. J. Smyth* (Geography Publications, 2012); and author of numerous book chapters, most recently '"Through American Eyes": A Hundred Years of Ireland in the *National Geographic Magazine*' in *Spacing Ireland: Place, Society and Culture in a Post-boom Era* (Manchester University Press, 2013).

**John Elder** taught at Middlebury College in Vermont (from 1973 to 2010), where his interests in the classroom included American nature writing, Romantic poetry and Japan's haiku tradition. His most recent books – *Reading the Mountains of Home* (Harvard University Press, 1999), *The Frog Run* (Milkweed Editions, 2002) and *Pilgrimage to Vallombrosa* (University of Virginia Press, 2006) – combine discussions of literature, descriptions of the Vermont landscape and memoir. Since retirement from full-time teaching, he has devoted much of his time to learning about and playing Irish traditional music. A multi-media website reflecting on that musical project, on pastoral literature and on the writing of Tim Robinson in particular may be viewed at www.pickingupthelute.com.

**Eóin Flannery** is Reader in Irish Literature at Oxford Brookes University, where he is the Director of the Oxford Brookes Poetry Centre. He is the author of three books: *Colum McCann and the Aesthetics of Redemption* (Irish Academic Press, 2011); *Ireland and Postcolonial Studies: Theory, Discourse, Utopia* (Palgrave, 2009); *Versions of Ireland: Empire, Modernity and Resistance in Irish Culture* (Cambridge Scholars Press, 2006). His next book, *Ireland and Ecocriticism: Literature, History, and Social Justice* will be published in 2015 by Routledge. He also recently edited a special issue of *The Journal of Ecocriticism* entitled 'Ireland and Ecocriticism' (2013).

**Derek Gladwin** is a Postdoctoral Research Fellow, through the Social Sciences and Humanities Research Council of Canada, in the Department of English at the University of British Columbia. He is author of the forthcoming book, *Contentious Terrains: Boglands and the Irish Postcolonial Gothic* (Cork University Press, 2016), and co-editor of the collection *Eco-Joyce: The Environmental Imagination of James Joyce* (with Rob Brazeau; Cork University Press, 2014). His next book, *Spatial Injustice in the North Atlantic Environmental Humanities*, will be published with Routledge in 2016.

Contributors

**Robert Macfarlane** is the author of several award-winning books on the relations of landscape, memory and imagination, among them *Mountains of the Mind* (Vintage, 2003), *The Wild Places* (Penguin, 2007), *The Old Ways* (Penguin, 2012) and *Holloway* (Faber & Faber, 2013). He is a Fellow of Emmanuel College, University of Cambridge.

**Catherine Marshall** is joint editor of the Royal Irish Academy's series Art and Architecture of Ireland. She was the first Head of Collections at the Irish Museum of Modern Art, and is a founding editor of Comcol, the international newsletter of ICOM's Collection. She chairs the steering committee of Bealtaine, the largest annual arts event in Ireland, and sits on the boards of the Butler Gallery and the Kilkenny Centre for Arts Talent. She has curated exhibitions in North America, Europe and Asia. She has lectured at Trinity College, Dublin, University College Dublin, National College of Art and Design and the National Gallery.

**Andrew McNeillie**'s latest of six collections of poems, *Winter Moorings* (Carcanet Press), came out in 2014. Lilliput Press published his memoir, *An Aran Keening*, in 2001. He is the founding editor of the literary magazine *Archipelago* and runs the Clutag Press. Born and brought up in North Wales, he read English at Magdalen College, Oxford, after working as a news journalist and spending a year living on Inis Mór in the Aran Islands. He is Emeritus Professor of English at Exeter University and was for a period literature editor at Oxford University Press.

**Gerry Smyth** is Reader in Cultural History at Liverpool John Moores University. He has published widely on Irish fiction, cultural theory and popular music. His latest book is *The Judas Kiss: Treason and Betrayal in Six Modern Irish Novels* (Manchester University Press, 2015). He is currently working on a new book entitled *Celtic Tiger Blues: Music and Irish Identity from Joyce to Jedward*, due for publication by Ashgate in 2016. He is also a musician and actor.

**Kelly Sullivan** is Assistant Professor/Faculty Fellow in Irish Studies at New York University. She has published essays on Irish artists Harry Clarke and Gerard Dillon, and researches and writes about twentieth-century British and Irish literature. Her novel, *Winter Bayou* (Lilliput Press), appeared in 2005, and her current book project, *Epistolary Modernism: Letters in Late Modernist Literature*, explores the importance of correspondence in novels and poetry by Elizabeth Bowen, Graham Greene, Louis MacNeice and others.

**Moynagh Sullivan** lectures in the Department of English at Maynooth University where she is the Director of the MA in Gender and Sexuality in Writing and Culture. Her research interests include psychoanalysis, Irish Studies, postmodernism, Gender Studies and popular culture, and she has published widely in these fields. She is co-editor (with Borbola Farago) of *Facing the Other: Interdisciplinary*

*Studies in Race, Gender and Social Justice in Ireland* (Cambridge Scholars Publishing, 2009); (with Wanda Balzano) of 'Irish Feminisms', special issue of *The Irish Review* (2007); and (with Wanda Balzano and Anne Mulhall) of *Irish Postmodernisms and Popular Culture* (Palgrave, 2007).

**Eamonn Wall** is the author of *Writing the Irish West: Ecologies and Traditions* (Notre Dame, 2011), *From the Sin-e Café to the Black Hills: Notes on the New Irish* (Wisconsin, 2000) and *Junction City: Selected Poems 1990–2014* (Salmon Poetry, 2015). His articles, essays and reviews have appeared in *The Irish Times*, *Irish Literary Supplement*, *Washington Post* and other publications. He lives in St Louis, Missouri, where he is employed as a professor of International Studies and English at the University of Missouri-St Louis, teaching courses in Irish and English Literature, and serving as director of UMSL's Irish Studies program.

**Jerry White** is Canada Research Chair in European Studies at Dalhousie University, and the former editor of the *Canadian Journal of Irish Studies*. His most recent books are *Two Bicycles: The Work of Jean-Luc Godard and Anne-Marie Miéville* (Wilfrid Laurier University Press, 2013) and *Revisioning Europe: The Films of John Berger and Alain Tanner* (University of Calgary Press, 2011).

# Acknowledgements

The community that has made these pages possible extends across oceans and disciplines, into the depths of archives and artwork, and has, from our perspective, resulted in a study that bears witness to the promise of authentic academic dialogue and collaboration.

We owe abundant thanks to the contributors of this collection, without whom this study could not exist. From the inception of this project, we received enthusiastic and professional responses from every contributor in this volume. We offer a special debt of gratitude to Nessa Cronin, who in addition to offering a brilliant essay was integral in connecting us to the necessary resources at the National University of Ireland, Galway.

The visual component of this book would have been a daunting prospect without the assistance and generosity of many people and organisations. We thank John Cox, Kieran Hoare and Aisling Keane at the James Hardiman Library at NUIG for granting permissions and supplying us with archival stills and maps from the Tim Robinson Archive; Angie Shanahan for creating wonderful drawings for the collection; and Pat Collins for providing the film stills, shot by Colm Hogan. We also want to thank the Department of Geography at NUI Maynooth for paying reproduction rights for the *National Geographic* photograph accompanying Patrick Duffy's essay, and contributors Kelly Sullivan, Nessa Cronin and Andrew McNeillie for allowing us to use their photography. We thank Carcanet Press for granting permission to reprint Andrew McNeillie's poem from *Winter Moorings* and to Andrew McNeillie, editor of the journal *Archipelago*, for permitting us to reprint John Elder's essay 'Catchments' from their Winter 2012 volume. Lastly, we want to thank the following organisations for allowing reproductions of maps and photos: the two Ordinance Survey maps, reproduced with the permission of the Board of Trinity College Dublin; the map of Little Otter Creek in Bristol, reproduced with open access permission from the United States Geological Survey; and the photo of Aran by Cotton Coulson (Keen Press/National Geographic Creative), reproduced with permission of *National Geographic* magazine.

We owe our commissioning editor, Tony Mason, along with his staff at Manchester University Press, a huge debt of gratitude for supporting this project from the beginning. He has guided the production of this collection with confidence and enthusiasm, and we are most grateful for the time and care he has taken to see the completion of this book.

Finally, we thank Tim Robinson, whose work makes these conversations possible and whose generosity in permissions and humble encouragement maintained a graceful balance of remove and embrace.

# Foreword: Tim Robinson

## *Robert Macfarlane*

In 'Lob', his great long poem of 1917, Edward Thomas conjures the figure of a walker who goes by several names, whose knowledge of the landscape and its habits is so deep that it seems, eerily, to be born of more than one lifetime's experience, and who is most often seen ahead on the road – glimpsed at the turn of a corner or through a gap in the hedge – but who can never quite be caught up with or pinned down, despite his patient tread.

I have read 'Lob' perhaps a dozen times over the course of my life, and each time have found new strangeness in it. It is an uncanny poem, and an aspect of its uncanniness, I have come to understand, lies in its ability to shift shape and change texture between readings, just as Lob himself does between encounters. When I read it most recently, the poem was again freshly strange to me, for on that occasion I found myself suddenly seeing in Lob an unmistakable foreshadow of Tim Robinson (formerly known as Timothy Drever): there was the same subtle liquidity of identity, the same astonishing acquaintance with place and its lore, the same commitment to 'keep[ing] clear old paths that no one uses',[1] as Thomas puts it – and above all the same sense that Tim, like Lob, has always been ahead of us on the road, quietly showing the way.

I was introduced to Tim's work eleven years ago, and I cannot now imagine being innocent of it. That introduction was made by Andrew McNeillie (whose fine poem in praise of Robinson ends this volume of essays), at a time when I was beginning to write a book called *The Wild Places*, exploring ideas and geographies of wildness in the 'archipelago' of Britain and Ireland, as I was then thinking of it. I still remember Andrew's astonishment on discovering that I had not yet read Tim. The implication was clear: to embark on a book about landscape in ignorance of Robinson would be like trying to write a book on memory in ignorance of Proust, or a book on architecture in ignorance of Ruskin. So began my belated journey into Tim's books, more or less in the order in which they were published: first, in a dizzying rush over the course of a summer, the *Stones of Aran* diptych and *Setting Foot on the Shores of Connemara*, then *My Time in Space*, then

*Tales and Imaginings*, and then – greedily, impatiently – the *Connemara* trilogy as it appeared in 2006, 2008 and 2011.

Meanwhile, I collected the maps and ephemera, and I came to see that while Tim's writing necessarily took the form of separate publications in various forms and media (essay, paper, diptych, trilogy, map, artwork), it was best imagined as a single sustained project of exceptional scope, dispersed in terms of material texts, but continuous over forty years in its modes of attention, concentration and value. Not an oeuvre, then, but a terrain. And even as I sought – with the completionist avidity of the devoted fan – to read and own everything that Tim had ever published, I also became aware of the more-than-usual silliness of such an urge, given Tim's preoccupation with what he calls 'the delusion of a comprehensible totality'.[2]

For if anything so rigid as a lesson might be said to be imparted by Tim's work, that lesson would address the limits of knowledge with regard to any apparently limited domain, and the virtues of uncertainty as a desirable state of mind. In this respect, as in others, Tim's writing recalls that of Nan Shepherd, whose slender masterpiece *The Living Mountain* (written in the 1940s but not published until 1977) took the Cairngorm mountains of north-east Scotland as their region of focus. Shepherd figures the Cairngorm massif – like Tim figures the Aran Islands – not as a crossword to be cracked, full of encrypted ups and downs, but rather as an endlessly re-fathomable volume of knots, flows, webs and tides. In the work of both writers, greater understanding of these landscapes' geometries and activities serves only to finesse them into further marvellousness – and to reveal other realms of incomprehension. 'The more one learns of this intricate interplay of soil, altitude, weather, and the living tissues of plant and insect', wrote Shepherd of the Cairngorms, 'the more the mystery deepens.' 'Knowing another is endless,' she continued – though the sentences might be taken from Tim's own prose – 'the thing to be known grows with the knowing ... Slowly I have found my way in, [but] if I had other senses, there are other things I should know.'[3] Slowly, like so many others, I have found my way into Tim's work – and as I have done so it has grown with the knowing.

The decade or so since I first set foot on the shores of Tim Robinson have coincided with a period of intense cultural activity in terms of British and Irish landscape writing. A loose group of what might tentatively be called 'new topographers' has emerged, among them Peter Davidson, Tim Dee, Kirsty Gunn, Kathleen Jamie, Richard Mabey, Sara Maitland, Philip Marsden, Autumn Richardson, Jean Sprackland and Richard Skelton, producing writing that is often first-person led, lyrically inflected, ethically alert and animated by the desire to map landscapes and their consequences for personal imagination, cultural memory and community relations. I say 'tentatively' not only due to a wariness with regard to school-making, but also because a common experience among these writers, myself included, has been the unnerving sense that much of what might be 'newly' done in this area has already been long since anticipated by Tim.

In the 2011 issue of *Archipelago* – a landscape journal edited by Andrew McNeillie and permeated by Tim's influence – Tim Dee set out to analyse the qualities of this emergent body of 'new' topographic writing. While acknowledging its wide diversity of tones and terrains, Dee felt able to identify a number of distinctive shared traits. I remember, while reading Dee's essay, mentally ticking these traits off against Robinson's existing work. Dee's list included hybridity ('several ... forms [combined] within one book ... memoirs, anthologies, essays, anthropologies, cultural geographies, travelogues, and natural histories').[4] *Tick.* A modesty with regard to the presence of the first person and a prudence with regard to the authority of his or her perceptions ('sustainable ... modest ... cautious but knowing ... personal and intimate, and the opposite of the aggrandizing big game hunter or summitteer').[5] *Tick.* Braced by science but alive to the imagination's needs and discoveries. *Tick.* Inclined to fine granulations of perception. *Tick.* A wariness, even a testiness, with regard to anthropomorphism and animism. *Tick.* A commitment to what might be called the mundane, in both the everyday and the 'worldly' senses of that word. *Tick, tick.* Elegiac at times in tone, with a sadness at loss flickering around the edges, and exhibiting a consequent devotion to what remains, but all the while fiercely eschewing sentimentality. *Tick, tick, tick!*

It is not only topographically minded prose writers who find themselves catching up with Tim Robinson. The cultural geographer John Wylie, in a fine 2012 article on Tim entitled 'Dwelling and Displacement', observed that 'geographers and others interested in questions of dwelling and selfhood, landscape and life, walking and writing, have potentially much to learn from Tim Robinson's work', and expressed his confidence that Robinson's writing would become the subject of 'extensive academic discussion ... in the years ahead'.[6] Innovative research by such phenomenologically inclined geographers as Wylie himself, Caitlin DeSilvey, Hayden Lorimer and Fraser MacDonald – all interested in what Wylie calls the 'enlacings and distanciations of self and landscape',[7] and in what DeSilvey calls 'telling stories with mutable things'[8] – demonstrates obvious overlaps with, and at times explicit debt to, Tim's writing. Comparable overlaps are increasingly apparent in other academic disciplines – history, Irish Studies, literary criticism, performance studies and anthropology among them – as is amply attested to within the essays that follow here, but evidence for which can also be found in the work of, say, the influential anthropologist Tim Ingold, or that of the interdisciplinary AARP/Atlantic Archipelago Research Project, jointly founded in 2010 by the universities of Exeter and Galway with a view to a Robinsonian 'deep-mapping' of the cultures and histories of Atlantic-facing European landscapes and peoples in the great coastal arc that runs from northern Scandinavia down to southern Spain. Film-makers (Pat Collins), composers (Susan Stenger) and visual artists (Norman Ackroyd) have also responded to Tim's work, finding their own arts informed and extended by its example. Watching the spread and reach of Tim's influence, I am put at times in mind of W. G. Sebald, whose unclassifiable prose fictions, written between 1988 and 2001, have come to exert such huge fascination upon scholars and artists in dozens of disciplines and media.

Two aspects of Tim's timeliness, or rather his ahead-of-his-timeliness, seem worth further remark. The first concerns his interest in the region – or better perhaps to say the bioregion, for it has been the islands, coast and immediate hinterland of the mid-west of Ireland that has compelled Tim's interest for forty years. Tim is a chorographer, and as such his work seems both in step with, and predictive of, the devolutionary spirit of Britain and to a lesser degree Ireland of the past few decades. The rise of the region in England, the precariousness of the Union, independence movements in Scotland and Wales, the micro-nationalisms of Cornwall and Shetland, efforts to preserve local distinctiveness in terms of dialect and custom, new forms of attachment to landscape that underlie or supersede that of nationhood – yes, it has been a seething period with regard to questions of belonging and place-identity, and out of the seethe have arisen exciting versions of what might be called progressive parochialism, as well as sourer and sharper versions of old patriotisms. I have heard it said that modern Irish politics are not openly confronted in Tim's writing, but aside from the basic inaccuracy of the proposition, it is clear to me that 'politics' operates principally as ethos rather than as dogma in Tim's writing: embodied as a style of particularised seeing, and a care of attention to the ways in which people inhabit places, and places inhabit people.

The second of Tim's anticipations concerns, for want of a preferable phrase, the 'environmental movement'. Brilliantly, Tim's work has picked a green path between an allegiance to 'dwelling' in the outmoded sense of living only and forever in a single place, which in Adorno's words, 'in the proper sense, is now impossible'; and the late modern drive to disembodiment and abstraction brought about by technology and global flows.[9] Place matters deeply but never naïvely in Tim's work. In an extraordinary passage from a lecture delivered in Cambridge in 2011, and later reprinted in *The Dublin Review*, he coins the term 'geophany' to indicate language that engages in 'a secular celebration of the Earth, with the height and power of the religious tradition but purged of supernaturalism'. Apologising for the unwieldy welding of the word, and the echo of 'theophany' it carries, he continues:

> If I have to call on the terminology of religion it is because that is the language evolved to address the highest; and the highest is what lies under our feet and bears us up. Geophany, then, the showing forth of the Earth through all the geophanic arts and sciences, should be our means toward a reformation of values.
>
> On such occasions the basic act of attention that creates a place out of a location would be renewed, enhanced by whatever systems of understanding we can muster, from the mathematical to the mythological, by the passion of poetry, or by simple enjoyment of the play of light on it. Here is a gateway to a land without shortcuts, where each place is bathed in the sunlight of our contemplation and all its particularities brought forth, like those mountainside potato plots gilded by midwinter sunset in the valley of the stone alignment.[10]

This is Robinson's Pisgah moment: a glimpse through into a realm conceivable but unreachable, the 'land without shortcuts', which can only be approached by an impossible sequence of 'good steps'.

It is a passage that seized me when I first read it, and that remains to me inspirational in its beautiful ambition of vision, combined with its acknowledgement of the impossibility of realising that vision. It is, indeed, one of countless passages of his prose that have grained their ways into the textures of my own style and my own thought to the degree that I am barely now aware of their presence. Were I to seek to pick out Tim's other gifts to me, though, they would certainly include his commitment to walking as a means of knowing; his need to get what he calls 'the feeling of [a] place, its obdurate reality, into my bones'[11] by means of bodily action; his 'neighbourly acquaintance' with the dead of a place as well as the living;[12] his interest in landscape not as a painterly backdrop against which human activity is played out, but as a dynamic, enfolding medium or flux, that burns *into* us and plays *through* us at all times; and lastly his commitment to 'the texture of immediate experience',[13] which is to say his belief that the disaggregation of phenomena into their constituent elements or qualia is a form of marvel-making rather than of reduction – for he practises what might be called a democracy of astonishment at the material world, whereby the glint of light off a strand of barbed-wire fence is as compelling as a coccolith, the tile of a church roof, a child's song or a gull's quick cry.

Three weeks before writing this introduction, I went to visit Tim and Máiréad at their flat in north-west London. I went for many reasons, but perhaps chief among them was a wish to hear from them both about Tim's decision to bring his writing and mapping of the west of Ireland to an end. News of this decision had reached me from several directions over the previous two years – coupled with various wild surmises as to what Tim planned to do next (the wildest of these surmises concerned a three-volume SF space-opera, which seemed exciting but implausible). Over lunch, Tim confirmed that what he had remarked in 2012 at a French cultural festival still held: 'I have an endless amount to say about Connemara. But the trilogy is finished, and with it a project that has taken me about forty years has come to an end. And at the moment I am trying to arrange my life so that I am facing a blank sheet of paper, once again – in physical reality, and in my mind.'[14] I felt, in part, a disappointment that this was indeed the end of the road for Tim's Irish writing, but I felt, in the main, a huge admiration for the bravery of Tim's choice, as he neared eighty, to abandon familiar terrain and strike out for new ground.

After we had eaten, we sat near the gas fire, and Tim read from some of his new work. The first eighteen months after first 'facing [the] blank sheet' had been intensely difficult, he said, but he was now working on a new series of essays. The piece he read out to me gyred – in recognisably Robinsonian fashion – from a shiny metal washer that as a young man he had picked up on a London street and glued to the wall of the studio he was then occupying, out to the missing washer that had supposedly hobbled the perceptive capacities of the Hubble Space Telescope. As I listened to Tim fly effortlessly from the foot-flurry of a city pavement, up to outer space and the stars' slow journeywork, I realised that this was – simply, amazingly – another version of Timothy Drever/Tim Robinson – a further remarkable renewal and refocusing of his unique vision. There he was, Lob again, always a turn of the path ahead of the rest of us – and never quite to be caught up with or pinned down.

## Notes

1 Edward Thomas, 'Lob', in Edna Longley (ed.), *Edward Thomas: The Annotated Collected Poems* (Newcastle: Bloodaxe, 2008), 77.
2 Tim Robinson, *Setting Foot on the Shores of Connemara and Other Writings* (Dublin: Lilliput Press, 1996), 1.
3 Nan Shepherd, *The Living Mountain* [1977] (Edinburgh: Canongate, 2011), 59 and 108.
4 Tim Dee, 'Nature Writing', *Archipelago* 5 (Spring 2011): 24.
5 Dee, 'Nature Writing', 24.
6 John Wylie, 'Dwelling and Displacement: Tim Robinson and the Questions of Landscape', *Cultural Geographies* 19:3 (July 2012): 367.
7 See John Wylie's research summary on his University of Exeter webpage: https://geography.exeter.ac.uk/staff/index.php?web_id=John_Wylie&tab=research.
8 Cailtin DeSilvey, 'Observed Decay: Telling Stories with Mutable Things', *Journal of Material Culture* 11:3 (November 2006): 318–38.
9 Theodor Adorno, *Minima Moralia: Reflections from a Damaged Life*, trans. E. F. N. Jephcott (London: Verso, 1974), 38.
10 Robinson, 'A Land without Shortcuts', *The Dublin Review* 46 (Spring 2012): 43.
11 Robinson, *Setting Foot*, 79.
12 Robinson, *Connemara: A Little Gaelic Kingdom* (Dublin: Penguin, 2011), 301.
13 Robinson, *Setting Foot*, 77.
14 'William Fiennes, Tim Robinson: La beaute du monde', Etonnants Voyageurs: Festival International du Livre et du Film, 7–9 June 2014, accessed 10 January 2015, http://vimeo.com/97608775.

# Map

Map of Aran Islands and neighbouring coasts by Robinson. Reproduced with the permission of Tim Robinson.

# Introduction: Ireland's 'ABC of earth wonders'

*Derek Gladwin and Christine Cusick*

In my face, the Atlantic wind, brining walls of rain, low ceilings of cloud, dazzling windows of sunshine, the endless transformation scenes of the far west ... The hill is Errisbeg, which shelters the little fishing-village of Roundstone from the west wind, in Connemara; ... it has been my wonderful and wearying privilege to explore in detail over the last fifteen years, the Burren uplands in County Clare, the Aran Islands, and Connemara itself.[1]

– Tim Robinson

**Setting foot on the west of Ireland**

Tim Robinson's work continues to garner significant cultural and critical attention both in Ireland and abroad. Over the last forty years, his maps and writings have incisively documented the geography of what he refers to as the 'ABC of earth wonders' – the Aran Islands, the Burren, and Connemara.[2] During this process of detailing specific places, Robinson has addressed the historical and geographical tensions that suffuse the Irish western landscape, one that the epigraph to this introduction aptly suggests brings an 'endless transformation' of scenes.[3] While there have been several selected reviews, articles and book chapters devoted to Robinson's contribution to Irish Studies, there has yet to be a sustained cultural and critical study exploring the complex web of his identities as a cartographer, an ecologist, an environmentalist, a natural historian, a botanist, a mathematician, a geographer, an artist, a translator and a landscape or topographical writer. Reading Robinson with keen attention to these constitutive elements reflects what John Wilson Foster has referred to as Robinson's 'Autocartography' – describing the elusiveness of mapping not as a substantive practice of meaning but as a process of arbitrary signs, one that cannot (like the person himself) be reduced to 'the sum of his formidable parts'.[4] This collection – including fourteen essays and one poem, along with maps, artwork and photographs – offers an engaging and fruitful opportunity to examine the dense, stimulating and imaginative world that comprises Robinson's work.[5]

Robinson literally set foot on the shores of western Ireland in the summer of 1972. What began as a whim from watching Robert Flaherty's mostly fictional 'documentary' *Man of Aran* (1934) soon became an obsession once he and his Irish-born partner, Máiréad, arrived on the Aran Islands.[6] The desolate and intricate island landscapes were so captivating that they both 'returned to live in Aran' as soon as they could leave London.[7] In the opening chapter of *Stones of Aran: Pilgrimage*, Robinson recalls what initially captivated and drew him to the west of Ireland were 'the immensities in which this little place is wrapped'.[8] While focused on the Aran Islands, this statement is indicative of landscapes as a whole and why, in addition to his work in the west of Ireland, Robinson's distinctive methods of map-making and topographical writing capture the geographical and cultural consciousness not only of Ireland, but also of the entire North Atlantic archipelago – an epistemology that is characterised by the discursive spaces associated with real and imagined areas of land and sea through cartography, culture and ecology, traversing many national borders.[9] Since arriving in Aran, Robinson has gone on to write several award-winning books (in addition to dozens of essays), create topographical maps of the Aran Islands, the Burren and Connemara, and deliver numerous public talks in Ireland, England, France and the United States. *Unfolding Irish Landscapes* – derived from the name of Robinson's own map-making company, Folding Landscapes – seeks to explore Robinson's place in Irish Studies, as well as in North Atlantic studies more generally.[10]

Attempting to label Robinson presents the largest challenge in a collection of work devoted to his writings and maps. Our aim, then, is not to define Robinson in some absolute or compulsory way. Rather, this collection examines the geographical places that his work has centred on, as well as the methods he uses to explain these places, and how these geographies explain his growing influence in Ireland. By the same token, Robinson's childhood in Northern England (Yorkshire) and time later spent in London and Europe provide a broader lens for examining his relationship with time and space in his map-making, and in the growing genre of what he refers to as topographical writing. His work, exhibiting both real and imaginative spaces, is not exclusively connected to Ireland. It is difficult to separate Robinson the figure (a proponent of protecting ecological diversity and the local concerns of his long-time home in Connemara) from his work (maps and writings); indeed, the two are intertextual, interconnected and inextricably intertwined, and this volume attempts to unfold some of these characteristics for both general and expert readers alike to further understand the landscapes he examines through cartography and writing. In a purposeful gesture of design, and an approach to Robinson's work that reaches beyond prose, the original essays in this volume are in conversation with many previously unpublished visual texts. Mindful of the importance of visual mapping, the images allow us to create another layer of interpretation, including frames of Robinson's earlier visual art, corners of the home he and his partner have created in Roundstone, and of course the maps and photos of the terrain and Robinson's place within it. The images draw from the abundantly rich resources of the newly created Tim Robinson Archive (James Hardiman Library, National University of Ireland, Galway), photographic records from some of our contributors, still images from Pat Collins's exemplary documentary (photographed by Colm Hogan) and original artwork from Angie Shanahan, an established artist from Cork whose work has been partly inspired by Robinson, and who created these pieces solely for this collection.[11]

The larger question that *Unfolding Irish Landscapes* attempts to explore, one that each contributor investigates through cross- and interdisciplinary approaches, is how has Robinson's work influenced the field of Irish Studies in the last forty years? In order to answer this question we feel Robinson's important works must be considered in a critical and creative volume of this scope and depth, through an array of writers and disciplines and across several countries. Many people within and outside of Ireland still have yet to recognise the significance of Robinson's work and the impact it has had, and continues to have, in the development and understanding of what is often considered a vitally important cultural and geographical part of western Ireland.

In addition, the variety and intensity of Robinson's approach underscores the international appeal of this collection. There remains a growing interest in Robinson's work throughout the United Kingdom, Ireland, Nordic countries, Europe and North America – as evidenced by the range of presenters at an international symposium dedicated to his work hosted by the Moore Institute for Research in the Humanities and the Social Studies in partnership with University of Exeter's Atlantic Archipelagos Research Project (AARP), at the National University of Ireland, Galway, in September 2011. Robinson was also elected as the 2011 Parnell Visiting Fellow at Magdalene College, University of Cambridge. He additionally had a residency with the Centre Culturel Irlandais in Paris, France, in 2012. Robinson and the photographer Nicolas Fève have recently collaborated on a book, edited by Jane Conroy with an introduction by John Elder, of photographic essays titled *Connemara and Elsewhere* (Royal Irish Academy, 2014). There is also a forthcoming Robinson publication: a translation, with Liam Mac Con Iomaire, of Máirtín Ó Cadhain's noteworthy novel *Cré na Cille* entitled *Graveyard Clay* (Yale University Press, 2015). Although this list is not exhaustive, it demonstrates the increasing international appeal of Robinson's work.

There have been many reviews over the years that critically engage with Robinson's work. One notable example is Seamus Deane's 1989 review of *Pilgrimage* in the *London Review of Books*, where he claims that Robinson's attempt at documenting the elusive geography of Aran demonstrates the 'demands to be represented' and how it ultimately 'cannot be'.[12] Deane's earlier review forecasts the trajectory of Robinson's concerns throughout his entire oeuvre, namely that his map-making and writing exist in the paradoxical 'dimension of time as well as space', while also functioning as 'a meditation on and a journey around …

the west coast of Ireland'.[13] Robert Macfarlane – a British landscape writer of acclaimed books including *Mountains of the Mind* (2003), *The Wild Places* (2007) and *The Old Ways* (2012), and who wrote the Foreword to this collection – emphatically states in a 2005 review in *The Guardian* that Robinson's writings are 'one of the most sustained, intensive and imaginative studies of a landscape that has ever been carried out'.[14]

In between Deane's review in 1989 and Macfarlane's review in 2005, much of Robinson's attention focused on writing and researching his largest prose undertaking to date, perhaps considered to be the culmination of all his work in Ireland, which is his highly celebrated *Connemara* trilogy – *Listening to the Wind* (2006), *Last Pool of Darkness* (2008) and *A Little Gaelic Kingdom* (2011). Robinson did, however, publish an autobiographical essay collection titled *My Time in Space* (2001), which includes many excerpts from previously published essays, and a book of creative short stories titled *Tales and Imaginings* (2002). Both of these works were preceded by a collection of his previously published essays, *Setting Foot on the Shores of Connemara and Other Writings* (1996), and *Stones of Aran: Labyrinth* (1996), which serves as a companion volume to *Pilgrimage*. But writing the *Connemara* trilogy literally demanded step-by-step attention from decades spent empirically researching while walking across Connemara – hence the term Robinson refers to as topographical writing, where a writer literally traverses the topos of a geographical region and mirrors that sensation and experience in the prose. Robinson succinctly states that this 'irreducible nub of topographicity is my emblem as map-maker', but we might argue that this process also informs his unique style of topographical prose as well.[15] At the same time, the energy that accompanied the the seriousness and attention to detail that Robinson brings to his process of knowing a place speaks to the result of such care, as well as the entrenched human need for his approach. Through both topographic prose and map-making Robinson undertakes one of the greatest explorations of the Irish landscape by a single person in recent history, paralleling, if not surpassing, Robert Lloyd Praeger's extensive catalogue of writings and natural histories of western Ireland.

Since 2005, paralleling Robinson's own unfolding of the *Connemara* trilogy, there have been about a dozen reviews and full-length published articles that recognise the critical importance of Robinson's work. There are essays that take Robinson's work as their primary focus, ranging from scholars such as Eamonn Wall, John Wilson Foster, Karen Babine, John Wylie, Christine Cusick, Daniel Sack and Derek Gladwin; but his work also subtly surfaces in other important scholarship, such as the work of Michael Cronin, Terry Gifford, John Elder, J'aime Morrison and Gerry Smyth.[16] Because many of the essays in this collection contain extensive engagements with and surveys about these critical works, we have decided in the space remaining to focus instead on explaining some of the influential characteristics that comprise Robinson's overall appeal and work through his two most recognisable contributions: cartography and writing.

Introduction: Ireland's 'ABC of earth wonders'  5

## The 'storings and sortings' of cartography and topographical writing

Robinson's initial move to Aran in 1972, while initially based upon what he refers to as 'an occasional brusque and even arbitrary change in mode of life',[17] quickly turned into an adventure in map-making because of a random suggestion to make a map of the Aran Islands for future visitors from the postmistress Máire Bean Uí Chonghaile in the village of Cill Mhuirbhigh (located on the largest island, Árainn).[18] The task Uí Chonghaile proposed was simultaneously simple and impossible, but the idea appealed to Robinson so much that he immediately began planning the project. It was at this moment over forty years ago that Robinson transitioned from visual artist to cartographer – two modes of personal and public exploration that share many similarities despite their inherent disciplinary differences – and in this transition he would somehow reconcile the false chasm between the creative and the rational impulse. After completing a map of Aran three years later in 1975 (revised and expanded in 1980), and finishing a map of the large limestone rock formation in Co. Clare known as the Burren in 1977 (see Figure 1), Robinson went on to map the immensely complicated and intricate topography of Connemara in the following decade, a project that finally resulted in a map by 1990.[19]

Mapping projects have profound effects on human perceptions of geography and material conditions of the terrain throughout culture and history. The practice of cartography has been associated with appropriating identity in Eurocentric colonial projects. This part of European history dates back to the so-called 'Age of Discovery' in the fifteenth and sixteenth centuries, when empires, although couched in rhetoric of exploration and discovery, adopted maps as agents of power and control. In Ireland, the Ordnance Survey, which was conducted by the British military in the mid-nineteenth century from 1824 to 1846, exemplifies the tensions surrounding cartographic projects in colonised zones. The Ordnance Survey mapping project attempted to, in part, standardise maps so that landowners could document the wealth of their landowning and Parliament could more accurately collect taxes.[20] One of the lasting consequences of the surveying process was the replacement of Irish place names with anglicised versions. The Ordnance Survey could be considered one of the most influential intellectual endeavours in nineteenth-century Ireland, and one that had lasting effects on nationality, language and colonial history.[21] In this regard, maps are cultural constructs shaped not only by those who create them, but also by those who perceive them, and such perceptions filter through an unreliable medium of language. Maps, then, are metaphoric ways of organising, orientating and then controlling all types of experiences, whether they are personal, social or cultural.[22]

Robinson recognised early in his cartographic practice that the 'making of a map ... is many things as well as a work of art, and among others it is a political, or more exactly an ideological, act'.[23] He maintains that the 'old Ordnance Survey shows this clearly'.[24] Mapping projects not only claim territory, but they also

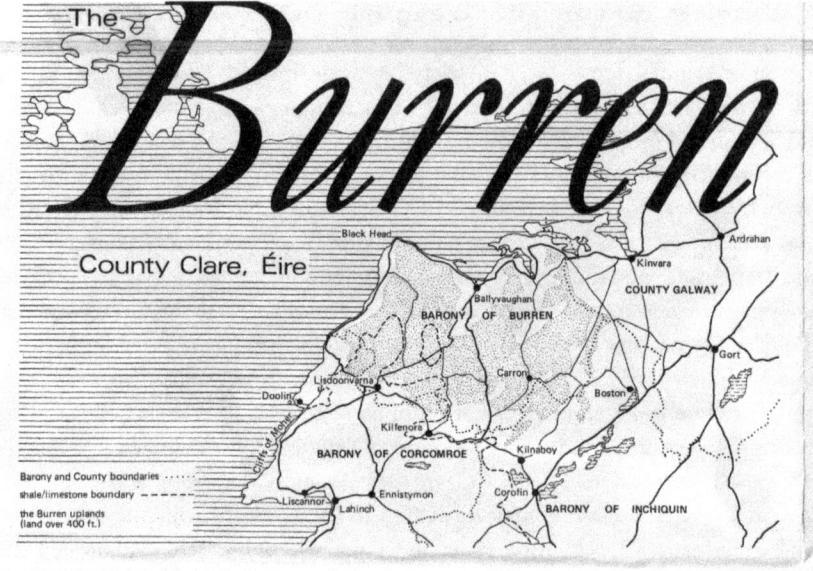

*Figure 1* Map of the Burren by Robinson (cropped left corner overview).

make certain claims on culture and history. Robinson diverges from this ideological approach of practising cartography as a way to claim territory and establish ownership. In *My Time in Space*, he clearly articulates the dangers associated with map-making in Ireland:

> I am acutely aware of the fact that cartography has historically been associated with conquest, colonization, control. The Ordnance Survey was a function of the army. Therefore I have taken care that the mapping I have been essaying for the last quarter-century or so in the west of Ireland be one that returns the territory mapped to itself, to its inhabitants.[25]

The inhabitants' embrace of Robinson's work is testimony that he has fulfilled his own prescriptive. Instead of maintaining the tradition of conquest and control, his accounts of previously mapped and contested landscapes provide an expanded sense of agency and plurality within the culture, folklore and geographical spaces. Robinson remains more concerned with listening to the landscapes than foisting meaning upon them. Or, as he puts it, 'a map is a sustained attempt upon an unattainable goal, the complete comprehension by an individual of a tract of space that will be individualized into a place by that attempt'.[26] In other words, complete understanding of a mapped place remains incomprehensible and even undesirable.

Robinson's attraction to mapping and map-making could have stemmed from his early studies in mathematics and physics at the University of Cambridge. And yet, he claims not to be 'very interested in maps from the technical point of view'.[27]

Introduction: Ireland's 'ABC of earth wonders'    7

*Figure 2*   Robinson's mapping process of Aill na nGlasóg, Árainn (April 1982).

Over time his own approach to mapping has become less technical simply because he now mostly 'maps' through writing. He came to the practice of topographical writing through both mapping, the process of surveying the landscapes on foot, and map-making – the process of designing and drawing the maps (see Figure 2). It is through these two forms of cartographic practice – mapping and map-making – that we can begin to understand Robinson's relationship, as well as our own, to the landscape. Robinson committed himself from the beginning, despite the time it took, to draw each map by hand, and he concedes that this tedious process 'is a scratching of the surface of the paper, and makes no claims to comprehensiveness'.[28] Even though he now appears to be less interested in the technical side of maps, his attention to this dimension is an important part of his success as a prose mapper. In each map he carefully takes the time to detail the lines and dots, as well as highlighting the importance of the open spaces. For Robinson, mapping is an art form that encourages and benefits from the process of collecting and organising, particularly 'large amounts of fact into an expressive whole'.[29]

Although there are clear distinctions between the process of mapping and the act of map-making, they are mutually dependent endeavours for Robinson. What draws him to this process is the 'one-to-one encounter between a person and a terrain'.[30] He once said in an interview with Christine Cusick that theoretical practices, such as critical approaches to environmental criticism, 'should come into the academy with mud on its boots'.[31] Robinson's geographically place-based practices must understand the technical, imaginative and material elements of examination. The most important part, however, is the experiential, where the map-maker, and writer in many cases, has muddy boots from literally traversing landscapes. This is why he insists on walking the terrain as part of his own process, one that Eamonn Wall, drawing from the Native American writer William Least Heat-Moon, refers to as 'deep-mapping' – a way of capturing the memories of the landscape through embodied experiences, such as walking, in addition to the technical form of cartographic documentation.[32] Robinson puts it this way: 'my map of Connemara is a record of a long walk, an intricate, knotted, itinerary that visits every place within its territory'.[33] Drawing the map, in other words, is a means of making sense of what happens when he puts foot to soil as an exploration of the landscape.

Another way of understanding Robinson's approach to cartography is to study how he views the places he maps. Robinson articulates his relationship to the landscape in what he calls the 'Echosphere', or the 'sphere of interaction and qualified autonomies of echoes'. He goes on to remark that these echoes reverberating through various landscapes 'have names and histories' where 'placelore knows a gradation of familiarity from town centre to lonely upland'.[34] Robinson's idea of the Echosphere represents a way to speak back to the people who have existed on the land like that of an echo ringing off of the material and cultural landscapes. According to Graham Huggan's foregrounding study of literary cartography entitled *Territorial Disputes*, the map is both a 'product and process: it represents both an encoded document of a specific environment and a network of perpetually recoded messages passing between the various mapmakers and map readers who participate in the event of cartographic communication'.[35] These echoes, drawing from Huggan's analysis, are messages that pass between the mapper and reader as an exchange of product and process. The initial sound of an echo begins at one point, in this case that point is the mapper, and then resonates through the landscape. These nodes of meaning, sound and experience pass between message and recorder in a more generative process of cartography and thereby depart from traditional forms used in colonial projects. In *Pilgrimage*, Robinson writes,

> But I find that in a map such points and the energy that accomplishes such fusions (which is that of poetry, not some vague 'interdisciplinary' fervour) can, at the most, be invisible guides, benevolent ghosts, through the tangles of the explicit; they cannot themselves be shown or named. So, chastened in my expectations of them, I now regard the Aran maps as preliminary *storings and sortings* of material for another art, the world-hungry art of words.[36]

This dynamic relationship between the echoes of the names and histories and the 'storings and sortings of material' is one that Robinson attempts to achieve in his own cartographic process and one that often emerges throughout the essays and images in this book.

In addition to his work in cartography, Robinson can be viewed as a landscape writer, producing what he calls topographic writing, which is a way to 'map' the

*Introduction: Ireland's 'ABC of earth wonders'* 9

*Figure 3* Going to Inis Meáin (April 1982).

material and cultural terrains of a specific geography through prose. Some scholars have described this approach as literary cartography, or a way of mapping though prose and narrative.[37] Kent Ryden suggests that when humans ask for stories we are essentially asking for 'memory over the land', thus implying a poignant connection between narrative, human recollection and topography, and the materiality of this connection manifesting itself in the processes of interpretation that connect person and place.[38] Through both process and prose product Robinson enlivens the interface between story and place. Central to Robinson's creative process is the very act of asking for stories. At times throughout his non-fiction he tells the reader that he has invited knowledge, but through his descriptions of encounters it becomes clear that in some cases his mere presence becomes a signal to the inhabitants of this place that he is in fact listening and valuing their collective and individual encounters with the liminal topography of island and coastal life (see Figure 3).

In J. H. Andrew's historical genealogy of Irish map-making, he remarks that Robinson is the only cartographer to 'achieve equal success' with both 'maps and text' and that in addition to this particular achievement Robinson 'added for good measure a poetic appreciation of landscape all too rare among geographical communicators in any country'.[39] One viable explanation for Robinson's acuity in this expression is the simultaneity of his processes. With each meticulous strike of dark line on the page, his maps are embedded with stories, and with every translated place name and human breath of the prose his narratives are inscribed by his spatial, social and historical interpretations of the terrain. This confluence is present throughout Robinson's writings, his listening to inhabitant stories unraveling into his processing of space and place. In *Labyrinth*, for instance, the seemingly anecdotal record of his roadside encounters with Dara Mullen, the island postman who Robinson 'often used to meet … on his rounds' when he was 'mapping the island', demonstrates the embodiment of process and product. Robinson recalls,

> Sometimes he would stop beside me on the road, wind down his window and tell me tales that seemed absurd but that turned out to have something in them. Once he kept me bowed to listen so long that when I straightened up I had a twinge in the back, as if I had shouldered a too-heavy post-bag of messages from the past.[40]

In this prose description of a walking encounter, the materiality of the past, of story, and of step, are at once part of Robinson as listener, mapper, writer, all bundled into the weight and physical burden of his task to give voice to his experience. The process of narrative, or storytelling, permits Robinson to bring these seemingly disparate tasks into the whole of his words on the page, and through his mingling of process and product he invites the reader to step with him into this confluence of space and time.

This is, essentially, what he means when he writes about the 'good step', the step that carries in its weight all the messiness that comes with human consciousness and an awareness of human inadequacy. He never attempts to use language or cartographic scratch as an endeavour of mastery. On the contrary, each act of representation pulls him down into the impossibilities of an encounter with the material space. It is telling that a recurring theme in *Unfolding Irish Landscapes* is an authorial exploration of Robinson's embrace of such uncertainty. Kelly Sullivan and Patrick Duffy, for instance, offer incisive explorations of Robinson's work as a philosophical encounter with the tentative, while Karen Babine declares that uncertainty is endemic to Robinson's chosen essay form. By situating these critical and creative discernments of Robinson's work against and often in dialogue with one another, this collection begins to uncover that the not knowing of Robinson's work is essentially the point, and the beginning, though never the end, of the knowing.

## Surveying critical and creative approaches

*Unfolding Irish Landscapes* builds upon the evolving fields of environmental criticism, geography, visual culture and landscape writing, and moves them forward in both content and approach through other diverse disciplinary lenses that include film, gender and postcolonial studies. These fields largely seek to make academic work more relevant for a general audience, thereby speaking to social groups beyond the academy, which is also our intent in this volume by acknowledging the creative and the critical components in not only Robinson's work, but also in the essays about it.[41] Many of the following contributors demonstrate the range and pervasiveness of Robinson's effect in various disciplines through both critical and creative approaches.

More importantly, these approaches are invoked because the simultaneous depth and comprehensiveness of Robinson's work and methodology to place-based studies call for such expanse. The collection has evolved organically out of the necessities of its subject and therefore responds to the impulses of a wide range of academic fields. This is partly why the award-winning Irish documentary film-maker, Pat Collins, recently made a film titled *Tim Robinson: Connemara* (2011). Robinson's popularity and appeal has grown internationally because of his ability to expand the prescribed disciplinary restrictions. Expanding the boundaries produces many effects, but perhaps the most notable is the ability to reach wide and varied audiences that continue to engage with Robinson's multifaceted sides. As editors, we are mindful that a collection attempting to synthesise the variance of the critical responses that Robinson's work evokes situates this study on the margins of academic discourse, which still largely relies on disciplinary categories and historical periodisation. And yet, in negotiating this challenge we hope to reveal what Robinson's work makes possible for a scholarship that seeks to purposefully connect the academy to our communities, to our social integration with the places we inhabit. With this in mind, we have divided this collection into three broad sections on cartography and geography, writing and narrative, and place and Irish culture. These categories encompass several viewpoints through which a reader can examine Robinson's critical, creative and cultural output in relation to many other disciplines and specific geographical spaces.

The esteemed landscape writer and critic from the UK, Robert Macfarlane, begins the conversation of this collection in the Foreword by explaining Robinson's influence on not only his own writing, but also those who are among the contemporary tradition of landscape writing (such as Peter Davidson, Tim Dee, Roger Deakin, Kirsty Gunn and Kathleen Jamie). Indeed, Macfarlane's own writing and approach bear striking resemblance to Robinson's. Such intersections are perhaps best captured in Macfarlane's *The Old Ways: A Journey on Foot* when he situates Robinson among those writers who 'had been animated at first by the delusion of a comprehensible totality, the belief that they might come to know their chosen place utterly because of its boundedness', but who 'after long acquaintance, at last understood that familiarity with a place will lead not to absolute knowledge but only ever to further enquiry'.[42] Both writers find their words by putting one foot after the other, content that this conflation of process and product will always be a perpetual method of seeking.

Section I, 'Explorations in cartography and geography', examines Robinson's innovative deployment of fields such as natural history, geography and cartography. The first essay in this section, by the historical geographer Patrick Duffy, examines Robinson's approach to representing the sense of place encountered in the landscapes in and around the Atlantic. In a world of collapsing distances and faster, wider-ranging travel, Duffy argues in his essay, '*Genius loci*: the geographical imagination of Tim Robinson', that Robinson's work in map and text illustrate the potential and possibilities of reversing to 'slow' landscapes. Next, the American environmental writer John Elder explores in his essay the way Robinson extends the definition of a 'catchment' area – a unit of the earth's surface bound by higher edges and within which springs, rainfall and smaller tributaries converge – to include its human metaphor as expressed through environmental care. Through a hybrid style of criticism and narrative, Elder explains that Robinson has not only come to know the catchments in intimate detail, but he has also tracked their confluence with the geology, language and history of western Ireland.

Beyond traditional modes of geography or geology, the cultural geographer Nessa Cronin traces the prehistory of images and concerns that would preoccupy Robinson during two distinct phases of his career: as a visual artist in London and Europe (under the name Timothy Drever) and as a cartographer of the west of Ireland. In '"The fineness of things": the deep mapping projects of Tim Robinson's art and writing, 1969–72', Cronin argues that Robinson's overall work can be regarded as a practice where a deep map is understood as being an attempt 'to record and represent the grain and patina of place', outlining an attempt to say 'everything you might ever want to say about a place'.[43] With a focus on two sets of artwork from his early career (*Moonfield* and *To the Centre*), as well as short prose writings and essays, Cronin's essay explores the fragmentary connection that Robinson draws between these two stages of his life and career. Ending this section on geography and cartography through another visual crossover with film-making, Derek Gladwin investigates how Pat Collins – who is considered one of the most articulate contemporary documentarians in Ireland – depicts Robinson as a mediator between landscape and culture through his own mapping enterprise, which resembles an equally comparable form of film-making. In 'Documentary map-making and film-making in Pat Collins's *Tim Robinson: Connemara*', Gladwin suggests that Collins and Robinson share a similar desire in their own forms of documentation to examine the subject of Connemara in order to create a place-based art form that magnifies the landscape while it also reduces the primacy of the 'maker' in the process.

Section II, 'Topographic writing and narrative', explores the writings and prose style of Robinson, as well as his relationship to writing more generally. Building upon his reputation for map-making, Robinson is now lauded as a prominent writer, a two-time Irish Book Award winner for *Listening to the Wind* and *A Little Gaelic Kingdom*. In a 2010 interview, when asked if he would describe his non-fiction as a form of 'narrative scholarship', Robinson responded, 'I try to present any scholarship there may be in my work as a personal experience, that of a learner ... All experience, bodily or mental, becomes the matter of a book.'[44] His answer expresses a paradigm shift where academic discourse can create space for personal narrative, a theme explored in this second section.

Christine Cusick opens this section with her essay, '"And now intellect, discovering its own effects": Tim Robinson as narrative scholar', by arguing that scholarship rooted in both experience and academic discourse requires that we examine the epistemological and hermeneutic assumptions that we bring to our chosen subject. To this end, Cusick offers close readings of Robinson's writing as a way to interpret his praxis of narrative scholarship. Kelly Sullivan, both literary critic and published novelist, looks into Robinson's monumental work *Stones of Aran* and asks whether it is possible to collect all the contradictions of the human world – geology, biology, personal history, myth and politics – into 'a state of consciousness even fleetingly worthy of its ground'. In 'Not knowing as aesthetic imperative in Tim Robinson's *Stones of Aran*', Sullivan contends that the driving imperative of both *Pilgrimage* and *Labyrinth* form an aesthetics of 'not knowing' coupled with authorial self-doubt about representing place. Shifting away from narrative technique, the award-winning Irish poet Moya Cannon offers a reading of 'Orion the Hunter', a work of short fiction Robinson dedicated to his late friend John Moriarty. Rather than formulate an argument about Robinson, in 'Thirteen ways of looking at a landscape: the poetic in the work of Tim Robinson' Cannon's own poetic sensibilities push her exploration, by way of 'Orion', into the ways in which Robinson developed as a cartographer, writer and cultural figure in Ireland. Concluding this second section on narrative and writing, Karen Babine argues that the genre of 'creative non-fiction', or the Montaignian essay, is largely missing in the Irish context. In 'Tim Robinson and Chris Arthur: in defence of the Irish essay', Babine maintains that Robinson and Arthur represent two exceptions of creative non-fiction writers who are still thriving, and who both operate almost exclusively in the non-fiction genre.

Section III, 'Place and the Irish cultural imagination', attempts to situate Robinson in the textured fabric of Irish culture, and in so doing offers the widest disciplinary application with the largest number of essays. Although originally from Yorkshire, England, Robinson has nevertheless made an indelible mark in Ireland; however, his presence in the field of Irish Studies continues to be a paradox. Throughout his career Robinson has worked in a thoroughly interdisciplinary manner, as writer, cartographer, linguist, natural scientist, folklorist, visual artist, ethnologist and cultural figure. Beyond this list of disciplinary taxonomies underlies the main focus of his concerns: Irish historical geography. In many respects, Robinson's work significantly contributes to the larger field of Irish Studies by exploring the ruptures, intersections and triumphs of tumultuous histories through a sense of place.

Eamonn Wall, a literary scholar and poet who has written the most on Robinson to date, begins the last section by exploring the methodology and reach of Robinson's work. Even though Robinson is not connected to the academy, his work exemplifies the idea of interdisciplinarity.[45] In '"But his study is out of doors": Tim Robinson's place in Irish Studies', Wall argues that Robinson has moved slowly and respectfully, taking 'the adequate step', allowing him to undertake many avenues of inquiry to great effect. In 'Maps, movements and migrants: reading Tim Robinson though Gluaiseacht Chearta Sibhialta na Gaeltachta', Jerry

White explores some of the possible connections between Robinson and the debates in the 1970s about the Irish language movement. White examines the beginnings of Robinson's mapping career, drawing on both the historical narrative of Gluaiseacht Chearta Sibhialta na Gaeltachta (the Gaeltacht Civil Rights Movement) and early editions of Robinson's work and documents pertaining to this Irish language movement through figures such as the film-maker Bob Quinn and the political journalist Desmond Fennell.

Critical approaches to Robinson's work typically converge on written or visual interpretations, whether they are focused on maps or prose. Gerry Smyth, however, considers the question of 'listening' as it relates to two philosophical systems: the phenomenology of listening associated with Jean-Luc Nancy and the existentialist listening associated with Martin Heidegger. In '"About nothing, about everything": listening in / to Tim Robinson', Smyth argues that each of these systems connotes metaphysical and ethical approaches to listening, which are of particular relevance to Robinson in his various roles as cartographer, environmentalist, scientist, folklorist and dweller in the landscape. As discussed in connection with Cronin's essay, Drever/Robinson left a vibrant career in the visual arts scene in England when he came to Ireland. Drawing on this specific history, Catherine Marshall, who served as the Head of Collections at the Irish Museum of Modern Art in Dublin for many years, investigates Robinson's relationship with other visual cultures in Ireland. In '"Another half-humanized boulder lying on unprofitable ground": the visual art of Tim Robinson/Timothy Drever', Marshall places Robinson and his earlier persona, Drever, in a visual context of the west of Ireland, alongside other Irish artists such as Paul Henry and Seán Keating, inviting speculation on the artist as voyeur or social activist, on the relationship between images and words, and between art and power. Although Robinson's maps and writings serve as typical entry points into his work, Marshall explores how they also function as art work.

In addition to Irish Studies, the Irish language movement in the 1970s, sound and visual art, Robinson's work also contains subtle invocations of gender analysis. In '"An ear to the earth": matrixial gazing in Tim Robinson's walk-art-text practice', Moynagh Sullivan argues that Robinson's powerful literary mapping of Connemara avoids gendering the Irish landscape as feminine, resisting the dominant trope in twentieth-century Irish writing and film in which the countryside stands in for woman and, often, mother. Sullivan investigates Robinson's mapping of Connemara and the Aran Islands alongside the work of artist, philosopher and psychoanalyst Bracha L. Ettinger – who also, similar to Robinson, maps psychic dimensions at the edge of consciousness – in order to illuminate the central encounter at the heart of Robinson's map-making: a walk-art-text practice. In addition to gender, the idea of place in Ireland can conjure colonial associations. In 'Essayist of place: postcolonialism and ecology in the work of Tim Robinson', Eóin Flannery situates Robinson's visual and verbal works within contemporary environmental and postcolonial contexts by arguing that his career and body of work are exemplary engagements with the diverse scales of environmental change, degradation and belonging across Irish history. Making reference to each of Robinson's publications on Aran and Connemara, as well as to his essays, Flannery highlights how Robinson's work restores a sustainable and ethical relationship with place in the Irish context. The collection ends appropriately with a poem by Andrew McNeillie titled 'On the rocks road' that he wrote about Robinson. McNeillie, literary critic, poet and creative writer, diverges from the critical essay form and offers a creative reflection of Robinson's relationship with the landscape and mapping upon his arrival to Ireland.

*Unfolding Irish Landscapes* serves a purpose beyond its primary goal to critically and creatively recognise the work of Robinson. This collection also attempts to spotlight western Ireland, as does Robinson's own work, in order to explore the histories and cultures that exist not only in connection to Connemara, the Aran Islands and the Burren, but also in the idea and practice of landscape, ecology and geography. Sharing the impetus behind Robinson's process and work, this collection aims to participate in the story of Ireland's west, allowing, within the limitations of human knowledge and expression, the place to remain at the centre of study. And through this simultaneous focus on the micro and macro function of place and practice, our hope is that the collection will encourage more authentic conversations across the gap that ostensibly divides academic and public discourse. As editors, we are mindful of the inadequacies of our human-bound tools of expression, but in dialogue with one another, our hope is that the voices of each writer in this collection may contribute to the incomplete ways of knowing a place. Robinson's landscape writing is like an intimate letter to the land – not an overly romanticised approach, but a letter that honours the ragged as well as the pristine edges of intimacy through the unity of geography, prose, history and poetry. This book investigates the many ways in which these personal letters to the Irish landscape have been imagined and applied. In the spirit of Robinson's step, we do not desire nor claim to adequately introduce or completely understand the enigmatic figure that he remains. We do, however, hope to entice readers and invite them into the variegated worlds of Tim Robinson and the west of Ireland.

## Notes

1 Tim Robinson, *Setting Foot on the Shores of Connemara and Other Writings* (Dublin: Lilliput Press, 1996), 30.
2 Robinson, *Setting Foot*, vi.
3 We will not be capitalising 'west' or 'western' Ireland in this introduction unless we are referring to a specific region/nomenclature/context, a strategy also deployed by Robinson in his own writings. In many cases, however, the 'West of Ireland' is the preferred usage, and individual writers may reflect this desired effect in essays throughout the collection.
4 John Wilson Foster, *Between Shadows: Modern Irish Writing and Culture* (Dublin: Irish Academic Press, 2009), 121.

5 The only previously published essay reprinted here is John Elder, 'Catchments', *Archipelago* 6 (Winter 2011): 1–22.
6 The Aran Islands consist of three islands located at the mouth of the Galway Bay on the central west coast of Ireland. The largest one, which is the location of *Stones of Aran*, is called *Árainn* and *Inis Mór* in Irish and anglicised as Inishmore. The medium-sized (or middle) island is called *Inis Meáin*/Inishmaan and the smallest is *Inis Oírr*/Inisheer. The collective name for the islands is *Oileáin Árann* or the Islands of Aran. Mirroring Robinson's own usage, we will be using the Aran Islands, or Aran for short, as a general reference to these three islands. See Tim Robinson, *Stones of Aran: Pilgrimage* [1986] (New York: New York Review of Books, 2008), 8.
7 Robinson, *Setting Foot*, 1.
8 Robinson, *Pilgrimage*, 17.
9 For more on Robinson's relationship to and the idea of the archipelago, see the website Atlantic Archipelago Researchers Consortium at http://aarco.org/. The ASRC site contains many events and projects, along with specific members, that explore the taxonomy of work emerging from Robinson's own oeuvre, in part, while also expanding the study of culture, mapping and ecologies of the Atlantic Archipelago more broadly.
10 Folding Landscapes was founded in 1984 by Tim and Máiréad Robinson in order to sell Tim's maps. For more, go to the Folding Landscapes website at www.foldinglandscapes.com/.
11 Robinson has donated the contents of his personal records to The James Hardiman Library, NUIG. We are indebted to Nessa Cronin, John Cox and archivists Aisling Keane and Kieran Hoare at NUIG for making these resources available to us.
12 Seamus Deane, 'Ultimate Place: Review of Tim Robinson's *Stones of Aran: Pilgrimage*', *London Review of Books* 11:6 (16 March 1989), 9.
13 Deane, 'Ultimate Place', 9.
14 Robert Macfarlane, 'Rock of Ages', *The Guardian* (14 May 2005). Accessed on 9 September 2014, www.theguardian.com/books/2005/may/14/featuresreviews.guardianreview34.
15 Robinson, *Setting Foot*, 106.
16 Karen Babine, '"All the Sky Were Paper and All the Sea Were Ink": Tim Robinson's Linguistic Ecology', *New Hibernia Review* 15:4 (Geimhreadh/Winter 2011), 95–110; Michael Cronin, *The Expanding World: Towards a Politics of Microspection* (London: Zero Books, 2012); Christine Cusick, 'Mapping Placelore: Tim Robinson's Ambulation and Articulation of Connemara as Bioregion', in Tom Lynch and Cheryll Glotfelty (eds), *The Bioregional Imagination: Literature, Ecology, and Place* (Athens, GA: University of Georgia Press, 2012), 135–49; John Wilson Foster, 'Tim Robinson's Variegated World: *My Time in Space* by Tim Robinson; Tales and Imaginings', *The Irish Review* 30 (Spring–Summer 2003): 105–13; Terry Gifford, *Reconnecting with John Muir: Essays in Post-Pastoral Practice* (Athens: University of Georgia Press, 2006); Derek Gladwin, 'The Literary Cartographic Impulse: Imagined Island Topographies in Ireland and Newfoundland', *Canadian Journal of Irish Studies* 38:1–2 (2014): 158–83; J'aime Morrison, '"Tapping Secrecies of Stone": Irish Roads as Performances of Movement, Measurement, and Memory', in Sara Brady and Fintan Walsh (eds), *Crossroads: Performance Studies and Irish Culture* (Basingstoke, UK: Palgrave Macmillan, 2009), 73–85; Daniel Sack, 'Walking in and out of Place: the Pedestrian Performances of Tim Robinson', in Mary P. Caulfield and Christopher Collins (eds), *Ireland, Performance, and the Historical Imagination* (New York: Palgrave Macmillan, 2014), 19; Gerry Smyth, *Space and the Irish Cultural Imagination* (London: Palgrave Macmillan, 2001); Eamonn Wall, 'Deep Maps: Reading Tim Robinson's Maps of Aran', *Terrain. org: A Journal of the Built & Natural Environments* 29 (Spring/Summer 2012): n.p.,

'Digging into the West: Tim Robinson's Deep Landscapes', in Pascale Guibert (ed.), *Reflective Landscapes of the Anglophone Countries* (Amsterdam and New York: Rodopi, 2011), 133–45, 'Walking: Tim Robinson's Stones of Aran', *New Hibernia Review/Iris Éireannach Nua* 12.3 (autumn/fómhar 2008): 66–79, and *Writing the Irish West: Ecologies and Traditions* (Notre Dame, IN: University of Notre Dame Press, 2011); John Wylie, 'Dwelling and Displacement: Tim Robinson and the Questions of Landscape', *Cultural Geographies* 19:3 (2012): 365–83.
17 Robinson, *Pilgrimage*, 17.
18 Robinson, *Pilgrimage*, 18.
19 Oileáin Árann, or the map of the Aran Islands, Co. Galway, was originally published in 1975; it was updated in 1980 and 1996. There is a two-inch map of the Burren that shows the uplands of north-west Co. Clare; the original was published in 1977 and an updated version came out in 1999. The third of Connemara, and perhaps the most labour-intensive map, taking almost ten years to complete, was published in 1990. This one-inch map is accompanied by an introduction and gazetteer explaining some of the local history and translations of the Irish place names.
20 Gillian M. Doherty, *The Irish Ordnance Survey: History, Culture and Memory* (Dublin: Four Courts Press, 2004), 13. For a comprehensive overview, see J. H. Andrews, *A Paper Landscape: The Ordnance Survey in Nineteenth-Century Ireland* [1975] (Dublin: Four Courts, 2001). Other literary projects, such as Brian Friel's play *Translations* (1980), have also confronted this moment of history.
21 Doherty, *The Irish Ordnance Survey*, 11.
22 Graham Huggan, *Territorial Disputes: Maps and Mapping Strategies in Contemporary Canadian and Australian Fiction* (Toronto: University of Toronto Press, 1994), 14.
23 Robinson, *Setting Foot*, 3.
24 Huggan, *Territorial Disputes*, 3.
25 Tim Robinson, *My Time in Space* (Dublin: Lilliput, 2001), 99.
26 Robinson, *Setting Foot*, 72.
27 Robinson, *Setting Foot*, 79.
28 Robinson, *Setting Foot*, 102.
29 Robinson, *Setting Foot*, 76.
30 Robinson, *Setting Foot*, 76.
31 Christine Cusick, 'Mindful Paths: An Interview with Tim Robinson', in Christine Cusick (ed.), *Out of the Earth: Ecocritical Readings of Irish Texts* (Cork: Cork University Press, 2010), 210.
32 For more on deep-mapping and walking in Robinson's work, see chapter 1 in Eamonn Wall, *Writing the Irish West: Ecologies and Traditions* (South Bend, IN: Notre Dame University Press, 2011), 2–16.
33 Robinson, *Setting Foot*, 81.
34 Tim Robinson, 'The Irish Echosphere', *New Hibernia Review* 7:3 (Fómhar/Autumn 2003), 9–22.
35 Huggan, *Territorial Disputes*, 4.
36 Robinson, *Pilgrimage*, 18–19. Our emphasis.
37 For more on literary cartography, see Huggan, *Territorial Disputes*; Gladwin, 'The Literary Cartographic Impulse'; and Robert T. Tally Jr. (ed.), *Literary Cartographies: Spatiality, Representation, and Narrative* (New York: Palgrave Macmillan, 2014).
38 Kent C. Ryden, *Mapping the Invisible Landscape: Folklore, Writing and the Sense of Place* (Iowa City: University of Iowa Press, 1993), 54.
39 J. H. Andrews, 'Paper Landscapes: Mapping Ireland's Physical Geography', in John Wilson Foster (ed.), *Nature in Ireland: A Scientific and Cultural History* (Dublin: Lilliput Press, 1997), 203.

40 Tim Robinson, *Stones of Aran: Labyrinth* (Dublin: The Lilliput Press, 1995), 98–9.
41 Recent publications in these fields, principally in relation to Irish Studies, include works such as Wall, *Writing the Irish West*; Donna Potts, *Contemporary Irish Poetry and the Pastoral Tradition* (Minneapolis: University of Minnesota Press, 2011); Maureen O'Connor, *The Female and the Species: The Animal in Irish Women's Writing* (New York: Peter Lang, 2010); Christine Cusick (ed.), *Out of the Earth: Ecocritical Readings of Irish Texts* (Cork: Cork University Press, 2010); F. H. A. Aalen, Kevin Whelan and Matthew Stout (eds), *Atlas of the Irish Rural Landscape*, 2nd edn (Cork: Cork University Press, 2011); Robert Brazeau and Derek Gladwin (eds), *Eco-Joyce: The Environmental Imagination of James Joyce* (Cork: Cork University Press, 2014); and Caroline Crowley and Denis Linehan (eds), *Spacing Ireland: Place, Society and Culture in a Post-Boom Era* (Manchester: Manchester University Press, 2013).
42 Robert Macfarlane, *The Old Ways: A Journey on Foot* (New York: Penguin, 2012), 111.
43 Mike Pearson and Michael Shanks, *Theatre/Archaeology* (London: Routledge, 2001), 65.
44 Cusick, 'Mindful Paths', 210.
45 See note 16 for all of Wall's previous scholarship on Robinson.

# I
# Explorations in cartography and geography

# 1

# *Genius loci*: the geographical imagination of Tim Robinson

## Patrick Duffy

It was as if he had walked under the millimeter of haze just above the inked fibres of a map, that pure zone between land and chart between distances and legend between nature and storyteller.[1]

– Michael Ondaatje

## Introduction

For forty years Tim Robinson has been engaged in a uniquely detailed exploration of the rocky outposts of the Aran Islands, Connemara and the Burren – ancient environments deeply incised with the marks of human occupation for more than two thousand years. His maps and close-grained narratives of their human relationships have been accompanied by a self-reflection on his own process of understanding these places: the process of excavation and discovery, the meaning of his cartographic and literary experiments, and how his testimony will be received by his readers – what he has called 'moments of writing' and 'moments of reading'.[2] He sees his geographical explorations in map and prose as an endless and progressively detailed, but ultimately inadequate, local revelation of the scenes, settings and sounds of landscape, place and people. In a world of collapsing distances and faster, more wide-ranging travel, Robinson's works in map and text illustrate the potential in 'slow' landscapes where his 'twisty journey' on foot and bicycle has permitted a more intimate engagement with nature, environment and community.

Although Robinson has resisted being labelled ('environmentalist' or 'cartographer'), much of his work reflects trends in geographical thinking on landscape over the past half century. Robinson is a geographer by instinct, if not by training. Geographers have approached the study of landscape broadly in three ways – emphasising the tangible, material expression of landscape-as-object; exploring the manner in which landscape is perceived and represented (usually by the external gaze); and more recently considering landscape as a space for the imagination and senses. Ted Relph's *Place and Placelessness* (1976) was an influential early work focusing on the human experience of place, manifested in a sense of place and identifying with place as being central to our existence as humans. In the 1960s, J. B. Jackson also highlighted the manner in which our relations with the landscape have been shaped by the nature of our participation in it, and modified by the technologies of speed with landscapes designed for seeing at 40 mph.

In the past twenty years, academic interest has shifted from elite landscapes of power to ordinary landscapes 'that people inhabit and work in ... landscapes that people produce through routine practice in an everyday sense'.[3] And recently interest has turned to the landscape as a lived in, embodied, 'traversed, and *felt*' phenomenon, sometimes articulated in a performative approach to its study.[4] Some of this has been manifested from the 1990s in the concept of 'deep mapping', where local, rural places are explored intensively through a conflation of literature, documentary and story, folklore, legend and oral testimony, archaeology, natural and local history and memory – an accumulation of multifarious 'data' and 'texts' which in toto produce a textured 'deep map' capturing essential senses of place: 'landscape becomes a palimpsest – a stratigraphy of practices and texts'.[5]

In many ways, Robinson has articulated this geographical quest in singular fashion: 'anomalies of bedrock and boulders ... with a patchy cloak of humanity thrown over ... roads, villages (dozens of them), old churches, trades (kelp, boats, poitín), ill-remembered history, well-remembered song'.[6] Reading Robinson's maps and books is to vicariously experience the senses of place of Connemara and the Aran Islands and the texture of their landscapes.

## Exploring Robinson's maps

Surveying and map-making today are the collective works of teams of people. Robinson is unique in harking back to an earlier practice of the pioneering individualist as explorer and map-maker combined. His maps have been hand-crafted from the official Ordnance Survey (OS) maps of the nineteenth century – the work of sappers and surveyors in the 1820s and 1830s – and John O'Donovan's 'heroic but hurried foray' into place names of the area in 1839.[7] The OS ignored many of the particularities of local landscapes, emptying the landscape of colour, texture, details of habitation of grass and crops, reducing the landscape to a 'condition of nameless uniformity' to make it visible to state administration.[8] In focusing on the more limited canvas of the Galway Bay area, Robinson restores the local detail that was omitted by the Ordnance Survey (see Figures 4a and 4b).

In contrast to the (monochrome) richness of Robinson's maps, the OS map is an empty vessel. The earlier nineteenth-century editions of the survey are largely blank landscapes, though the six-inch scale details the 'field-boundaries as an eye-tormenting tangle of fine lines'.[9] They also include a small selection of archaeological remains – and Robinson takes pleasure in rediscovering numbers of cultural markers that were missed by the nineteenth-century survey. During

*Figure 4a*    Ordnance Survey, first edition six-inch map of Aran (1840s).

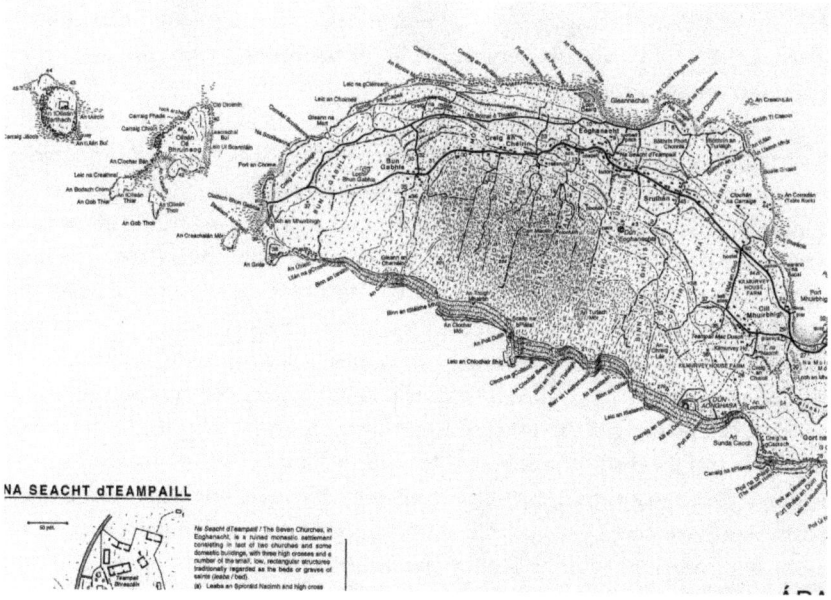

*Figure 4b*    Map of Árainn by Robinson.

fieldwork in Connemara, for example, he mapped many previously unknown megalithic sites, shell middens, as well as hundreds of children's burial grounds (cillíns), holy wells, mass rocks and famine graves. One of the comparatively few holy wells marked on the OS map was in the wrong place, he noted, twenty yards above the high water mark![10] His megalithic tombs are individually identified by type. There is an organic 'feel' to Robinson's map, in the manner in which place names have been rescued and restored in authentic Irish form and inscribed on the map. The scale and scope of his undertaking is shown by the approximate areas covered by his maps: the Burren is $c.500$ sq km, Connemara $c.1,200$ sq km, Aran Islands $c.42$ sq km. As a measure of his enthusiasm, he found that 'the smaller the place the more there is to be said about it', though in the smallest area he surveyed, Aran, he 'knew he would never get to the bottom of it'.[11]

The scale differences are an important consideration too, of course: Aran is on a scale of 2.25 inches per mile; Burren, 2 inches per mile; and Connemara is a 1 inch map, so in Aran proportionately more detail is possible. His early ambition was that through walking in what he called an 'endless proliferation of detours',[12] going out in all weathers, he would get the

> feeling of the place, its hard, resistant reality, into my bones ... I used to hope that the intensity of my physical experience of the place would persist through all these layers of time and space and tracing paper and burn through into the final drawing, so that it would not just be a factual record of the place, but an expression of a feeling.

He goes on to say, 'Perhaps if I vividly remembered walking along a certain shinglebank, I would be able to put some echo of my footsteps into the dots representing it on my map.'[13]

The notion of fractals and zooming in to add layers of depth and detail to the bare bones of the OS map is a recurring trope in all of Robinson's cartographic endeavours, where every nook and cranny was marked by its inhabitants over the centuries. This quest for greater detail highlights the challenging nature of the landscapes of Aran and south Connemara – where Ros a Mhíl, for instance, is a mere twenty miles from Roundstone, 'but the coastline in between is so complex that even as estimated from a small-scale map it is at least 250 miles long'.[14] He set out at one stage to mark in all the new houses on his map and to 'cross off' those that were abandoned (detail that was then unavailable on published OS maps) (see Figures 5a and 5b). But he stopped this exercise as 'the absences' on his map 'blackened it with the shadow of the past' – the blighted landscape of famine.[15] No doubt also the realisation dawned that such detail was beyond the capacity of one man. Data on new settlements have since been available in OS map revisions over the past twenty years.

The cartographic style of Robinson's maps, 'personal in its methods and subjective in its motivation', is a dense layering of inky symbology – pen strokes and hachuring replicate the rocky clefts and fissures (grykes) in the Aran Islands and the Burren, and an extensive array of strokes, tiny dots and dashes are used to represent sandy or rocky shores, shingle, dunes, scarps, forests, lakes and turloughs, drumlins, hills, cliffs and storm beaches.[16] Because he thinks the most revealing features of a map are its blank areas, he has preferred building up his maps with

*Figure 5a* Map of south Connemara by Robinson.

little dots and dashes 'standing separately' on blank white paper, 'rather than layers of colour that cover whole areas and implicitly claim to say something about every part within them'.[17] His Connemara and Burren maps do not have fixed landward boundaries – they are demarcated by roads and zonal shading, harking back to a premodern Gaelic conception of space. His survey of the Aran Islands resurrects as far as possible their historic territorial structure of thirty-six quarters, villages and townlands; in Connemara and Burren he uses townlands – into which he fills his survey details.

The central significance of Robinson's surveys is his restoration of the largely forgotten harvest of place names which carpet the area. Around the coast of Árainn (Inishmore) he recorded over two hundred place names, with sixty-six

*Figure 5b* Ordnance Survey, first edition map of south Connemara.

on Inis Meáin and forty-five on Inis Oírr. In the 1830s edition of the OS maps, a sparse spattering of minor names along the south cliffs of Árainn are anglicised renditions of some Irish names which endure to the present in Robinson's map.[18] And almost as many he rescued in the interior – scores of names of territorial units, early church or monastic sites, cairns, duns and other megalithic remains.

There are hundreds of names recorded for Connemara. In An Ceathrú Rua, a peninsula of about ten square miles, there are up to one hundred names located. Many mentioned in the *Gazetteer* do not appear on his map (or are alternatives or extensions of a name on the map), which possibly reflects the slippery usage of these local names by the community – *Sruthán an Bheannaithe, An tOileáinín Beannaithe, Trá Cnoc Carrach, Cladach an Ghleanna, An Caisleán Cruinn, Na Caisle, Poll na nGabhar, Tobar Charraig Mhóir, An Trá Choiréalach, Trá na bPáistí, Bothar na Ronna, Tobar an Mhulláin, An Tráighín idir Dhá Ing, Tobar Naomh Thomáis, Cuan na Loinge, Loch na Croimine, Oileán an Phríosúin, Cora na bPáistí, Caladh na Siongán, Aill an Tobac, Barr an Chrompáin, An Chara.*[19] *An Bóthar Buí* (the yellow road) is marked on the map but unmentioned in the *Gazetteer*.[20] In *Connemara: A Little Gaelic Kingdom*, numbers of additional shore names are not marked on the map: *An Crompán Sliogánach, An Meall Glas, Barr na Cora* (*Cora na Roinne* is marked on the map), *Na Muiltíní, Caladh an Uisce*.[21] Alternative or additional names for the same place are listed in the *Gazetteer*, such as *Loch Raithneachán* in Dolan townland aka *Loch Cara an Oileáin*; *Loch Tanaí* and/or *Loch na Croimine* (in *An Cheathrú Rua Thiar*); *Crompán* with *Aill an Tobac*.

Because of the restricted scale of his maps, the intricate network of stone walls which is draped over the human landscape is too dense in detail for representation, and in his attempt to be comprehensive many of his names, too long for inclusion on the map, are referenced by number and listed in the accompanying gazetteers.[22] In subsequent years, he accumulated great numbers of additional names for the myriad offshore rocks and tiny islets in south Connemara, which he was unable to record on his map.[23] Details on the history of townland boundaries, which are universally important features on maps of rural Ireland, are rare. 'The Moot', for instance, is a two-mile straight-line earthen bank defining the boundary of Letterdife townland near Roundstone. As a loosely remembered oral boundary, the OS followed its usual practice in the 1830s of imposing 'a rectilinear abstraction on such a recalcitrant reality'.[24]

Underneath these names must be hundreds more field names which, like the walled enclosures, are not detailed in Robinson's map. He has estimated that there are c.14,000 little fields in Aran, comprising c.1,500 miles of drystone walls – a monument to the labours mainly of nineteenth-century inhabitants (see Figure 6).[25] The geography of this labyrinth of walls mirrors the underlying geology and topography, displaying a rectangularity in the overall pattern of fields and boreens shaping the community's relationship with farmland and sea for fishing and seaweed rights – as well as influencing the nature of Robinson's traverses through the landscape.

The countless numbers of little fields in both Aran and Connemara defeated Robinson's fractal-like mission to record as much as possible, especially at the 1 inch scale: 'Thousands of names must have been given them over the centuries, most of them forgotten; I have only recorded a few dozen.'[26] Many of the meanings of these field names were lost to local memory, such as *Creigeán na*

*Figure 6*   Aran: 'a wilderness of stones and stone walls', photo by Cotton Coulson.

*Banríona* (the queen's crag-cum-field); *Scrios Buaile na bhFeadóg* (the open tract of the pasture of the lapwings).[27] As elsewhere in the west of Ireland, where changing farming practices and falling population have broken the link between people and land, and where Irish has disappeared as a daily form of speech, there is a loss of memory and meaning: 'A farmer with no Irish is a stranger in his own land.'[28] Field names in Irish and English (named by the families who owned and worked them) represent what might be characterised as familial or domestic space, as opposed to the more general communal space of other local names, and have been recorded throughout Ireland in recent years in better resourced surveys and at larger scales than Robinson's maps.[29]

Throughout Ireland, local place names have a poetry that rhymes with local identity and sense of belonging – names 'that sigh like a pressed melodeon' across the landscape in the words of poet John Montague. The names on Robinson's maps ring with lyrical euphony too, and reflect an acute awareness of the local environment: *Nead an Iolra, Log an Fhia, Meall an Fhathaigh, Gob an Damha, Caladh na hInse, Cuainín an Róin, Poll an Phíobaire, Aill na Síog, Loch na nUilleann, Sruthán na Seilide, Fuaigh na gCacannaí, Sruth na Rón, Carraig na gCapall*.[30] The names in Irish have a certain exoticism and mystique which would have been true also of the thousands of townland names in Ireland before they were phonetically rendered in English: *Trá na hAdhairce, Tuarán na gCaorach, Carraig an tSagairt, Carraig na bPortán, Loch an Mhongáin, Cloch Leathbhealaigh, Loch an Fhraoigh, Loch Pholl an Mhaide Giúise, Gleann na gCoileach, Sruthair na Mioltóg, Duirling na Roilleachaí, Loch na mBreac Caoch*.[31] The most common names refer to natural features of land and shore – coves, caves/holes, lakes, inlets or pools, harbours, wells, cliffs, headlands and rocks, often with associated characteristics of flora or fauna.

During fieldwork Robinson was alert to the changes which were underway in the landscape – in the Burren, for instance in the 1960s and 1970s, there had been extensive consolidation of fields, and field walls had been removed with many known Stone Age tombs. When he was updating the map, he had to delete a holy well, a children's burial ground, a 'saint's chair' and other artefacts.[32] Undoubtedly many artefacts were protected by the stories and traditions circling around them, and frequently the power of the priest had an important role, as was noted in 1929 by Liam Price in Wicklow where a prehistoric mound was removed by a farmer on being told by the priest 'he might take it down as it was only a pagan who was buried in it'![33] Mám Éan pattern day (patron day) in Connemara was condemned by the clergy in 1920s, and revived in the 1970s with the imprimatur of the Church – though locals persist with more ancient practices at the neighbouring holy well.

**The storied landscape**

There are two ways of addressing Robinson's recourse to narrative texts on Aran and Connemara – a simple one of his need to amplify the detail of the mapped landscape, especially exemplified in the gazetteers accompanying the maps, but also in the dense factuality of the books; and secondly, his more philosophical and reflective exploration of the deeper dimensions of the landscapes to engage with what he called their 'placelore' in his 1996 *Companion* to the Aran map.

Like cartographers in the past, Robinson very early on realised the limitations of the orthodox map as a mode of communication. From the seventeenth century onwards map-makers and surveyors regularly supplemented their works with 'terriers' and 'memoirs' containing written descriptions of the areas represented on their maps. Explanatory memoirs accompanied the maps of the Geological Survey, for example. The OS Memoirs, Letters and Name Books in the 1830s provided comprehensive details on the topography, settlement, economy, customs and traditions of the parishes surveyed.[34] Robinson's gazetteers and his books might be viewed in this tradition. His landscape narratives also find echoes in earlier work, like Robert Lloyd Praeger's *The Way That I Went* (1937), concentrating on 'land, landscape and wildlife, with customs, economy and history taking second place', and the notebooks of Liam Price on his Wicklow field trips focused especially on landscape archaeology, folklore and place names for the period 1928–66.[35] As Pat Sheerin claims, 'Land and place are made up of language as much as, if not more than, they are made of earth and buildings.'[36]

More recently, such texts have been seen as a form of literary mapping in the sense that 'all writing is in one way or another cartographic … storytelling is an essential form of mapping, of orienting oneself and one's readers in space'.[37] Exploring the Aran, Burren and Connemara maps is unsatisfactory without some parallel reference to the books. Robinson himself saw his maps as a preparation for 'the world-hungry art of words' which would more thoroughly encompass the breadth and depth of his geographical imagination.[38] In 1994 he reflected that he had lost interest in maps and cartographic techniques as a means of representing the landscapes through which he moved, though he forewarned that his first book with the Aran map (*Pilgrimage*) would not accommodate its 'condensed labyrinth … so there will have to be another book someday and it is going to take a lifetime and still be unfinished'.[39] *Bóthar an Screigín*, for example, was as much as could be marked on the poor fields of Cill Éinne in Aran; writing afforded more scope to detail the backstory to the place.[40] Ultimately, in his view, 'only prose, and prose at length, recursive and excursive, can begin to map it and act out the building-up of the overarching, underpinning, encircling, realities of sky, land and sea out of uncountable glints of detail'.[41]

Consequently, his strategy as it developed has been to commence with field-work and maps, then to compile an accompanying more detailed gazetteer, and finally to produce much more comprehensive narratives on the particularities of place and space and people. His books and gazetteers are classic reflections of his wandering approach – territorially, imaginatively and intellectually – containing a reservoir of riches on place names and *dinnseanchas*, which reach well beyond the scope of his maps. The *Connemara Gazetteer* was reprinted with a few changes in 2005. *Listening to the Wind* was published sixteen years after his map of Connemara and contains a lot of additional data in its peripatetic meanderings.[42] In other words, the map is quite inadequate to capture the wealth and richness of detail of names in what he called 'the topographic web of Connemara'.[43] *Scailp*, the object of one of his early explorations in the hinterland of Roundstone, is not marked on the map. It seems to afford a panorama of Roundstone Bog. But his answer to his query 'what is this place?' hints at an elusive ethereality – perhaps it couldn't be fixed on the map?

> [W]hat is this place? A steep little comb of the hillside, tumbling down to the lakeside; a roughly built stone hut, roofless and full of bracken, crouching in the shelter of a lumpy rock outcrop; a patch of green grass on top of a knoll where sheep sometimes lie beside a hollowed-out stone-heap, the remains of a pen for lambs; a silvery tree trunk, almost branchless rising out of a split rock; a wind off the lake and the bog beyond it stretching to the horizon, a presence of pure space.[44]

His text therefore provides an extended commentary on the geographical patterns and shapes displayed on his maps by filling in the environmental, historical, folkloric or simply the contemporary social contexts reflecting an intimate, well-known, well-trodden landscape of sea and shore. There is a local intimacy in the commentary, summoning up the texture of landscapes reflected in events, or personalities whether saint, celebrity or local inhabitant, or in eccentricities and distinctive features of local topography. In some cases the human imprints echo geological timescales – the pattern of *bóithríní* (boreens) and stone walls in Aran mirror the fissures caused by tectonic forces which separated Europe from America millions of years ago. And like many other infrastructural works throughout Connemara, roads such as *Bóthar na Scráthóg* in An Cheathrú Rua were built as relief works during local famines in the later nineteenth century.[45]

His account of *Ceathrú an Chnoic* (the Quarter of the hill) in the east end of Árainn is a typical example of his approach to expanding the bare bones of the map:

> Bothar an Chnoic / the road of the hill, built as a relief work in the [nineteenth] century, now disused and interrupted by field-walls, crosses *Ceathrú an Chnoic* ... On its east are traces of a very delapidated ringfort, and *An Carn Buí* / the yellow cairn, a prominent hilltop regarded as a fairy dwelling. The road ends at *Poll na Scailpe* / the inlet of the cleft. *Cloch Liam* / Liam's stone. *Poll Leathan* / broad inlet. *Poll Dorcha* / dark inlet. *Leac na gCarrachán* / the flagstone of *Na Carracháin* / the stone heaps, a rocky shoal and well-known fishing ground offshore here; an angle of the cliff above it is *Coirnéal na gCarrachán*, the inlet at its west end is *Poll na gCarrachán*. *Na Clocha Móra* / the big stones, a group of huge boulders on the clifftop and leaning against the cliff which together with Teampall Bheanáin furnish a mark for Na Carracháin. *Dabhach an tSnamha* / the tub of the swimming, a rock ledge with many smooth rockpools.[46]

Placelore and place names are clearly central to his geographical mission to rescue significant elements of the cultural landscape. In An Cheathrú Rua, fieldwork involved traipsing down boreens, with local informants Joe Tommy and Máirtín 'closely inspecting every bullock in the little fields and at the same time calling out placenames which I tried to scribble down as I tussled with maps in the breeze'.[47] There are hundreds of minor names around the hugely complex coastal zone of Connemara, with an equally dense array in the interior. Anglicised names like Roundstone are expressions of the Irish sounds – *Cloch na Rón* (Seals' stone), a rendition for over three hundred years, or *Cromaill* meaning 'bent cliff' (leaning over a narrow sound), known in English as Cromwell's sound. Slackport is probably *Sleabhach* (an edible seaweed), but a local story was invented to explain it in English, about visiting sailors looking for shops and pub, finding none, dismissing it as a 'slack port'! Clifden, originally *An Clochán*, was 'fashionably anglicized' as Clifden in the early nineteenth century. Atticlogh contains a bog nicknamed The Congo from turf cutting in the early 1960s when Irish United Nations troops were killed in the Congo.[48]

The gazetteers include summary descriptions contextualising many names that have endured for three or more centuries – presaging the more reflective writing in the books on Aran and Connemara, which in many ways might be described as reflections on a tale of meaning.

> A placename ... is perpetually gathering and shedding meanings. It comes down to us as a loose bundle which may or may not still contain that kernel, the initial grain of sense that set it rolling through time ... If read as a mnemonic for a history of the mind's responses to a mysterious marriage of sound and place, the placename can be a world of power.[49]

His gazetteer to the Aran map contains explanatory details on landscape remains and associated names, as for example in the townland of Eochaill: *Dún Eochla* is a double-ramparted Iron Age stone cashel – its outer rampart enclosing a circular area approximately 100 yards across. Nearby is a disused lighthouse built in 1818 and replaced in 1857 by lights on the other islands and a signal tower from 1805 to defend against a French invasion which could communicate with similar towers on Inis Oírr (Inisheer) and the Clare and Connemara coasts by means of flags and black discs hoisted on a mast. The Irish–English names, *Bóithrín an Tower* and *Bóithrín an Lighthouse*, reflect the fact that both were manned by English-speakers.[50]

Robinson refers to what he calls colonialist attempts to render the sounds of one language in the spelling of another – Bunowen for *Bun na Abhann* was classically used in Brian Friel's play *Translations* – 'Irish placenames dry out when anglicized, like twigs snapped off from a tree. And frequently the places too are degraded, left open to exploitation, for lack of a comprehensible name to point out their natures or recall their histories.'[51] The legacies of some OS renditions are aesthetically displeasing, if not downright comical – for example, Sruffaunoughterluggatoora and Muckanaghederdauhaulia in Connemara, and Bullaunancheathrairaluinn in the Aran Islands. His task as topographer is to rescue place memory: without the recording and rehearsing of meaning 'place itself founders into shapelessness, and time, the great amnesiac, forgets all'.[52] From island names, to townlands, through villages, to quarters and local micro names, there is a dazzling array of place names in the Aran Islands: 'the landscape here speaks Irish'.[53] Robinson observes that in 'recording, largely from the lips of Aran itself, as best I can these place-words, and expressing them in … map and book', he wants to resist the inevitable loss of meaning of the spoken Irish in its native environment.[54]

Lanes, ledges, rocks, points, islands and inlets, shores, caves, pools, blowholes, clefts, cliffs and many others features are named, many having obsolete historical references rescued from oblivion by Robinson: *Port Bhéal an Dúin*, the port of the fort's mouth (Dún Aonghusa): 'Every feature of the bay has its name, though many of them are almost forgotten, now that a detailed knowledge of its accesses and obstacles is no longer a matter of survival to the boatmen.'[55] *Leic an Níocháin* is a flagstone where women used to wash clothes in seawater, but today there is no need with piped water in each house, though the name survives to commemorate an earlier age.[56] Some show long memories of events from generations ago: off the Aran shore is

> *Ancaireacht Sheáin Uí Mháile* / Sean O'Malley's anchorage; he is said to have been a landowner in Mainistir whose steamers came with timber from Europe and anchored there … *Poll na Marbh* / the hole of the dead, an area named from graves said to be those of some O'Flaherty men killed in a feud between two branches of the sept over ownership of the castle of Ballynahinch in Connemara, in 1584.[57]

*Meall an tSaighdiúra* is said to have been called after an English soldier who profaned Tobar Chonaill holy well and died. There is also *Binn an tSaighdiúra*, which may be connected with an OS sapper who fell to his death while surveying the area in the 1830s. *Cúgla* (*Cúige Uladh*) is called after migrants from Armagh ejected from Ulster in the 1795 sectarian strife. *Duirling na Spáinneach* (rockbank of the Spaniards) remembers the armada ship *Concepcion* wrecked off this point whose survivors were executed in Galway.[58]

The Aran Islands, the focus of Robinson's first engagement with the Irish landscape, was an attractive scenic location for elite visitors, artists and tourists throughout the nineteenth and twentieth centuries, and part of Robinson's narrative is to reference local associations with notable personages. For example, the Zetland hotel was originally The Viceroy's Rest and was named in honour of the Lord Lieutenant of Ireland, the Earl of Zetland, who holidayed in the 1880s and 1890s. Munga Lodge was owned by the Frewins, one of whom was Moreton Frewin who 'made and lost a fortune in Wyoming and married Clara Jerome', one of the Jerome sisters of New York City. Clifden Castle, built by John D'Arcy in 1815, was used as a holiday home by Thomas Eyre of Bath who had bought the estate. Thomas Hazell was a Scotsman involved in the kelp trade who bought Doon and Cashel houses. Daniel Bowden Smith, retired governor of Bengal, moved into the Roundstone area in the 1840s. The Duchesse de Stacpoole, who lived in Errisbeg house and died in the 1950s, was one of the three terrors for drivers in Connemara ('the ditches, the donkeys, and the Duchesse').[59]

Connemara's poverty in the nineteenth century also attracted proselytising members of the Irish Church Mission, and during the famine Quakers set up workshops and self-help enterprises. Details of local landownership history and the construction of landed estates, houses and demesnes from the seventeenth century provide an important layer of meaning in the *dinnseanchas* or storytelling about places, picked up (as he says himself) from locals encountered on his journeys, but also it is clear from follow-up research undertaken by Robinson. The region's remoteness also afforded many hiding places in caves and bogs during troubled times, which are commemorated in names.[60] The Aran Islands became an iconic destination for the *National Geographic* magazine through the twentieth century, in 1931 for example, captivated by its 'simplicity, harmony ... and theatrical panorama of sea and sky, the medieval homeliness of speech and hearth and tool'.[61] Indeed it was Robert O'Flaherty's *Man of Aran* film (1934) which first attracted Robinson to the islands and his paean to the islands echoes the romanticism of magazine and film: 'one of civilisation's loftiest windows onto its own origins in the past and the natural world ... To live on Aran is a rare and demanding privilege.'[62]

His maps which show the complex interdigitation of land and sea and lake and bog, islands, peninsulas and cliffs, overlain with the intricate material and intangible filaments of human settlement, present a 'chaos of visual impressions'.[63] For someone with Robinson's sense of the fractal immensity of landscape there was nowhere else like the labyrinthine recesses of Aran or the 'bewilderment of forking seaways' in Connemara.[64] And there was no other way to approach its understanding than through an extended commentary of revelation.

The lake landscapes of south Connemara, a baffling entanglement of land and water to Praeger in the 1930s, had a frightening sense of place for Robinson that is beyond the scope of his map: 'The outline of each lake bristles with projections, every one of which is itself spiny; they stab at one another blindly. There is a fractal torment energizing the scene, which is even more marked in aerial photographs, in which the lakes seem to fly apart like shrapnel.'[65] Elsewhere he refers to the messy confusion of the islands and peninsulas that comprise *Na hOileán* at *Garomna*, where one is drawn into increasingly complex layers of engagement, as with a *glasoileán*, an island off an island that is accessible only at low tide, off an island connected by causeway to the mainland.[66] The cliffs on Aran (and the Burren) were perpendicular landscapes which were extensions to the fieldscapes of the inhabitants and were virtually impossible to represent effectively on a map: 'for generations a hunting-ground for fowlers, the cliff has its named and familiar paths, its exits and its entrances … haunted spots and favourite nooks … the cliff face was a wide province of the islanders' mental landscape'.[67] Around the westernmost quarters of Árainn, Robinson got lost in a maze of interminable boreens, 'all a long way from anywhere', fields, walls, 'rigmaroles … the unnerving landscape of Na Craga, the familiar mesh of countless stony fields, all clenched and gnashed together like the cogs and ratchets of an antique clock long ticked to a stop, its key lost'.[68]

Robinson's reflections on the complexity of local landscapes illuminate the furthest extremities of his mission impossible to explore the intangible reality of place encountered in south Connemara:

> A fish the size of the *donnánach* [small rockfish found inhabiting a holy well] has access to holes and crevices the surface area of which is many times greater than that of the foreshore as we would measure it, in acres, say. Within those retreats are smaller ones, home to little creeping things that enjoy still larger areas, not just in inverse proportion to their own size, but absolutely. Extending this thought to the single-celled population of the seashore, it seems that our acre contains within it thousands of acres. Since thought has access everywhere, rightly considered the nondescript rocky shore is cavernous, labyrinthine, unfathomable. Add to these fractal dimensions the echoing linguistic space opened up by placenames and oral lore, and the mythic realms inhabited by prophetic fish and wonder-working saints, and it is clear that my account of it, even if I wrote volumes on this one dog-eared corner of land, would be a mere footnote to reality.[69]

This is probably the most apposite and emblematic manifesto of Robinson's pilgrimages through the landscape. His twisty journey into the recesses of place, space and the mind, in a hopeless fractal quest, is adverted to in Derek Gladwin's discussion of the fundamental 'unknowability' of place and landscape.[70]

## Reflections on slow landscapes

Robinson's map of Connemara is 'the record of a long walk. An intricate, knotted, web-like walk that visits every place within its territory.'[71] Satellite navigation in a GPS-coded landscape today is the extreme opposite of Robinson's organic engagement with local places and landscapes. With the collapse of distance, and faster, more wide-ranging travel, there is growing interest in a return to slow landscapes, which allow more intimate relationship with nature and environment, explored at walking speed, for example. Perhaps this is why the walks and works of Robinson are so appealing, connecting with the scenes, settings and sounds of nature (wind and rain, streams and surf, birdsong) and people (songs, story and conversation).

For most people today, landscape through a car window is a faster, less tactile experience, insulated from wind and rain, mediated perhaps by car radio/music, with much of the passing landscape a largely unknown blur. There is no embodied connection between traveller and landscape, especially in rapid-transit motorwayed space. Relph argued that through the twentieth century a less authentic sense of place that he called *placelessness* was dominant: 'the casual eradication of distinctive places and the making of standardized landscapes that results from an insensitivity to the significance of place … and the casual replacement of the diverse and significant places of the world with anonymous spaces and exchangeable environments.'[72]

My father's unpublished diaries of daily life in south Ulster record the intensity of social interaction in the largely pedestrian (and bicycling) world of the 1930s and 1940s. This is evident, for example, in its more leisurely landscapes of walking/strolling/cycling, chatting on roadsides, and in places that were textured with local names and points of reference, loaded with internal significance for the community's inhabitants.[73] The poet Patrick Kavanagh's celebration of his Monaghan landscape was one embedded with intimate ordinary details, which he celebrated as 'important places and great events' – its banks and stones, fields and lanes walked and cycled, or trudged on horsecart or behind the plough, local places of people, stories, voices, animals.[74]

Praeger, who (mostly) walked around Ireland, bemoaned the replacement by the motor car of the horse-drawn jaunting car,

> so ideal for viewing the country: you saw everything by the road-side and far away; you could hop on and off as you liked. Many a rare plant did I 'spot' … which I would never have detected from a saloon car even if going at a staid eight miles an hour instead of the usual thirty or forty.[75]

'Spending unnumbered days' tramping south Connemara 'prising placenames like winkles' out the crannied stones, as Robinson found, was facilitated by bicycle and walking which allowed him to stop and chat, make notes, look at the landscape – seeing, feeling, sensing: 'turning right into Farrell's Road, the lane that climbs gently between meadows half-lost to whitethorn and furze bushes ... The low drystone walls and ditches of the wayside are overgrown with fuchsia and willow and brambles; I know them foot by foot from blackberrying every autumn.'[76] A similar close-focus encounter is evident in the American writer Rebecca Solnit's walk through west Cork in the early 1990s:

> I crossed the bridge over the Bawnaknockane River, where it enters Roaringwater Bay, passed Knockroe, Knockaphukeen, Coosane, Ballybane, Barnaghgeehy, Letterlicky Bridge, across the Durrus, Hollyhill, Cappanaloha, to Bantry Bay – a fine collection of names for a ten-mile walk over a range of modest hills ... It was lined with low stone walls and hedgerows full of flowers profoundly familiar and subtly different in equal measure ... Cows bellowed, sheep bleated, birds chirped, and occasionally cars roared, all under a sky that seem distended and sagging with its burden of water.[77]

With speeding up of travel and expanded horizons of living, therefore, such local landscape detail has been edited out, or bypassed as people took to cars and reached further beyond the local, across regional and national landscapes. Today most travellers' experience is of 'fast landscapes' of house-high hedges whizzing past, or empty fields and placeless space. As Solnit has observed, people now live in a series of interiors – home, car, gym, office, shops – disconnected from each other. 'On foot everything stays connected, for while walking one occupies the spaces between those interiors in the same way one occupies those interiors.'[78]

While Robinson would instinctively empathise with journalist Frank McDonald's shock at the bungalow blight evident in a drive through south Connemara in the 1980s – 'a ruined landscape, bristling with telegraph poles and banjaxed by bungalows' – he takes a more positive view of these developments from his interactive vantage point on foot, as being part of a disorderly communal vitality, order in its apparent *dis*order, that was imperceptible 'from the window of a car'. An Cheathrú Rua: 'lots of houses and lots of hydrangeas ... higgledy-piggledy, assembled as if by successive throws of dice rather than according to a plan ... It is only as contradictory and untidy as life.'[79] Walking is Robinson's main modus operandi and he is constantly aware of his footsteps following earlier well-trodden paths – surveyors and sappers slogging through fields with chains and theodolites in the early nineteenth century, John O'Donovan and other OS personnel collecting data on names and antiquities, not to mention the daily and seasonal itineraries of thousands of local inhabitants and their livestock. He quotes with approval Rev. Henry McManus on the big country of the western landscape in the mid-nineteenth century, no doubt experienced at walking pace: 'Instead of hedges bounding your view, as elsewhere, with only a narrow strip of sky visible above you, here you behold at one glance hundreds of square miles, spanned at their extremities by the whole conclave of heaven, filled of a fine day with a blaze of living light, indescribably glorious.'[80]

'*Stones of Aran* is all made up of steps.'[81] Robinson's meandering pilgrimage through the landscapes of Aran, Connemara and the Burren are classic examples of slow landscapes in practice where the intimate details of place have been experienced and expressed by him in maps and words. On bike and foot he could engage with the landscape at a pace and in a way which matched the time-space reality of its original makers and shapers, revealing layers of knowledge and imagination laid down in a pedestrian world which have limped on in local memory to the present. Robinson's singular achievement has been to rescue, record and restore this legacy.

## Notes

1 Michael Ondaatje, *The English Patient* (London: Bloomsbury, 1992), 246.
2 Tim Robinson, *Connemara: Listening to the Wind* (Dublin: Penguin Ireland, 2006), 3.
3 See Tim Cresswell, 'Landscape and the Obliteration of Practice', in Kay Anderson, Mona Domosh, Steve Pile and Nigel Thrift (eds), *Handbook of Cultural Geography* (London: Sage, 2003), 274.
4 James D. Sidaway, 'Shadows on the Path: Negotiating Geopolitics on an Urban Section of Britain's South West Coast Path', *Environment and Planning D: Society and Space* 27 (2009): 1091–116. John Wylie, 'Dwelling and Displacement: Tim Robinson and the Questions of Landscape', *Cultural Geographies* 19:3 (2012): 365–83.
5 Cresswell, 'Landscape', 278. See also Michael Pearson and Michael Shanks, *Theatre/Archaeology* (London: Routledge, 2000) and Eamonn Wall, who considers Robinson's work an exercise in 'deep-mapping', in 'Walking: Tim Robinson's Stones of Aran', *New Hibernia Review/Iris Éireannach Nua* 12:3 (autumn/*fómhar* 2008): 66–79.
6 Tim Robinson, *Connemara: A Little Gaelic Kingdom* (Dublin: Penguin Ireland, 2011), 317.
7 Tim Robinson, *Connemara: Introduction and Gazetteer* (Roundstone: Folding Landscapes, 2005), 1. In Robinson, 'A Connemara Fractal', 12–13 (see note 13), he describes his mapping technique. Eamonn Wall has referred to him as undertaking 'a one-man Clare Island survey' in the Aran Islands. See Wall, 'Walking', 69.
8 J. H. Andrews, *Plantation Acres: An Historical Study of the Irish Land Surveyor and his Maps* (Belfast: Ulster Historical Foundation, 1985), 393.
9 Tim Robinson, *Stones of Aran: Labyrinth* (Dublin: The Lilliput Press, 1995), 11.
10 Robinson, *A Little Gaelic Kingdom*, 299. See, for example, *Connemara Gazetteer*, 33, 79, 87. He restores *Cnoc na gCorrbhéal* to an unnamed hill on the Ordnance Survey map near Roundstone in *Connemara: Listening to the Wind* (Dublin: Penguin Ireland, 2006), 10.
11 Robinson, *A Little Gaelic Kingdom*, 368; *Labyrinth*, 440.
12 Tim Robinson, *Stones of Aran: Pilgrimage* (London: Penguin, 1986), 12.
13 Tim Robinson, 'A Connemara Fractal' in Timothy Collins (ed.), *Decoding the Landscape* (Galway: Centre for Landscape Studies, UCG, 1994), 13–14.
14 Robinson, 'Connemara Fractal', 14.
15 Robinson, *A Little Gaelic Kingdom*, 313.
16 Robinson, *Listening to the Wind*, 143.
17 Robinson, 'Connemara Fractal', 29.
18 Portdeha for *Port Daibhche*, Culoogarrow for *An Colbha Garbh*, or Illaunanaur (mis-recording *Gleann na nDeor* according to Robinson in *Pilgrimage*, 24), Scalpaluig (*Binn an Loig*), Benaharnaun (*Binn an Charnáin*).

38  Patrick Duffy

19  Robinson, *Gazetteer*, 119–23. Respectively translated as, the stream of the blessing, blessed little island, stony hill beach, the valley shore, the round castle, the water courses, the goats' hollow, the big rock well, the coral strand, the beach of the children, the road of the division, the well of the boulder, the little beach between the two notches, St Thomas's well, the ships' bay, the lake of the crooked-horned cow, the island of the prison, the rocky point of the children, the harbour of the ants, tobacco rock, the head of the creek, the ford. In some cases, as with many place names, forms are not always grammatically correct, as noted by Robinson.
20  However, it is described in *A Little Gaelic Kingdom*, 294.
21  Robinson, *A Little Gaelic Kingdom*, 301. Respectively translated as, the shelly creek, the green knob, the tip of the rocky point, the small boulders, the harbour of the water. *Cloch na nDaoine Móra* (the big people's/'big shots' stone), where visitors would sit and watch the locals fishing, is not marked on the Aran map, but mentioned in *Pilgrimage*, 24.
22  Robinson's Aran map was first published in 1975, with a second version in 1980, and a further updated version with an accompanying 'Companion' (or gazetteer) in 1996. See Tim Robinson, *Oileáin Árann: A Companion to the Map of the Aran Islands* (Roundstone: Folding Landscapes, 1996).
23  See Robinson, *A Little Gaelic Kingdom*, 321–2. These are part of an archive he is presenting to National University of Ireland, Galway.
24  Robinson, *Listening to the Wind*, 11.
25  Robinson, *Labyrinth*, 11.
26  Robinson, *Labyrinth*, 438–9.
27  Robinson, *Labyrinth*, 438–9.
28  See Michael Viney, *A Year's Turning* (London: Penguin, 1998), 119.
29  Séamas Ó Catháin and Patrick O'Flanagan, *The Living Landscape: Kilgalligan, Erris, Co. Mayo* (Dublin: Comhairle Bhéaloideas Éireann, 1975); Nollaig Ó Muraíle, 'Placenames of Clare Island', *New Survey of Clare Island Newsletter* 4 (March 1998, RIA): 6; Uinsíonn Mac Graith agus Treasa Ní Ghearraigh, *The Placenames and Heritage of Dún Chaocháin* (Béal an Atha: Comhar Dún Chaocháin Teo, 2004); Éamonn Lankford, *Suirbhé Logainmneacha Chorcaí* (115 volumes, Cork, www.placenames.ie/). Frances Tallon (ed.), *The Field Names of County Meath* (Meath Field Names Project, 2013). From 1999 a field names project under a local community FAS scheme has been underway around Leitir Mealláin and Garumna in south Connemara.
30  Respectively translated as, the eagle's nest, the deer's hollow, the hummock of the giant, the point of the sandbank, the harbour of the island, the little harbour of the seal, the piper's hole, the cliff of the fairies, the lake of the elbow/angle, the stream of the snails, the cove of the cormorants, the stream of the seals, the horse rock. See Robinson, *Connemara: Gazetteer*, 33, 39, 48, 60.
31  Respectively translated as, horn beach, small field of the sheep, the priest's rock, the rock of the crabs, the lake of the overgrown swamp, the half-way stone, the lake of the heather, the lake of the bog-deal hole, the glen of the (wood) cocks, the (tidal) flow of the midges, the rock bank of the oyster-catchers, the lake of the blind trout. See Robinson, *Connemara: Gazetteer*, 64, 65, 72, 73, 93.
32  Introduction to the Burren map (1999).
33  Christiaan Corlett and Mairéad Weaver (eds), *The Liam Price Notebooks: The Placenames, Antiquities, and Topography of County Wicklow*, Vol. 1 (Dublin: Dúchas, 2002), 37.
34  See the forty-volume printed edition of Ordnance Survey memoirs edited by Angélique Day and Patrick McWilliams (Belfast: Institute of Irish Studies, QUB). OS letters have been published by a variety of publishers: Christiaan Corlett and John

35 Medlycott (eds), *Wicklow* (Roundwood Historical Society, 2000) and Michael Herity (ed.), *Meath* (Dublin: Four Masters Press, 2001).
35 Seán Lysaght, *Robert Lloyd Praeger: The Life of a Naturalist* (Dublin: Four Courts, 1998), 151; Corlett and Weaver, *The Liam Price Notebooks*, 37.
36 Pat Sheerin, 'The Narrative Creation of Place: the Example of Yeats', in Timothy Collins (ed.), *Decoding the Landscape*, 2nd edn. (Galway: Centre for Landscape Studies, 1997), 150.
37 Robert T. Tally Jr, 'On Literary Cartography: Narrative as a Spatially Symbolic Act', *NANO: New American Notes Online* 1 (January 2011): no pag.
38 Robinson, *Pilgrimage*, 11–12. As Eamonn Wall has noted, there is more to a place than its roads, buildings and hills. Each road was more nuanced and detailed than a mere line drawn between settlements.
39 Robinson, 'A Connemara Fractal', 12, 29: 'Such knots of significant place can hardly be conveyed in a map alone. That is why I have combined this map with a book.'
40 Robinson, *Labyrinth*, 89–90.
41 Cusick, Christine, 'Mindful Paths: An Interview with Tim Robinson', in Christine Cusick (ed.), *Out of the Earth: Ecocritical Readings of Irish Texts* (Cork: Cork University Press, 2010), 208. See also Derek Gladwin, 'The Literary Cartographic Impulse: Imagined Island Topographies in Ireland and Newfoundland', *Canadian Journal of Irish Studies*, 38:1–2 (2014): 158–83, who suggests that by incorporating a wide range of literary genres Robinson constructs a new kind of spatial analysis that facilitates greater landscape insights.
42 Such as, *An Gleann Mór*, Path of Afola, 8, *Oileán na nUan* [Lamb Island, p. 31], which was a small island on *Loch na bhFaoileann* (which means the lake of the seagulls because so many cormorants used to nest on it that farmers used their droppings as fertiliser).
43 Robinson, *Gaelic Kingdom*, 7. A documentary film, *Tim Robinson: Connemara* (Harvest Films, 2011) provided Robinson with an opportunity to add further layers of interpretation and representation to his mapped and written landscapes. For more on the film, see Gladwin's essay in Chapter 4 of this volume.
44 Robinson, *Listening*, 12. See also Tim Robinson, 'The View from Errisbeg: Connemara and the Aran Islands', in Frank Mitchell (ed.), *The Book of the Irish Countryside* (Belfast: Blackstaff Press, 1987), 42–52.
45 Robinson, *Connemara: Gazetteer*, 121. See also Kathleen Villiers-Tuthill, *Alexander Nimmo and the Western District* (Clifden: Connemara Girl, 2006).
46 Robinson, *Companion to the Map of the Aran Islands*, 27.
47 Robinson, *Gaelic Kingdom*, 301. Local people were his main source of information: 'Sean was at the door of his house as I came by, and I turned in to thank him and to drink my tea. I asked him to check all the placenames I had scribbled on my map and make sure I had them all in the right positions, which he did most scrupulously'. See *Pilgrimage*, 119.
48 Robinson, *Connemara Gazetteer*, 43, 60, 67, 78. Elsewhere in Ireland naming has marked military setbacks involving Irish regiments, such as the battles of Spion Kop (Boer War), or Balaclava (Crimea) or Gallipoli (WW1). A marshy field in Tipperary was named Falklands by the farmer in 1982.
49 Robinson, *Pilgrimage*, 116.
50 Robinson, *Companion to the Map of the Aran Islands*, 38.
51 Robinson, *Listening*, 81.
52 Robinson, *Gaelic Kingdom*, 69.
53 Robinson, *Companion to the Map of the Aran Islands*, 19.

54 Robinson, *Companion to the Map of the Aran Islands*, 19.
55 Robinson, *Pilgrimage*, 60–1.
56 Robinson, *Pilgrimage*, 60–1.
57 Robinson, *Companion to the Map of the Aran Islands*, 34.
58 Robinson, *Connemara: Gazetteer*, 74, 75, 78, 81, 96.
59 Robinson, *Listening*, 100.
60 See, for example, *Fr Miley's Den* (1798), *Scailp Val* (1798) and incidents from the War of Independence and the Civil War. See also *Connemara Gazetteer*, 47, 48, 78.
61 Robert Cushman Murphy, 'The Timeless Arans: The Workaday World Lies Beyond the Horizon of Three Rocky Islets off the Irish Coast', *National Geographic Magazine* 59:6 (June 1931): 747–8.
62 Robinson, *Companion to the Map of the Aran Islands*, 84, 85.
63 Robinson, *Listening*, 26
64 Robinson, *Gaelic Kingdom*, 320.
65 Robinson, *Listening*, 74. In *Gaelic Kingdom*, 61, he refers to the 'lovely desolation of lakes and bog' in south Connemara.
66 Robinson, *Gaelic Kingdom*, 316.
67 Robinson, *Pilgrimage*, 47.
68 Robinson, *Labyrinth*, 437, 10.
69 Robinson, *Gaelic Kingdom*, 300–1. See also *Pilgrimage*, 12, on the endless dimensions of the step.
70 *A Twisty Journey: Mapping South Connemara* (Binn Éadair: Coiscéim, 2006) is a reprint of articles Robinson wrote for the *Connacht Tribune*. See Gladwin, 'The Literary Cartographic Impulse', 162 and 171.
71 Robinson, 'Connemara Fractal', 14.
72 Relph, *Place and Placelessness*, 143.
73 Diaries 1932–1995, Co. Monaghan, unpublished.
74 See Patrick Kavanagh's poem 'Epic', first published 1951: Peter Kavanagh (ed.), *The Complete Poems of Patrick Kavanagh* (New York: Peter Kavanagh Hand Press, 1972), 238.
75 Robert Lloyd Praeger, *The Way That I Went* (Dublin: Allen Figgis, 1969), 20.
76 Robinson, *A Twisty Journey*, x, and *Listening*, 21.
77 Rebecca Solnit, *A Book of Migrations: Some Passages in Ireland* (London: Verso, 1997), 61.
78 Rebecca Solnit, *Wanderlust: A History of Walking* (London: Verso, 2002), 9.
79 Robinson, *Gaelic Kingdom*, 297, 311, 294.
80 Robinson, *Listening*, 190
81 Robinson, *Pilgrimage*, 12.

# 2

# Catchments

## *John Elder*

Rain is general all over Connemara. In every month of the year and for parts, at least, of two days out of three, it swirls in over Roundstone Bog then drums down onto Errisbeg. Some of these spatters will wash due south, over the two-lobed peninsula of Goirtín and out to the sea, without relying on a fixed channel. Others, relinquishing their individual surface tension after being hung up in the sphagnum, will pool eastward over the sheep tracks that edge the bog, eventually arriving at the newly paved roads and driveways of an uninhabited housing estate; now they can hasten onward, past the studio of Folding Landscapes and into Roundstone Bay. But the drops trickling north and east from Errisbeg will gather into rivulets that trace a network of softly inscribed glens. These will become the tributaries for dozens of irregular lochs strewn across the bog's interior like puzzle pieces scattered on a table. Visible streams connect some of these bodies of water to one another, while arteries under the peat link the whole system into one soggy pulsation.

'Catchment' is the word by which Tim Robinson designates a unit of the Earth's surface bounded by higher edges and within which springs, rainfall, smaller tributaries converge, in most cases flowing onward to an outlet that joins it to a more broadly encompassing drainage. Every point on the Earth's surface is mapped in such a way by elevation and the movement of water. In *Listening to the Wind*, the first volume of his *Connemara* trilogy, Robinson further characterises a catchment as 'an open, self-renewing system, supporting and supported by a vast number of life-forms and all their interrelations. Even its basic topography, the most skeletal and reductive representation of its geometry, is profoundly suggestive of a way of looking at the world and caring for it.'[1] Across the seasons and over the decades, Robinson has walked the catchments near his home in Connemara (see Figure 7). Not only has he come to know them in intimate detail, but he has also tracked their confluence with the geology, language and history of western Ireland.

*Figure 7* Connemara topography (from Pat Collins's film *Tim Robinson: Connemara*, photo by Colm Hogan).

Since moving to Ireland almost four decades ago, Robinson has produced maps and gazetteers of the Aran Islands, the Burren's limestone labyrinth and Connemara. In 1986 and 1995 he published *Pilgrimage* and *Labyrinth*, respectively, the two parts of a project called *Stones of Aran* that has been widely acknowledged as a masterpiece. In 2011 he saw into print the third volume of his equally impressive and absorbing *Connemara*. Though Robinson is British (half Scottish, half English) by birth, his writing has long since been recognised as a treasure of contemporary Irish literature because of the questing intellect manifest within it, the vividness of his prose and the dignified, authentic and respectful manner in which he engages with the stories and landscape of his adopted home.

Beyond Robinson's specifically Irish impact, however, his extraordinarily localised work also holds great meaning for the inhabitants of other rural areas who are striving to discern and to convey the significance of their own home – landscapes. He both perpetuates and enlivens a tradition of Romantic revolt against centralised, mechanised and hierarchical views of humanity and the land. In an essay from *The Spirit of the Age*, William Hazlitt identified the genius of Wordsworth's early poetry as 'a proud humility', taking 'the commonest events and objects, as a test to prove that nature is always interesting from its inherent truth and beauty'.[2] In the rapt attention Robinson brings to 'the thousands of tiny trickles' in a catchment, he similarly asserts the significance of a boggy region located far from any capital, and at the extreme western edge not only of Ireland but also of Europe.

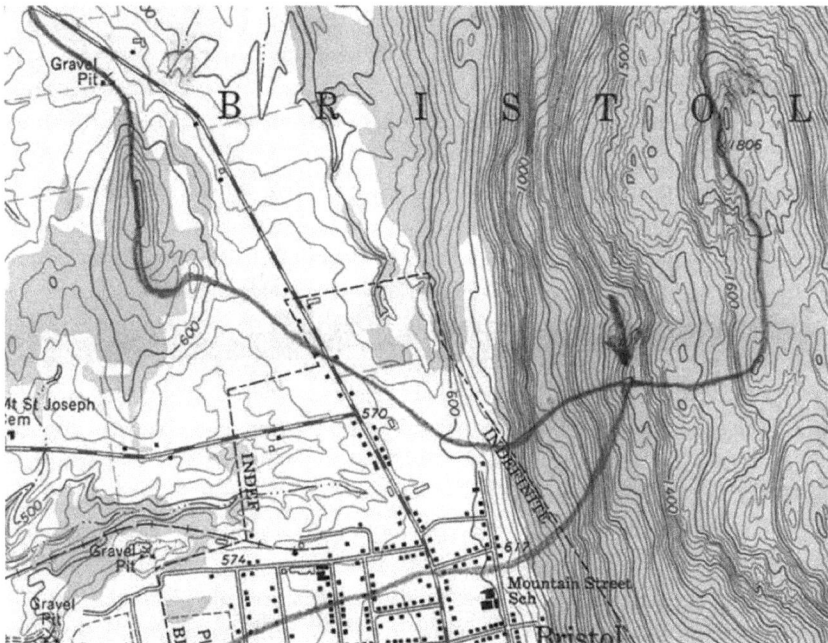

*Figure 8*  Map of Little Otter Creek in Bristol (map from USGS with markings by Elder).

As a fellow denizen of the hinterlands, though living far from Ireland, I feel personally grateful for Robinson's example of dogged and eccentric excellence at this ancient edge. By focusing on what the American poet and environmental activist Gary Snyder would call the lineaments of the land, rather than on political boundaries, Robinson invites his readers to lay their maps edge to edge with his, contributing to a collective effort of re-mapping and re-storying our homes on earth. The best way I can find to express my appreciation for Robinson's achievements is through a dialogue between my own landscape in the Green Mountains of Vermont in north-eastern United States and his in western Ireland. Specifically, I want to concentrate on the part of his Connemara map that depicts Roundstone Bog and on the corresponding portion of *Listening to the Wind*, relating them both to a landform near my home that is called the Hogback Ridge.

My US Geological Survey map of the Bristol Quadrangle is heavily seamed from so often being folded into backpacks or jacket pockets; its formerly white margin is soiled and splotched, its corners dog-eared (see Figure 8). The street grid of Bristol is shown in the south-eastern corner of the map. But its central feature is a dramatic and heavily forested upland called the Hogback Ridge running north to south and rising over a thousand feet above our village. Hogback is a finger of quartzite about ten miles long and between two and three miles across. Glaciation and erosion have stripped away the softer dolostones that once enclosed it within a higher plateau, sloughing them down to enrich the heavy clays to our west in the Champlain Valley. Hogback thus constitutes a resistant limb of bedrock and, although logged off several times over the past two centuries, has never been cultivated or settled. Our house on North Street stands less than a quarter mile from this roadless area that enlivens the village with its wildness.

In walking out our backdoor to an outlook called the Ledges, from which I can look straight down at our house as well as clear across Addison County to Lake Champlain, I pass under a band of broad-branched, open-grown trees where wild turkeys often roost. They venture out to glean on nearby farms in the months between harvest and the coming of the snow. In continuing my uphill hike, I enter thickening woods where evergreens are more prevalent. Deer trace their daily circuits here, bedding down under the hemlocks at night, while moose pass through on their way from the heights to the wetlands at the western edge of town. Farther up the slope comes a broken, rocky scramble of land where bobcats and black bears den.

Along with the various discolourations on its margin from tea and peanut-butter sandwiches held by fingers numb from the cold, there are three lines drawn onto my map in pencil. One of these traces the ridge of Hogback from the north to its southern terminus at a rugged cliff called Deer Leap, then angles sharply west through the village. All of the land east and south of this line drains into the New Haven River as it runs through Bristol toward its meeting with the northward flow of Otter Creek and its ultimate destination in Lake Champlain. Much of the water washing down to the north and west of the ridge will end up in Lewis Creek, travelling north through Hinesburg after its detour through Bristol Pond. On a cold afternoon in February, two and a half years ago, my neighbour David Brynn and I hiked up onto Hogback to figure out where the *rest* of the precipitation on the ridge ends up. We hoped to confirm that there was a fan-shaped wedge of land on Hogback's western slope that intervened between the New Haven and Lewis Creek drainages, ultimately conveying water due west along Plank Road into a tributary of the Little Otter Creek.

We climbed up onto Hogback from Mountain Street, passing through a tattery piece of wooded ground just to the south. Map in hand, we wanted to locate a particular little knoll we believed might be just above us to the east. While far short of the height of country, it seemed from the map to be a possible location for the headwater of the Little Otter. We had little success at first, since most of the surface water was frozen into taut skins of ice the size of drumheads, silent and motionless now until we crunched through them in our boots. At one point, though, as we stopped to listen to the world beyond our own motion we heard a bubbling. A patch of cobbles lay with the looseness of talus just below the place marked by an X on our map. In our part of Vermont, the temperature of groundwater holds at around 46° Fahrenheit (about 8° Celsius), so that springs can be discovered running in all seasons. This one spilled down into a descending seam of land, establishing a watercourse that did indeed run directly west.

Above that point of origin and division was a rugged granite erratic carpeted with common polypody, a robust and abundant fern in the Green Mountains. Kernels of snow from a recent storm nestled among its shrivelled but still green fronds. Just upslope from this landmark crossed an icy, treacherous road left over from a recent A. Johnson Company logging operation that had also left a hedge of slash along its track. The boulder we were concentrating on was about ten feet high and with steep sides. But we managed to scramble atop it. We'd tossed up a few rocks before making our grand ascent so that we could erect a commemorative cairn on a bare, level area of that ferny roof. Three large flat stones established a statuary base of narrowing circumference, on which I set a rounder and more massive rock. Then we saw that a long tooth of ledge was already waiting for us on the erratic. One side of this triangular piece was cut away in a curve that let it tuck up smoothly beside and over the top of that round stone. It held steady there, a slender tip flickering above the cairn like a grey candle flame.

After we had marvelled at this providential balancing act, we got out a jar of maple syrup we'd brought with us and dripped some of its dark viscous contents down the sides of that crowning flame. It collected onto the stones below but it also clung to and ambered the grainy surfaces of the vertical piece. We hoped that some rodent might happen upon this trove of calories and win the chance to survive for one more Vermont winter.

My Folding Landscapes map of Connemara, too, is far from pristine. Flecks of dried sphagnum tell the story of an outing I took into the Roundstone Bog when visiting Ireland last May. I wanted to pursue the topic of catchments through a comparison between Hogback and Roundstone Bog because these two largely roadless areas are similar in size as well as each being immediately adjacent to the centre of a village. One heavily forested, the other nearly treeless, they are alike in being easily explorable by classes from the local schools, by tourists who decide to set out for an afternoon hike after a slice of pizza at Cubber's in Bristol or a glass of cider and a crab sandwich at Roundstone House, or by writers coming back day after day in search of the big picture. One place has been preserved from development through its steepness and rockiness, the other through its standing water and mire, but both offer portals into the processes of self-renewal through which the meaning and potential of these rural communities too have evolved. I intended to identify the catchments circulating through Roundstone Bog, just as David and I had done when climbing up the eastern side of Hogback. What I found, however, was that so much open water, surrounded by immense stretches of saturated sphagnum, kept me from discerning any clear sequence of watersheds comparable to that three-part scheme of division in Vermont. At one point I attempted to find my bearings by reconnoitring away from a sheep trail I was following. Plunging up to my waist in the mire, I feared I'd meet the Tollund Man's Irish cousin face to face before finally managing to wallow back to firm ground. Though I reached no clarity about catchments that day, it was still a lovely outing, with harriers gliding overhead, the northern horizon crenellated by the range called the Twelve Pins, and graduated terraces cut into the trailside peat where turfs had been dug out in a way reminiscent of the staircase walls of Vermont's marble and granite quarries.

After many hours of poring over his map upon returning to Vermont, I finally resorted to emailing Tim Robinson to ask what *his* analysis of the Roundstone catchments was. He responded,

> Most of the area you mention drains through the Ballinaboy river to the NW, another good proportion of it through the Doohulla river to the SW, a marginal basin to the E is drained by a stream called Kelly's River (not named on the map) … Since the relief is so subdued over most of this area the margins of the catchments must be very vague; there are indeed substantial surface streams linking long series of lakes, but there are level boggy patches in which the direction of flow could be determined by tiny accidents and interventions, mere oozings, droolings and snivelings marshaled one way or the other by a cow's hoofmark or a snipe's egg.

This message was a useful reminder not to give way to droolings and snivelings myself the next time I ventured, map in hand, into bogland.

In comparison with a USGS topo map, Robinson's cartography has a strikingly open quality. There's no colour scheme of white for low ground, green for mountainous terrain and blue for water. And instead of the dark brown contour lines that swirl and gather when elevations quickly rise in the Green Mountains, Robinson indicates contours with fine broken lines. Rather than overlaying schematic information onto the landscape, his intention is to scour off the blurry surface of the Connemara palimpsest. He wants to rake away the topographic gravel deposited over the past century and a half so that the springs may run clear again and the holy wells may be revealed. One of the ways he manages this, having started out with the Ordnance Survey maps of Ireland, is to dispense with as many of the superimposed English place names as possible – substituting the Irish names which he has painstakingly recovered through a combination of bibliographic research and conversations with elderly, Gaelic-speaking neighbours. Names and stories, not simply rainfall, brings exactness to the map of Connemara's catchments.

A reader who unfolds a copy of Robinson's map while standing atop Errisbeg, the 917 foot high promontory that presides over the south-eastern edge of Roundstone Bog, will note that the map's primary identification for that peak is the much older name *Iorras Beag*, meaning 'small peninsula' in Irish. Such a return to the earlier name both establishes its historical primacy and suggests a characteristic and indigenous way of responding to the land. Mountains are not separate from and exalted over low places. They are instead simply the most visible manifestations of a larger area from which they derive a meaning and identity much more significant than the mere fact of elevation. Both the mountain and its southern skirts have been named, it seems, for the Goirtín Peninsula, which lies due south of Iorras Beag and south-east along the coast from Roundstone Bay.

In the *Gazetteer* Tim Robinson published to accompany his map, he elaborates on the peninsula's name, '*Goirtín*/small plot. A fine tombolo or neck of foraminiferous sand links this islet to the mainland; the deposits of seashells and the darker levels of soil visible in the faces of the eroded dunes here and on the tombolo represent ancient (perhaps Neolithic) settlement levels.'[3] For both the native Connemaran and the visitor just being introduced to Ireland and its ancient place names, the deeply historical character of Robinson's etymology has a strong appeal. As the dunes of imperial history erode, they also disclose. A midden of language is exposed beneath the sandy years, allowing a hiker who carries this map to imagine centuries of forerunners in the misty, windswept vastness of the bog, including some who made it their home long before all recorded histories of this turbulent place.

The Ordnance Survey maps with which Robinson began were highly accurate with respect to the scalloped indentation of coastlines, or to the location of towns and mountains. As Brian Friel's play *Translations* (1980) reveals, however, the creation of these maps in the early part of the nineteenth century saw the introduction of many misleading place names. Surveying parties were sent out under the leadership of officers in the Royal Engineers, to take stock of the empire's Hibernian real estate. But many of the names they came up with were either inept attempts to translate the Irish names' meanings or transliterations of their sound into English orthography – as understood by cartographers who generally knew no Gaelic. The overall effect was one of effacement, since the long-established Irish names had often carried with them mythical, historical and familial narratives that spun and wove a living web of stories in the land. In *Listening to the Wind*, Robinson never loses sight of the damage caused when such deep cultural continuity with the land is lost. As he says at one point, 'Irish placenames dry out when anglicised, like twigs snapped off from a tree.'[4] A place may become more vulnerable to exploitation when bereft of a deeply rooted name that pointed to its natural characteristics and recalled its history. Ecological and social withering can follow from linguistic desiccation.

Robinson's insistence on indigenous place names, in his writing and on his maps alike, has sent me back to the topo map for the Bristol Quadrangle with new eyes. There are in fact almost no place names of any kind associated with Hogback itself and its catchments. English names from the nearby towns of Bristol, Starksboro, Hinesburg and Monkton nibble around its edges, while the New Haven River is prominently labelled as it wraps around Deer Leap. But with the other exceptions of the Ledges and the tributaries flowing down into the Champlain Valley, the many plateaus, cliffs, boulder fields and vernal pools that are landmarks for hikers within Hogback's dramatically contorted terrain go nameless on the map.

Residents in this part of Vermont who are of European descent are impoverished by a general oblivion to the names and stories woven into the land over many centuries by the Western Abenaki, just as the original people's descendants here are disenfranchised by that ignorance. In contrast to other parts of North America where Indian names are prevalent and large local tribes persist, many Abenaki withdrew into Québec because of the frequent armed conflicts in the region during the latter part of the eighteenth century. Those who did remain, after European settlers arrived here around the beginning of the nineteenth century, encountered prejudice and hostility that often led them to disguise their identity.

I would not propose a complete parallel between the histories of indigenous naming in Vermont and Connemara, since the Irish have reclaimed control over their land and language while the Abenaki remain a politically marginalised group struggling for official recognition as a tribe by the state of Vermont. Still, Robinson's dedication to helping recover the earlier place names of Connemara convinces me that a deeper experience of affiliation with this north-eastern corner of Addison County will require learning as much as possible about the original Abenaki names for this landscape. A good place to start might be with the catchments on the slope above Bristol. In the work of an ethnographer named Gordon Day, I have recently discovered that Otter Creek was called *Pecunktuk*, or the Crooked River, in Abenaki, that the Little Otter was known as *Wonakakituk*, or the River of Otters, and that Lewis Creek was *Sungahnetuk*, the Fishing Place – or perhaps the River of Fish Weirs, with reference to stones placed there for use in such weirs. The name of Alaska's Mount McKinley, the highest peak in North America, has already reverted to its Native name of Denali, thus scraping away the detritus of a third-rate president's surname and restoring the grandeur of its original Koyukon title, the High One. But Robinson's example suggests the necessity for a sustained process of recovery extending even to more modest locales like Hogback – carried out with the patience required for tracing the thousands of rivulets in a catchment.

Such an effort of attentiveness to the testimony of elderly residents in suppressed communities, while essential to both greater groundedness and greater fairness, inevitably sounds an elegiac note. 'In noting such almost obliterated communal memories,' Robinson writes, 'I sometimes feel like a priest bending his ear to the mouth of a dying man to capture the profound and determinative sense of his last breath.'[5] The solemnity of this sentence is all the more striking coming from a writer who has, from as early on as his 1986 book *Pilgrimage*, asserted his identity as an atheist. As the self-comparison to a priest reveals, however, reverence is at the heart of Robinson's view of history and the land. Not reverence for a transcendental truth, but rather a compassionate and deeply respectful concern for how people worked and carried on in the face of a resistant environment and under the heel of rapacious invaders. When elegy reaches deep enough to touch the bedrock of history, it arrives at a tragic perspective on survival in the land and it comes to seem, in Robinson's words,

> wrong to treat the Famine as just a period. It was in fact the keystone in a triumphal arch of suffering ... There has never been a year in which it would have been appropriate to celebrate the end of the Famine; instead, it has been forgotten while the ragged edges of its shadow still lie around our feet.[6]

History, in a wounded landscape, can feel like a haunting – the embedded persistence of so much that has been lost.

The mapping of catchments and the tracking of tectonic plates as they drift across the mantle with their cargo of continents both require attention to the rocky surface of the Earth. Despite their striking differences of spatial and temporal scale, these are both physical phenomena whose evidence and effects are apparent everywhere. Robinson's knowledge of geology not only infuses his readings of the surface and history of our planet with drama but also provides a model for understanding shifts and continuities in the realms of ecology and culture, as well as in the pilgrimage he and his wife Máiréad have undertaken to Europe's ultimate headlands. While a number of environmental writers have integrated current geological models into their work, none, in my opinion, has done so more compellingly than Robinson. Here he is in a late chapter of *Listening to the Wind* entitled 'Walking the Skyline', as he describes a band of rocks near the Twelve Pins:

> These rocks are of schist and marble; that is, they are of clayey and limey materials originally deposited in layers on an ocean floor and later metamorphosed and upended by geological forces. The Dalradian ocean in which they were born once stretched from what is now the Shetlands to the west of Ireland, and it lay within a vast supercontinent that comprised most of the Earth's present land masses, for this was long before the Atlantic came into existence ... Eventually the stretching and rifting of the continental plate culminated in the birth of the Iapetus Ocean, the predecessor of the Atlantic, which lasted for some 100 million years and began to close up again around 510 million years ago. The Dalradian sedimentary rocks were caught up in the reunion of the continents; they were crumpled and torn, pushed down into the hot depths of the Earth and thrust up into mountain chains of Himalayan proportions.[7]

Just as every point on the surface of the Earth is situated within a specific local catchment, so too all parts of the landscape are borne along on tectonic plates that collide from epoch to epoch, raising new mountain ranges from deposits on the continental shelves that are collected, compressed and lifted up as one edge rises over another. In this regard, I was fascinated to discover another concrete connection between Ireland and Vermont when I set aside my Bristol topo and the Folding Landscapes *Connemara* in order to lay two bedrock maps of the regions side by side. They had been made up for me by Ray Coish, who teaches Geology at Middlebury College in Vermont. I took the rubber bands off the long, rolled-up maps and anchored them on my study floor with books at the corners so that I could stand above them and take a long view.

These maps revealed a remarkable correlation between the bedrocks of Connemara and the Green Mountains. The intricate mix of rocks in the Dalradian schist Robinson describes is closely replicated in the composition of schists in the mountains of Vermont. Such a similarity is neither a simple coincidence nor a common sort of parallel. Instead, it reflects a shared geological origin. Just to the east of present-day Lake Champlain (which defines the western boundary of Vermont for about half of our state's 159 mile length) there once lay a continental shelf. When the Iapetus Ocean began to close as Robinson describes, it folded together and compressed the sediments on that shelf, raising them into mountains that were in fact the much higher ancestors of our present-day Green Mountains and the Appalachian chain to which they belong. They rose near the middle of the great supercontinent called Pangaea.

When Pangaea began to break up and the Atlantic Ocean opened, approximately 130 million years ago, the new line of separation lay to the east of the earlier shore. The Appalachians were pulled apart, with one remaining ridge running from Newfoundland through New England and the Carolinas to Georgia. Other parts of the range migrated eastward with the European plate. Millions of years of erosion, including major episodes of glaciation, have effaced much of the mountain chain in Europe, while also lowering it significantly in North America. But the remaining bedrock shows remarkable similarities between Connemara and the Green Mountains, both in overall composition and in the specific, intricate sequence of mineralogical bands that organises the two landscapes.

Just before Pangaea began disassembling itself into separate continents and drifting apart, Newfoundland and Ireland were directly attached to each other. Their bedrock maps are thus absolutely identical. But the geological parallels between Vermont and Connemara also remain quite striking, except in one regard. On the geological map of Ireland, the schist in the centre of Connemara is composed of east–west bands of the various Dalradian Appin and Argyll groups, as well as of the Killary-Joyces Succession; these are represented on my map by a bracing shuffle of emerald, fuchsia, pink, brown and violet. But the bands in Vermont, which exactly align with the orientation of the Green Mountains, appear as north–south on the map. When the Iapetus was closing to form the Appalachians, the coastline of the North American, or Laurentian, plate curved dramatically, which led to a similar change of orientation in the mountain chain formed along its shelf. Thus, today's Appalachian bedrock, after heading north and then north-east through New England and Newfoundland, turns almost due east in parts of Ireland, Wales and England, before resuming a north-eastern course into Scotland and Norway. Portions of the original mountain system also ended up in the Netherlands, coastal France and north-western Spain.

One reason for my fascination with this family resemblance, geologically speaking, is a growing interest in Irish music over the past decade. My life, from birth in Kentucky to adulthood in Vermont, has involved a progression up the Appalachian spine. The fretless banjo eventually gave way for me to playing Irish flute. Even before observing those colourful Appalachian bedrocks splashed into the maps of Ireland and Vermont, I was struck by the strong continuities between Celtic musical traditions and those of the southern Appalachians and New England. (The subject has of course been much written about.) Tunes like 'Blackberry Blossom' and 'Langstrom's Pony' have been played on both sides of the Atlantic since the middle of the eighteenth century. In another intriguing connection to their identical bedrock, Newfoundland is often described as the most Irish place outside of Ireland, as well as being the only one with its own name in Irish – *Talamh an Éisc*. When I asked Ray Coish, a native of Newfoundland, if there was much Irish music played where his family came from he said, 'Well, we call it Newfoundland music.'

It's intriguing to contemplate such musical continuity within a tradition that, while not confined to any one country, is strongly associated with certain rocky landscapes that resist easy cultivation. A haunting heritage is transmitted through modal structures that can madden amiable guitar players who try to join in with their standard chord progressions. Right up through such Celtic offspring as bluegrass, there's a keening quality in the music, as melodies break across ancient droning shores. For the growing number of Celtic music's devotees who live far from the Appalachian arch, including many in Eastern Europe, Russia and Japan, it also seems to express a longing for values too often neglected in the modern conversation of nations. Though no one would settle in the rocky, depopulated reaches of Connemara or the Green Mountains out of a desire to get rich, many who have been 'in populous city pent' long for the strategic retreat to rootedness and away from consumerism that such places seem to represent. For them, tunes that don't fit meekly into a standard key, instead waywardly toggling between C natural and C sharp, offered a liberating contrast. The bedrocks of the Champlain Valley Belt, Green Mountain Belt and Rowe-Hawley Belt that flow from the chambered heart of Connemara down the length of Vermont have no causal relation to the fiddle, flute, banjo, concertina and pipe music connecting our two regions. But the similar tunes played in both are nonetheless tokens of something more essential to our sense of place than the products lining the shelves of big-box stores and the sizzling urgency of news flashes on the television or the computer.

In chasing this musical rabbit, I've once more veered off the page of Robinson's majestic and highly localised precision. Despite the melodic cadences and fricative rhythms that bring such music to his prose, there's no getting around this warning lodged near the beginning of the Preface: 'History has rhythms, tunes and even harmonies; but the sound of the past is an agonistic multiplicity. Sometimes, rarely, a scrap of a voice can be caught from the universal damage, but it may only be an artefact of the imagination, a confection of rumours.'[8] How easily a reflection can puff up into a confection. Indeed, my earlier remark that Robinson's writing invites inhabitants of other hinterlands to lay their maps edge to edge with his was not quite right either. The vitality of his book is like that of a catchment. It is a system to be observed, walked around and returned to; neither a doorway nor a staircase that might lead away from Connemara. Yet for all that, his fierce attention to overlooked places that have temporarily lost their names remains 'profoundly suggestive of a way of looking at the world and caring for it'.[9] Time to brush last spring's sphagnum off the map and climb back up Errisbeg.

## Notes

1. Tim Robinson, *Connemara: Listening to the Wind* (Dublin: Penguin, 2006), 273.
2. William Hazlitt, 'Mr. Wordsworth', *The Spirit of the Age* [1825] (Whitefish, MT: Kessinger Publishing LLC, 2004), 93.
3. Tim Robinson, *Connemara Part 1: Introduction and Gazetteer* (Roundstone, IRE: Folding Landscapes, 1990), 80–1.
4. Robinson, *Listening*, 81
5. Robinson, *Listening*, 120–1.
6. Robinson, *Listening*, 201.
7. Robinson, *Listening*, 364.
8. Robinson, *Listening*, 3.
9. Robinson, *Listening*, 273.

# 3
# 'The fineness of things': the deep mapping projects of Tim Robinson's art and writings, 1969–72

## *Nessa Cronin*

> But if it is true that Time began, it is clear that nothing else has begun since, that every apparent beginning is a stage in an elder process. The compass rose that unfurled about me in Aran, I now discover, had its stem in London.[1]
> 
> – Tim Robinson

Tim Robinson's work has become a touchstone for those interested in, and concerned with, the changing nature of the modern Irish landscape. In particular, the production of the maps of *The Burren* (1977; 1999), *Oileáin Árann/A Map of the Aran Islands, County Galway* (1980; 1996) and *Connemara* (1990), along with his books on the west of Ireland, have established Robinson as one of the foremost writers, cartographers and thinkers of the Irish landscape over the last forty years. While he is primarily known for his contributions to Ireland's cartographic and cultural heritages, his earlier work as the visual artist Timothy Drever is less known. In November 1972 Robinson famously left 'the visual for the verbal' and departed the London art scene for the west of Ireland with his partner Máiréad.[2] In his autobiographical essay collection *My Time in Space*, he notes,

> The move from city to island, from the visual arts to literature, from minimalist abstraction to the most scrupulous cartography of the grain of the actual – had even that drastic step not been enough to shake up my little store of conceptions? ... To the artist it is intolerable that one cannot climb to one's own horizon and look beyond.[3]

If one wished to proffer an initial assessment in terms of the trajectory of his work from the late 1960s to 2012, one could argue that while the conceptual vocabulary stayed the same (a vocabulary dominated by concerns of space, time, dwelling and belonging), all that changed was the vehicle for such concerns. While Robinson's non-figurative artworks and environmental installations dealt with mathematical and existential questions concerning the limits of knowledge, his cartographic and literary works retreat to the opposite end of the global scale with a fascination for the fractal geometries of the Polish-French mathematician Benoît Mandelbrot.[4] After 1972 Robinson's 'site of enquiry'[5] was no longer placed out in a Camden art gallery nor contested on London's Hampstead Heath, as after that point the west of Ireland would become the home for his maps and topographical writings.[6]

In 1997 Robinson revisited some of the artworks and installations from his previous life as a visual artist for a group exhibition of European artists entitled The Event Horizon, curated by Michael Tarantino of the Irish Museum of Modern Art (IMMA).[7] Robinson's installation for the show was entitled *The View from the Horizon*, a title which in many ways played with the impossibility of such a perspective, and which also marked, conversely, a view from nowhere. He remarked that the title therefore 'hinted at my unease as to whether any of this work would do at all'.[8] In revisiting the 'resurrected' artworks from the late 1960s and early 1970s, a 'network of imagery' become more obvious with images of 'gravity and rainbows, of compass roses and cardinal positions, of the birth of the universe and the moment of vision' dominating his creative imagination and persona.[9] The problem for Robinson was that he initially thought such images had originated in his map-making activities and writings in and on Ireland, from what he termed his 'geophanic years'.[10] However, with the vantage point of The Event Horizon he now realised that there was a much longer historical and biographical trajectory embedded within the 'suite of images', in that his earlier artworks contained themes and concerns that would later find expression in his Irish writings and map-making practice.[11]

This chapter is an *essai* in tracing the 'pre-history' of images and concerns that would preoccupy Robinson across the two distinct phases of his career: from his life as a visual artist in London and Europe, to the cartographic narratives of the west of Ireland. It argues that when looked at as an oeuvre, Robinson's work can be regarded as a 'deep mapping' practice concerned with 'the fineness of things',[12] where a deep map is understood as being an attempt 'to record and represent the grain and patina of place', outlining an attempt to say 'everything you might ever want to say about a place'.[13] With a focus on two sets of artworks from his early career (*Moonfield* and *To the Centre*) and short prose writings and essays in *The View from the Horizon, Setting Foot on the Shores of Connemara* and *My Time in Space*, this chapter explores the fragmentary connection that Robinson draws between these two stages of his life and career.

### Drawing, mapping, writing the landscape

When considering the range and depth of Robinson's oeuvre across these three art forms (visual arts, cartography, writing), and the profile of his work to both a national and international audience, surprisingly little has been published on Robinson by academic scholars, with almost all critical scholarship focused on his prose writings on Ireland.[14] In many ways, the literary quality of his work has drawn attention to his engagement with a distinct anglophone tradition of nature writing and topographical literature, which has arguably overshadowed other aspects of his work. Jos Smith's 'Step towards the Centre of the Earth: Interview with Tim Robinson' moves from a discussion of Robinson's visual artwork to a discussion about his cartography and writings in Ireland. While Smith's questions are perceptive and show a deep understanding of Robinson's work, as an interview that seeks to cover the range of Robinson's work for the general reader it does not focus on any one particular aspect of his work (which is understandable considering the format chosen).[15] Indeed, the only work that has attempted to merge the work of Timothy Drever and Tim Robinson is the booklet *The View from the Horizon*, which was written by Robinson himself to coincide with the 1997 IMMA exhibition.[16]

Eamonn Wall notes the multi-dimensional aspects of Robinson's work and states that 'all three arts are informed by his earlier practice as a visual artist', but his focus is primarily on Robinson's prose writings.[17] For Wall, literary language, 'for all of its limitations, is the mode best suited to an intense examination that seeks to reach for the totality of place'.[18] While Wall draws on his reading of travel writer and historian William Least Heat-Moon to briefly explore the concept of 'deep mapping', his attention is more on an analysis of the relationship between walking and writing in Robinson and is less focused on developing the concept of deep mapping per se.[19] Cultural geographer John Wylie examines the 'creative tension of land and life' in Robinson's earlier texts on Árainn and Connemara to argue that 'a displacement of land and life from each other, a displacement of dwelling, is in actuality the incessant precondition of landscape'.[20] While Wylie acknowledges Robinson's visual art origins and cartographic work, his primary focus remains on habits of dwelling and a phenomenological critique of landscape writing. Wylie's interest in Robinson's writing, however, does not extend to Robinson's mapping practice, which, as I argue later in this chapter, is a sustained but *provisional* exercise rather than what social scientists and cartographers call 'ground-truthing'.[21]

One critical commentator who is alert to the questions of culture and translation, language and colonialism is Michael Cronin in his brief (yet insightful) discussion of Robinson's work in his essay, 'Inside Out: Time and Place in Global Ireland'.[22] Cronin argues that Robinson seeks to 'restore the infinite complexity' of the places under scrutiny, and that his work is 'linked to the rehabilitation of dwelling as a creative or enabling way of engaging with places subject to the peripheralising dismissal of velocity'.[23] While Cronin does not allude to Mandelbrot here, a connection between the project of restoring the 'infinite complexity' of the world is worth considering alongside Robinson's fascination with fractals. In fractal geometry, small spaces replay out their particular geometric composition time and again so that measurement becomes potentially infinite. Cronin focuses on the question of time, speed and the rhythm of the *pace* of Robinson's work. He argues that Robinson's writing forces the reader *to slow down*, and to engage with his 'decelerated practice of walking the fields of Inishmore'.[24] The danger seems to be that with speed, one can slip off the surface and lose one's foothold and grip on the world all too easily.[25] This interpretation of Robinson is later included and framed within a larger context in *The Expanding World: Towards a Politics of Microspection*, which argues for the need (ecologically and politically) for an in-depth analysis and understanding of the local in terms of reframing our relationship with global space and time, in terms of a 'micro-modernity' of the global village.[26] The relationship between speed (and to be more accurate, slowness), knowledge and cartographic writing will be explored later in my discussion of what 'makes' a deep map.

Cronin's argument in *The Expanding World*, as described by Eóin Flannery, is for a 'renewed ecological ethic', an ethic which does not privilege legal ownership but 'is underwritten by the longer-term human and non-human histories of local places'.[27] The micro-modernity and micro-politics of Cronin are also evidenced in Robinson's previous claim for 'a local micro-geography' enabled in part through the capturing of placelore.[28] The challenge felt by both Cronin and Robinson is the question of how to counter the perceived collapse of space and the fetishisation of speed when we are embedded within the geopolitics of transnational capital and locked within the logic of late modernity.[29] Flannery's own discussion of Robinson repays close rereading as it points to a significant ecocritical shift in an Irish context. Flannery recognises that Robinson's writings elucidate how 'our very modes of perception, our temperaments, are remoulded and distorted by the white heat of economic modernization and financial profiteering'.[30] He notes, again commenting on the impact of Robinson's work, that 'the actual physical assault on the landscape is merely a symptom and is not the underlying disease'.[31] One theoretical gap recognised here is the significance of a particular colonial legacy (Flannery's 'underlying disease'?) and its next generation postcolonial 'hangover' that in many ways still shapes and informs landscape management and related public policy in Ireland today. While there are rhetorical gestures toward the concept of a 'sense of place' as being critically important to Irish culture in geographical and literary writings on Ireland (with examples of the role of the Gaelic Athletic Association often abounding in such analyses), the question as to the cultivation and care of the Irish landscape is one that is less easily posed, let alone answered.

When one examines the intellectual history and conceptual arc of the Drever/Robinson corpus of images, maps and texts as a whole, a number of key issues begin to emerge – concerns that can be traced back to his early interest in mathematics and his subsequent engagement with the art worlds of Istanbul, Vienna and London, and from there to his later career as a map-maker, observer and writer of the Irish landscape. As noted, the question of the relationship between Robinson's different work (as artist, cartographer, writer) evolves into a discussion of an 'ecological ethic' over the course of his work on the Irish landscape. What draws these two elements together I would argue is what has been termed elsewhere as a 'deep map', where a deep map is both process and product of a deep engagement with a particular place over a period of time. In the following section, I unpack the concept of deep mapping as a means to understand all three of Robinson's artistic projects.

## Deep mapping: 'a conversation not a statement'

The concept of a deep map is often attributed to Wallace Stegner's *Wolf Willow: A History, A Story* (1955), and is more fully developed in William Least Heat-Moon's *PrairyErth: A Deep Map* (1991), which is an experiential account of the people and place of Chase County, Kansas. In *Desert Notes: Reflections in the Eye of a Raven* (1976), Barry Lopez warns against creating 'the wrong sort of map', of maps that are too 'thin':

> Your confidence in these finely etched maps is understandable, for at first glance they seem excellent, the best a man is capable of; but your confidence is misplaced. Throw them out. They are the wrong sort of map. They are too thin. They are not the sort of map that can be followed by a man who knows what he is doing – the coyote, even the crow, would regard them with suspicion.[32]

The 'thinness' attributed to the positivist 'finely etched maps' of traditional cartographic practice also has echoes of Clifford Geertz's argument for critical acknowledgement of the 'thick descriptions' of culture and society in anthropological contexts and settings.[33] Lopez foregrounds the varied and contested use-values for which maps are made and commandeered: notwithstanding the implicit and problematic gendering of such cartographic endeavours, maps that 'work' are ones that 'can be followed by a man who knows what he is doing'. While the concept is associated with Stegner and Least Heat-Moon, the development of a critical definition of deep mapping as a critical tool of analysis is claimed by archaeologist Michael Shanks and theatre practitioner Mike Pearson. In their co-authored work, *Theatre/Archaeology*, Shanks and Pearson argue that they regard the eighteenth-century gazetteer to be an early instance of their understanding of deep mapping:

> the deep map attempts to record and represent the grain and patina of place through juxtapositions and interpenetrations of the historical and the contemporary, the political and the poetic, the discursive and the sensual; the conflation of oral testimony, anthology, memoir, biography, natural history and everything you might ever want to say about a place.[34]

A more recent and comprehensive account of the history of deep mapping is offered by the visual artist and scholar Iain Biggs, where he notes that the term deep mapping with regards to its origins and praxis has now become associated with two distinct types of place-based practice. In North America (and in environmental circles) deep mapping usually refers to 'an environmentally oriented literature (which may extend into radio and photo essay) dealing exhaustively with a local or regional site and often linked to "vertical" or "deep" travel writing'.[35] In Britain, and in performance and archaeological circles internationally, Biggs observes that the term primarily refers to 'a site-based performance practice – known as "theatre/archaeology" or "performance archaeology" – originating with Mike Pearson, Michael Shanks, Clifford McLucas, and the radical Welsh performance group *Brith Gof*.[36] McLucas (1945–2002) would go on to develop a working 'manifesto' as to what he called 'deep maps' that emerged from his working practice as an artist, scenographer and film-maker in Wales and the US from the 1980s until his untimely death in 2002.[37] 'There are ten things that I can say about these deep maps,' McLucas states, adding that these ten necessary components shape and inform each other in the overall imagining, process and 'production' of the work/map. Therefore, deep maps are: big; slow; sumptuous; encompass a variety of media; require an engagement of the 'insider and the outsider'; bring together the amateur and the professional; may only be possible/imaginable now; are 'politicized, passionate, and partisan'; will be 'unstable, fragile and temporary'. In elaborating on the last point he notes that deep maps 'will be a conversation and not a statement'.[38]

As Biggs notes, deep mapping is ultimately concerned with the 'specifics of place'.[39] In outlining a critique of the origins and ideological impulse behind current deep mapping practice in the UK and North America, Biggs is also concerned that institutional (e.g. university) and art-curatorial settings may disable and close down deep mapping conversations if disciplinary expertise and privilege, institutional surveillance and the vagaries of national and international funding become the priorities of these processes. He argues for the need to 'preserve "open" deep mapping' in order to build 'strategic alliances between institutionally based and freelance practitioners'. Such an open practice will have an 'expanded account of deep mapping that stresses its *varied, provisional*, and *inclusive* (indeed protean) nature to help negotiate such alliances'.[40] In many ways, this is an argument to extend the academic understanding of what 'research' and 'practice' actually are as distinct/co-joined categories, and to move to what cultural geographer and ethnographer Karen E. Till describes as 'more socially responsible research practices' in the arts and social sciences in general.[41] This is, in essence, an argument for a deep mapping practice to be kept 'open', an argument for a reflexive and transdisciplinary research praxis,[42] where such research takes place across and beyond disciplinary and institutional borders to incorporate expertise from non-academic settings, groups and individuals in a socially responsive and responsible way.[43] Deep mapping is therefore a subjective and partisan project in counter-cartography, a 'multi-constituency' rather than 'multi-disciplinary'[44] approach that may do much to undo the often arbitrary silos of knowledge and institutional culture in academic settings.[45] Such an imperative to maintain openness closely aligns itself to McLucas's recommendation that deep maps are about conversations, not statements.

In summary, deep mapping is simultaneously about recording and representing, complementarity and contestation, process and product, and is primarily concerned about understanding our place in the world through the lens of personal experience. It is not just about the uncovering of the *longue durée* of archaeological meaning or geological time, but the act of understanding how all those layers connect (horizontally and laterally) and create a meaningful engagement with, and experience of, place. While deep mapping as a concept can aid the cartographic and creative imagination in terms of different ways of representing place (and as such it has proved to be a powerful creative platform for socially engaged artists in particular), as a critical tool it offers an alternative spatial vocabulary, as an enabling narrative that can create spaces of resistance that disrupt and challenge our received ways of thinking about space, place and belonging.[46] Robinson's Irish work in particular can therefore be read as a slow, longitudinal project in the deep mapping of 'the fineness of things'. It is also concerned with what I have described elsewhere as 'lived and learned landscapes',[47] and the intersections between 'official', state-produced or sanctioned, knowledge of place (e.g. the institutional 'Archive', scholarly texts, translations) and the knowledge associated with the complex and fractal-like lifeworlds of 'ordinary', everyday experience (e.g. fragments of auto-ethnography, folklore, song).

## 'A Career in Art' (1964–69)

In 'A Career in Art', published in the essay collection *My Time in Space*, Robinson writes of his first encounter with art through his teacher Miss Heaps, 'a sweet-tempered bun-shaped Ilkley lady'. It was, however, his later experience in grammar school (shaped by his teacher Tommy Walker) that let him to read Mathematics at university while dedicating the rest of his time, untutored, to painting.[48] After completing his degree at Cambridge, Robinson settled as a teacher in Istanbul for three years with his partner Máiréad, and it was during this time that he assumed his mother's maiden name, and adopted Timothy Drever (rather than the more 'prosaic Robinson') as his artistic pseudonym.[49] A summer travelling through Dalmatia and northern Italy led the Robinsons to settle in Vienna, and the artwork from this period took on the sombre mood of the city, which was then a 'Cold War city unstably encamped on the ruins of its recent past'. His works 'took a turn into nightmare' where 'atomic bombers flew in at the window, skeletalised birds fell through a lethal sky, monstrous creatures crawled in the sewers of towering "cities in a vacuum"'.[50] Robinson's first exhibition as Timothy Drever was in a gallery owned by the 'fantastic realist' Ernst Fuchs, who was according to current legend 'subject to visitations from angels who periodically commanded him to chastity, and his minutely executed painting revelled in a sex-haunted religiosity'.[51] A review of this exhibition in the *Wiener Zeitung* praised Drever's 'sleepwalking surefootedness', a phrase he writes that he has 'found comforting since', and is another element of the 'suite of images' that would later resonate in his Irish writings.[52]

In 1964 the Robinsons then moved to London where Robinson commenced in earnest his London art career, financially aided by his work as a freelance illustrator of technical diagrams. One of his ongoing series *The Dreams of Euclid* was accepted for the competitive John Moores Biennial in Liverpool, a positive review of which lead to his being invited by Guy Brett (Signals Gallery, London and art critic of *The Times*) to participate in a group show called Soundings. After this he began working on artworks as 'multiples', which involved the production of artworks in 'identical instances' that also bore the hallmarks of the 'numbered, limited edition'.[53] The concept was 'seen as a blow against the capitalist art world; economies of scale and industrial production processes would place these works in the financial reach of anyone who could afford a beer'.[54] Drever's multiples were two pieces in strip-cartoon format called 'The Theory and Practice of Dreams', which referenced the 'sleepwalking surefootedness' of his Vienna period.[55] During this time, he also got involved in public demonstrations relating to the arts, one of which was a protest campaign against a Conservative Party proposal to charge an entry fee into public galleries and museums.[56]

## Transitions I (1969–71): *Moonfield*

In 1969 Drever co-authored an article with Peter Joseph in *Studio International* entitled 'Outside the Gallery System'. The article was, he writes, partly a manifesto and partly a 'prelude to the installation of two outdoor works in the grounds of the Iveagh Bequest at Kenwood on Hampstead Heath'.[57] Both artists were moving away from traditional paintings on canvas at that time as 'we wanted to use wider dimensions that would implicate the viewer's or participant's own location and movement' while also foregrounding intentionality and accident.[58] This is the year in which the 'suite of images' that would prefigure much of his subsequent visual and literary works on Ireland would come to be shaped and formulated. Joseph's work consisted of three flat discs (approximately eight feet in diameter) propped up against trees in an open area across the landscape. They were, as he would later note, 'signals that conveyed nothing except their own position'.[59] Drever's contribution was entitled *Four-Colour Theorem*, after the then unsolved mathematical problem in topology. This was a 'collection of about a hundred pieces of hardboard, from eighteen inches to four feet across, of four different shapes and colours, laid out on a large lawn in a walled garden, to be rearranged at will by whomsoever chose to engage with them'.[60] The exhibition attracted a lot of attention, both by critics and the general public. However, after one weekend, they were to find that a somewhat less discerning audience had partly destroyed the installations and that 'one of Peter's discs had vanished from the glades and my theorem had been sacked by skinheads and thoroughly disproven'.[61]

Later that year, both artists retreated to 'the safety of "the gallery system"' and held a joint show called New Space at the Camden Arts Centre. Here, Drever exhibited *Moonfield*, which was based on the same shapes as *Four-Coloured Theorem*,

*Figure 9*   *Moonfield* (Timothy Drever, Camden Arts Centre, London, 1969).

but this time the shapes were now painted conversely black on one side and white on the other (see Figure 9). *Moonfield* was, he acknowledges, 'inspired by those almost indecipherable black-and-white TV images of the first moon-landing'.[62] The monochrome lens of his Viennese artworks was being revisited. While Joseph's artwork (a seventy-foot-long wall of yellow painted canvas, bisecting the room) achieved the rare honour of 'being denounced by George Steiner for its dumb hostility to language', Robinson retrospectively sees *Moonfield* as both the 'zenith and moonset of my brief arc across the skies of modern art'.[63] As a highpoint in his career it marked the realisation of many of his spatialised, mathematical ideas of space, time and agency, and as a 'moonset' it was to mark the end of one phase of his work, and the beginnings of a transitional period that would end with a move to the west of Ireland.

In February 1971 Drever published an article entitled 'Field Work 3: A Structured Arena' in the *Bulletin of the Computer Arts Society*.[64] The dual aims of the Computer Arts Society (based in London, with a European branch in Amsterdam) were to 'encourage the creative use of computers in the arts and allow the exchange of information in this area'.[65] In the opening paragraph to 'Field Work 3', Drever notes that upon considering the impact of two environmental works he had recently exhibited (which are here unreferenced, but most probably refer to *Four-Colour Theorem* and *Moonfield*) he 'became less interested in the transient results of their [his audience's] activities, and more aware of the way in which those activities themselves were patterned by the structure I had provided'. He goes on to clarify that he stopped thinking of these installation works as 'participational', and began instead to see 'the actions of those involved as the object of my activity as an artist'. This then led him to think of creating other spaces/contexts 'which would impose certain rhythms on anything taking place within them, and on the consciousness of anyone entering them'.[66] His concerns with the experiential nature of occupying or inhabiting space, and the question of spatial agency, were beginning to be choreographed and take shape.

One future project that was envisaged in this article was a 'concrete floor of regularly-spaced shallow waves', with the floor area enough for 'a specific rhythm to be generated by the act of walking across it'. In many ways this project would prefigure his own ideas concerning the ethics and aesthetics of the 'adequate step', harnessed from his experiences of walking across the landscape and developed (as I argue here) as a 'deep map' in his later writings.[67] In addition to the 'wavy floor', specific black-and-white rods would make an appearance over and on the floor area with 'a scattering of rigid, fragile, "measuring rods"' which 'would change its character'. He also suggested that footballs could be bounced on it 'unpredictably', and that musical, dance or theatre groups could 'let the rhythms of their own activities interact with its periodic structure'. The main interest in this piece lay, however, in the '"interference" of the floor's stable and coherent wave-structure, with the unstable and fluctuating forms of action super-imposed upon it'.[68] In conclusion, Drever considers that it would 'also be good to see the system set up in a wave-tank, or if possible as a computer graphic display'.[69] After the move to their Connemara home in Roundstone from Aran – called Nimmo House – Robinson recreated this idea of the 'rectangular array of waves' in a symmetrical pattern of bricks laid out, chessboard-like, in the back garden of their home and has since discussed the possibility of a modern dance work to be choreographed on the site (see Figure 10).[70]

### Transitions II (1971–72): *To the Centre*

The title page of *The View from the Horizon* foregrounds the dual 'authorship' of the book; it states that the book is comprised of 'constructions by Timothy Drever, 1972' and 'texts and maps by Tim Robinson, 1975–96'. In the Introduction he writes:

> The question became sharp for me recently as I approached the contemplation of the body of texts and maps I would have claimed had been inspired by my encounter with the west of Ireland, because in trying to forsee what I might do next I mentally revisited that earlier time of change, unwrapped some artworks stored away from my last year in London before the transition date of November 1972 – and discovered in them a concentrated abstract of the suite of images that has controlled my subsequent writing and is implicit in my cartography.[71]

The formative 'suite of images' which found a concentrated, abstracted form in his environmental installations would now find a home in the world of maps and

*Figure 10*   Garden at Nimmo House, Roundstone, Co. Galway (photo by Nessa Cronin).

literature. *Moonfield* was a monochrome, environmental art installation. This early series of wooden rods of differing lengths and monochrome markings would signify for Robinson a pull toward the centre of the earth, marking the first register of the concept of the step. This step, of connecting to the land, the past and the future, is a concept that would strongly figure in his future aesthetic of walking, and keeping a foothold, on the landscape in his subsequent writings.

In the three-part series collectively referred to here as *To the Centre* (which comprises the three works *To the Centre*, *Autobiography* and *Inchworm*), Robinson again used the medium of monochrome measuring rods. He describes them as 'a collection of wooden rods of lengths from about three to eight feet and of thicknesses from a quarter of an inch to two inches or so, painted in black and white bands in various combinations of width'. *To the Centre* was 'the principal piece of the three' and it comprised a 'slender, yard-long, white-painted wooden rod, suspended vertically from a large number of fine threads of many colours, which radiate from the centre of its upper end and are pinned almost randomly to the ceiling or the upper parts of the walls and furnishings of whatever space it is mounted in' (see Figure 11).[72] When it was housed in his studio in London, it became almost totemic in that he regarded it as 'a pace taken towards the centre of the earth', and as such it can be taken as a foundational image for Robinson's entire corpus of work.[73] In

*Figure 11* To the Centre (IMMA Installation, 1997, from *The View from the Horizon*, Robinson/Drever, 1997).

many ways these works were also strangely prescient of the surveyor's measuring rod that would inform his cartographic work and imagination in Ireland.

In an explanatory gesture to the reader/viewer as to the title of the second work, entitled *Autobiography*, Robinson notes that it reminded him of 'my grandmother's little set of ivory spillikins'. An art critic friend wondered if they were 'measure become organic', and Robinson goes on to surmise that they might be seen as 'growth-stages in the life of a measuring rod'.[74] Of the third work, *Inchworm*, Robinson writes:

> I have no memory of making it and was surprised when I found it bundled up with the others. It comprises thirty-five thin white wooden rods each thirty-five inches long. On each one a different inch-long segment is picked out in grey, so that if they are laid side by side in a certain order the grey inch appears to progress regularly from one end to the other, but of course this symmetry is broken when they are dropped and scattered. None of these works had names so far as I can remember, but for present purposes I have called them, respectively, 'To the Centre', 'Autobiography', and 'Inchworm'.[75]

*Inchworm* is the common name given to the *Geometridae* moth in North America, the caterpillars of which have an 'earth-measuring gait' with 'twiglike bodies with two pairs of claspers like short soft legs at the rear'.[76] He describes the caterpillars as exuding 'a viscous substance from spinnerets on their faces that instantly dries into a silk thread and acts as a safety line, at least when they are still small and light'. The caterpillar, so suspended, hangs 'before one's face like a little question- or exclamation-mark, before beginning to climb up again, laboriously twisting the front of its body from side to side and apparently packing the thread among its front legs'.[77] His observations of the caterpillars at his home in London became 'deeply curative' vigils but 'of what, I cannot quite say'.[78] As creatures marking a transitional state, it is fitting that Robinson should be drawn to the lifecycle of the caterpillar in this period in his life.[79] Significantly, it was during this period that the decision to leave London for Ireland came to take shape, as he writes, 'I believe it was also in this period that M and I formed the decision to abandon London and go to the Aran Islands to wait upon our own futures.'[80]

### Departure/arrival: Oileáin Árann, 1972

The exhibition at IMMA included *Moonfield* and *To the Centre*, but also included his maps of Connemara and Oileáin Árann, alongside short texts excerpted from *Stones of Aran* and *Setting Foot on the Shores of Connemara*. While the view from the horizon may be a physical and metaphysical impossibility, the nearest thing to that is a shift in modes of thought, a 'shake up' of 'one's deepest vocabulary'.[81] The move to Árainn 'precipitated' Robinson 'into a directionless state' in which he was 'prey to anxieties and obsessions'. But, as he notes, it was the *place* itself that suggested 'a way out'.[82] The 'lure' in terms of trying to understand the nature of this particular place emerged with the understanding that any account could only ever be an interim report, as it was a project that 'could never reach an end'. As he writes, 'Aran is extraordinary in so many ways … that I was soon lured into trying to understand the island, by its promise that this project could never reach an end. Accumulating impressions in a diary, I became a writer; and then, noting placenames and routes and locations on paper, a cartographer.'[83]

After Máire Bean Uí Chonghaile (postmistress of Cill Mhuirbhigh) suggested that he should make a map for tourists, Robinson produced a rough sketch map of the island that evening.[84] The project appealed to him as it involved 'all the things I liked doing: walking, drawing, asking questions'.[85] It was important that the map would not be the 'wrong sort of map' – the 'thin' map that Lopez admonished cartographers of earlier – but that it would be 'expressive as well as informative' as to the uniqueness of the islands.[86] The dual mandate of the map to be *expressive* and *informative* (with the emphasis largely on the former) meant that the aesthetic and affective nature of the map would be a quality that would be foregrounded and negotiated through a variety of draughtsmanship techniques and cartographic methods. After some time, what can now be termed as a deep mapping practice was beginning to take hold: 'Finally, a vague cloud of ideas about maps and their relationship to the place mapped, some of which I had half-realized in artworks before now, could perhaps be worked out in practice.'[87] This first mapping project per se was seen as a 'step in an interior evolution', taking its lead from the immediacy of the surroundings and context, but leaning on ideas and expertise from the past, 'half-realized in artworks before now'. The process of mapping the islands was noted by Robinson as being in one way a collaborative endeavour, with wide consultation and approval sought first from those inhabitants who had a deep knowledge of the islands (an ethnographic 'peer review' process) with its end goal and recording function clearly in sight. However, the 'execution' of the overall project itself (both the final 'product' and its process) Robinson viewed as being 'a private work of art'.[88]

Robinson's artwork moved towards a monochrome register in the late 1960s, culminating in the shapes of *Moonfield* and the rods of *To the Centre*. Now with the decision of map illustration (in terms of colours, tints and 'ornaments') Robinson decided that a coloured map 'could easily fall apart, visually and conceptually, into superimposed but otherwise separate layers' and 'looked for visual equivalents of their feel underfoot' for his map symbols and legends, devising his own symbology.[89] By representing the terrain underfoot onto the mediated two-dimensional terrain of the map, Robinson departs from traditional cartographic conventions. His goal was to 'preserve the texture of immediate experience' and thereby 'short circuit the polarities of objectivity and subjectivity, and help me keep faith with reality'.[90] The specificities of the 'textures of place' would determine his conventions, a cartographic practice that evolved in ways reminiscent of McLucas's deep mapping practice and that also resonates with theories of space and place as described by Edward Casey, Yi Fu Tuan and Tim Cresswell.[91] For Robinson, a map is 'a sustained attempt upon an unattainable goal, the complete comprehension by an individual of a tract of space that will be individualised into a place by that attempt', while also being a phenomenological register of the 'feel' of the landscape represented, under one's eyes and also as imagined, underfoot.[92]

Robinson's map of *Oileáin Árann* thus had a dual function; it was a public map that had a clear utility in terms of local tourism and in spatially recording the heritage of the islands, but it also served a private function in being the catalyst that prompted 'a step in an interior evolution' of what is interpreted here as Robinson's deep mapping practice. In many ways the later reproduction of the map of *Oileáin Árann* onto a large-scale vinyl format designed to be laid on the ground and walked across brought together the idea of cartography as art and, by extension, of art as being socially engaged. This map subsequently became known as the *Distressed Map of the Aran Islands*, and its 'afterlife' can also be read as being a re-articulation of his earlier environmental artworks in that 'viewers' were encouraged to walk, write and doodle on it when it was exhibited in Ireland and London.[93] The visual artworks from the London years and the cartographic narratives on Ireland (encompassing maps and various writings) were in many ways the opening up of various kinds of spatial conversations on the nature of space and our place in it. The spatialising imaginary that got breathing space on Hampstead

Heath, and was contained within a Camden gallery, would later find expression in Tim Robinson's deep mapping practice associated with the west of Ireland:

> This imagery of steps, walks, mazes, webs, is not entirely something I have freely chosen to elaborate, and it could become a knot-garden I have to cut my way out of. Perhaps I need to quit these worn ways and trodden shores, to test these ideas elsewhere, to travel in search of that impossibility, the view from the horizon. I know that the step, for instance, is not some poetic flower casually picked by the wayside of my west-of-Ireland life, because, looking back, I see it implicit in the work I was doing in London, such as the yard-long suspended rod, which now rhymes mysteriously with my trigonometric musings on the Connemara mountain.[94]

## Acknowledgements

I would like to thank Iain Biggs and Karen E. Till for introducing the concept of deep mapping to me during a symposium in Bristol in 2011, and members of the Mapping Spectral Traces International Collaborative who have shared their deep mapping expertise from the contexts of North American, Australian and British community mapping projects and socially engaged arts practice. I would also like to thank participants at the Know Your Place: Community Mapping Workshop, co-hosted by the Ómós Áite: Space/Place Research Network, Galway City Museum, March 2014, and those at the Conference of Irish Geographers, May 2014, where many of these ideas were discussed in detail. Thanks also to Tim Collins, Christine Cusick, Derek Gladwin, Jim Rogers and Eamonn Wall for ongoing conversations on Robinson's work and on Irish environmental and ecocritical literature in general. Finally, I would like to acknowledge official and unofficial Roundstone Conversations with Tim and Máiréad Robinson since 2009 that have added greatly to my understanding of the work of Tim Robinson and Folding Landscapes. This research was conducted with funding gained by a New Ideas Award, supported by the Irish Research Council.

## Notes

1 Tim Robinson, *The View from the Horizon: Constructions by Timothy Drever 1972, Texts and Maps by Tim Robinson, 1975–96* (Clonmel, Tipperary: Coracle Press, 1997), 23.
2 Tim Robinson, *My Time in Space* (Dublin: Lilliput Press, 2001), 53.
3 Robinson, *My Time*, 72.
4 Benoît Mandelbrot's theories of fractals and of the order of nature would very much inform Robinson's own minute and painstaking examination of 'the fineness of things' through his fieldwork and writings on Ireland.
5 I am grateful to Dr Méabh Ní Fhuartháin, Centre for Irish Studies, NUI Galway, for enhancing my critical understanding of this phrase, which has very much shaped my thinking on Robinson's work in recent times.
6 Robinson, *My Time*, 72.
7 The title for the exhibition was borrowed from an essay, 'The Event Horizon', by Italian film-maker Michaelangelo Antonioni in *That Bowling Alley on the Tiber: Tales of a Director*,

translated by William Arrowsmith (New York: Oxford University Press, 1987). On Antonioni see also, Matthew Gandy, 'Landscapes of Deliquescence in Michelangelo Antonioni's *Red Desert*', *Transactions of Institute of British Geographers* 28:2 (2003): 218–37. Details about the original 'Event Horizon' IMMA show are available at 'Event Horizon Season at the Irish Museum of Modern Art'. Accessed 10 December 2013, www.imma.ie/en/page_19249.htm.
8  Robinson, *My Time*, 72.
9  Robinson, *My Time*, 72.
10  Robinson, *My Time*, 72.
11  Robinson, *The View*, 10.
12  The phrase 'the fineness of things' is taken from a Robinson essay with that title, 'The Fineness of Things', in *My Time in Space*, 217–30.
13  Mike Pearson and Michael Shanks, *Theatre/Archaeology* (London: Routledge, 2001), 65. For the purposes of this essay, by the concept of the 'West' I am specifically referring to the geographical area of the Burren (Co. Clare), Oileáin Árann (Co. Galway) and Connemara (Co. Galway). However, in Robinson's writings he also draws comparisons between this 'west' and the other south-western and north-western regions of Ireland, particularly when discussing issues relating to Gaeltacht areas within those regions.
14  In particular, see Karen Babine, '"All the Sky Were Paper and All the Sea Were Ink": Tim Robinson's Linguistic Ecology', *New Hibernia Review* 15:4 (Geimhreadh/Winter 2011): 95–110; Christine Cusick, 'Mindful Paths: An Interview with Tim Robinson', in Christine Cusick (ed.), *Out of the Earth: Ecocritical Readings of Irish Texts* (Cork: Cork University Press, 2010), 205–11, and 'Mapping Placelore: Tim Robinson's Ambulation and Articulation of Connemara as Bioregion', in Tom Lynch and Cheryll Glotfelty (eds), *The Bioregional Imagination: Literature, Ecology, and Place* (Athens, GA: University of Georgia Press, 2012), 135–49; John Wilson Foster, 'Tim Robinson's Variegated World: *My Time in Space* by Tim Robinson; Tales and Imaginings', *The Irish Review* 30 (Spring–Summer 2003): 105–13; Derek Gladwin, 'The Literary Cartographic Impulse: Imagined Island Topographies in Ireland and Newfoundland', *Canadian Journal of Irish Studies* 38:1–2 (2014): 158–83; Eamonn Wall, 'Walking: Tim Robinson's Stones of Aran', *New Hibernia Review/Iris Éireannach Nua* 12:3 (autumn/*fómhar* 2008): 66–79, and *Writing the Irish West: Ecologies and Traditions* (Notre Dame, IN: University of Notre Dame Press, 2011); John Wylie, 'Dwelling and Displacement: Tim Robinson and the Questions of Landscape', *Cultural Geographies* 19:3 (2012): 365–83. However, other key aspects of Robinson's work that have remained relatively unexplored are the colonial and postcolonial contexts of his work (of which he himself is very explicitly aware and addresses in his writings), and the bilingual and Irish-language aspects of his work, both in terms of his use of Irish language sources for his cartographic mappings/writings and also his translation work with writer and scholar Liam Mac Con Iomaire. For a visual approach to Robinson's work, see other work in this collection: Derek Gladwin's essay 'Documentary Map-Making and Film-Making in Pat Collins's *Tim Robinson: Connemara*' (Chapter 4) and Catherine Marshall's essay '"Another Half-Humanized Boulder Lying on Unprofitable Ground"?: The Visual Art of Tim Robinson/Timothy Drever' (Chapter 12).
15  Jos Smith, 'A Step towards the Earth: Interview with Tim Robinson', *Politics of Place: A Journal for Postgraduates*, in special issue *Maps and Margins* 1 (2013): 4–11. Accessed 1 December 2013, http://blogs.exeter.ac.uk/politicsofplace/files/2013/08/POP_Issue01_Smith.pdf.

16 Catherine Marshall's contribution to the present essay collection in Chapter 12 also marks a critical departure in this cross-disciplinary conversation on Robinson's work as the visual artist Timothy Drever.
17 For Wall, the 'three arts' he refers to here are the arts of 'map-making (the actual work of cartography); mapping (the surveying and collecting of lore and data that precedes this); and writing', Wall, 'Walking', 68.
18 Wall, 'Walking', 68.
19 Wall, 'Walking', and 'Adequate Steps: Tim Robinson's *Stones of Aran*', in *Writing the Irish West*, 1–50.
20 Wylie, 'Dwelling and Displacement', 377.
21 In the field of remote sensing, and in social and cultural geography, 'ground truthing' refers to the practice of verifying a satellite image with data and information collected in the field. It more usually refers to a quantitative, rather than a qualitative, data-gathering and analytic practice.
22 Michael Cronin, 'Inside Out: Time and Place in Global Ireland', *New Hibernia Review* 13:3 (2009): 74–88. A later version of this essay was also published in Michael Cronin, *The Expanding World: Towards a Politics of Microspection* (London: Zero Books, 2012).
23 Cronin, 'Inside Out', 83.
24 Cronin, 'Inside Out', 83.
25 See also Kelly Sullivan's essay in this collection, 'Not Knowing as Aesthetic Imperative in Tim Robinson's *Stones of Aran*' (Chapter 6).
26 Cronin, *The Expanding World*, 7.
27 Eóin Flannery, 'Ireland and Ecocriticism: An Introduction', *Journal of Ecocriticism* 5:2 (July 2013): 4.
28 Tim Robinson, 'The Seanchaí and the Database', *Irish Pages* 2:1 (2003): 44.
29 In his essay 'A Land without Shortcuts' (based on his Parnell Lecture delivered at the University of Cambridge, 2011), Robinson goes some way to answering this question, with a shift in focus from the parameters of space to the concern for the changing nature of time in a digital age. Tim Robinson, 'A Land without Shortcuts', *The Dublin Review* 46 (Spring 2012): 25–44.
30 Flannery, 'Ireland and Ecocriticism', 5.
31 Flannery, 'Ireland and Ecocriticism', 5.
32 Barry Lopez, as cited in William Least Heat-Moon, *PrairyErth: A Deep Map* (Boston: Houghton Mifflin, 1991), 4.
33 Clifford Geertz, 'Thick Description: Toward an Interpretive Theory of Culture', in *The Interpretation of Cultures: Selected Essays*, edited by Clifford Geertz, 3–30 (New York: Basic Books, 1973).
34 Pearson and Shanks, *Theatre/Archaeology*, 64–5.
35 Iain Biggs, 'The Spaces of "Deep Mapping": A Partial Account', *Journal of Arts and Communities* 2:1 (2011): 7.
36 Biggs, 'The Spaces', 7.
37 Clifford McLucas (1945–2002) was an architect, visual theorist, artist and educator who developed a detailed programme of his own understanding of deep mapping through his work with the internationally renowned experimental performance theatre company *Brith Gof*, co-founded in 1981 by Mike Pearson and Lis Hughes Jones. He later collaborated with Mike Pearson and Michael Shanks.
38 All references to McLucas and his work on deep maps are from his webpage: http://cliffordmclucas.info/deep-mapping.html. I am grateful to Dr Iain Biggs for conversations on the 'history' of deep mapping during his time as an Irish Studies Scholar and Moore Institute Visiting Fellow at NUI Galway, Spring 2014.

39 Iain Biggs, '"Deep Mapping": A Brief Introduction', in *Mapping Spectral Traces 2010, Exhibition Catalogue*, ed. Karen E. Till, University of Virginia Tech College of Architecture and Urban Studies, 8. Accessed 1 December 2013, www.isce.vt.edu/files/MappingSpectralTracesCatalogFull.pdf.
40 Biggs, 'Deep Mapping', 9. Original emphasis.
41 Karen E. Till, 'Artistic and Activist Memory-Work: Approaching Place-based Practice', *Memory Studies* 1 (2008): 98.
42 The term 'transdisciplinarity' here denotes the work of engaging knowledge across multiple sites, particularly across and beyond traditional academic sites and boundaries of knowledge (i.e. universities and other institutes of education). Often it entails a counter-hegemonic practice in questioning and re-situating near-hegemonic knowledge, norms and practices relating to discrete disciplines. For further analysis see Karri A. Holley, *Understanding Interdisciplinary Challenges and Opportunities in Higher Education* 35:2 (ASHE Higher Education Report: 2002), particularly the chapter 'Defining Interdisciplinarity', 11–30.
43 Biggs argues that in many ways deep mapping practice should be an 'indisciplined' practice, questioning and challenging the limits of disciplinary knowledge and praxis, email message to author, 27 May 2014.
44 This point is made by Biggs (email message to author, 27 May 2014).
45 On the question of institutional disciplinarity in the context of Irish Studies, see Nessa Cronin, '"Disciplinary Ghettoes": Irish Studies and Interdisciplinary Negotiations', *Journal of Nordic Irish Studies* 6 (2007): 1–16.
46 A recent example of this was Know Your Place: A Community Mapping Workshop run in conjunction with Lifeworlds: Space, Place and Irish Culture International Conference, Galway City Museum, 29 March 2014. The workshop was centred on the devising of a community-based mapping 'toolkit', based on the mapping work and experiences of the X-PO Mapping Group in Kilnaboy, Co. Clare, Mná Fiontracha Mapping Group in Árainn and the Sliabh Aughty Community Group. The workshop was facilitated by Dr Ailbhe Murphy, Dr Deirdre O'Mahony and Ms Ann Lyons, and organised by Dr Nessa Cronin and Dr Tim Collins through the Ómós Áite: Space/Place Network, NUI Galway.
47 Nessa Cronin, 'Lived and Learned Landscapes: Literary Geographies and the Irish Topographical Tradition', in Marie Mianowski (ed.), *Irish Contemporary Landscapes in Literature and the Arts* (London: Palgrave Macmillan, 2011), 106–19.
48 Robinson, *My Time*, 31–2.
49 In this essay, for the sake of clarity with regard to the two distinct aspects of Robinson's career, 'Tim Robinson' refers to the works written/constructed by and under the name of Tim Robinson, while 'Timothy Drever' refers to the works written/constructed by and under the name of Timothy Drever.
50 Robinson, *My Time*, 41.
51 Robinson, *My Time*, 41.
52 Robinson, *My Time*, 42.
53 Robinson, *My Time*, 44–7.
54 Robinson, *My Time*, 46.
55 In many ways, as Iain Biggs has argued, one could read Drever's work from this period as being an engagement (consciously or otherwise) with the Surrealist tradition of the dream, as *Moonfield* even suggests 'a dream landscape rather than a mathematical/conceptual one'. Biggs email to author, 27 May 2014.
56 Robinson, *My Time*, 44–7.
57 Robinson, *My Time*, 51.

58 Robinson, *My Time*, 51.
59 Robinson, *My Time*, 52.
60 Robinson, *My Time*, 52.
61 Robinson, *My Time*, 52.
62 Tim Robinson, *Setting Foot on the Shores of Connemara and Other Writings* [1996] (Dublin: Lilliput Press, 2007), 214.
63 Robinson, *My Time*, 53.
64 Timothy Drever, 'Field Work 3: A Structured Arena', *Bulletin of the Computer Arts Society* (February 1971), n.p. I am grateful to Tim Robinson for this reference.
65 Drever, 'Field Work 3', n.p.
66 Drever, 'Field Work 3', n.p. This would also become part of his working practice in terms of the invitation for people to walk across, and engage with, what would later become known as the *Distressed Map of Aran*. This public artwork is briefly discussed at the end of this chapter.
67 On further explorations of the 'adequate step', see Robinson, *The View*, 27–9.
68 Drever, 'Field Work 3', n.p.
69 Drever, 'Field Work 3', n.p.
70 Robinson in conversation with the author during *Roundstone Conversations*, September 2013. The concept of a 'wavy floor' has already been architecturally designed in Ionad an Bhlascaoid Mhóir/The Blasket Island Centre, Dún Chaoin, Co. Kerry. The corridor floor of the visitor centre comprises an undulating slate floor, which is meant to reproduce the experience of the motion of the sea as the visitor walks along it. At the end of the corridor is a large floor-to-ceiling window framing a view of the islands – the intended final, symbolic destination of the visitor.
71 Robinson, *The View*, 9–10.
72 Robinson, *The View*, 11.
73 Robinson, *Setting Foot*, 214.
74 Robinson, *The View*, 11.
75 Robinson, *The View*, 13.
76 Robinson, *The View*, 57.
77 Robinson, *The View*, 57–8.
78 Robinson, *The View*, 59.
79 I am grateful to Dr Karen E. Till for this observation.
80 Robinson, *The View*, 59.
81 Robinson, *The View*, 16.
82 Robinson, *Setting Foot*, 211–12.
83 Robinson, *Setting Foot*, 212.
84 I am grateful to Dr Deirdre Ní Chonghaile (granddaughter of Máire Bean Uí Chonghaile) for further discussions about the history of this cartographic moment.
85 Robinson, *Setting Foot*, 2.
86 Robinson, *Setting Foot*, 2.
87 Robinson, *Setting Foot*, 2–3.
88 Robinson, *Setting Foot*, 2–3.
89 Robinson, *Setting Foot*, 76.
90 Robinson, *Setting Foot*, 77.
91 See, in particular, Edward S. Casey, *The Fate of Place: A Philosophical History* (Los Angeles and London: University of California Press, 1997) and *Getting Back into Place: Toward a Renewed Understanding of the Place-World* [1993] (Bloomington: Indiana University Press, 2009); Tim Cresswell, *In Place/Out of Place: Geography, Ideology and Transgression* (Minneapolis: University of Minnesota Press, 1996); Yi Fu Tuan, *Topophilia: A Study*

of *Environmental Perception, Attitudes, and Values* (Englewood Cliffs, NJ: Prentice-Hall, 1974) and *Cosmos and Hearth: a Cosmopolite's Viewpoint* (Minneapolis: University of Minnesota Press, 1996).
92  Robinson, *Setting Foot*, 77.
93  The *Distressed Map of the Aran Islands* was exhibited in many places, and was first shown in Vinyl, an exhibition curated by Simon Cutts as part of Cork's year as European City of Culture, 2005. In Cork it was placed in an old school playground, with a notice for the viewer: 'The original Aran map is published in a paper edition, and if a copy wears out one can get another. The present enlarged version on vinyl is singular, like the islands themselves, and what becomes of it remains to be seen. You are invited to walk on it, write on it, dance on it, treat it as you see fit.' On the Folding Landscapes website, the reader is told how this version of the map of the islands received its name: 'After a month of exposure to the elements, to skate-boarders and dancers, and to people wishing to commemorate their own experiences on the islands, it came back to Folding Landscapes rather worn and crumpled, and with several interesting graffiti, and we decided to call it the *Distressed Map of the Aran Islands*.' It was later exhibited in the Map Reading Room of the Royal Geographical Society as part of the Map Marathon curated by Hans Ulrich Obrist of the Serpentine Gallery in London, October 2010, where it acquired 'another interesting layer of damage'. Notes on the *Distressed Map* from www.foldinglandscapes.com/?page_id=21, accessed 1 December 2013.
94  Robinson, *The View*, 43.

# 4

# Documentary map-making and film-making in Pat Collins's *Tim Robinson: Connemara*

## Derek Gladwin

A map is a sustained attempt upon an unattainable goal, the complete comprehension by an individual of a tract of space that will be individualized into a place by that attempt.[1]

– Tim Robinson

In sum a film *is* a map, and ... its symbolic and political effectiveness is a function of its identity as a cartographic diagram.[2]

– Tom Conley

### Documenting through map-making and film-making

In the documentary film *Tim Robinson: Connemara* (2011), director Pat Collins spotlights a figure in Ireland known for map-making, nature writing and documenting parts of western Ireland with incisive detail. Collins, who is considered one of the most articulate contemporary documentary film-makers in Ireland (with over two dozen film credits to his name), depicts Robinson as a mediator between landscape and culture through his own mapping enterprise. The film focuses on the landscape of Connemara, which is a region in Co. Galway in the west of Ireland with topographies of granite mountains, an abundance of bogland and thousands of pools of water (see Figure 12). Collins's film-making and Robinson's map-making contain similarities in their structural process, particularly in what both ultimately attempt to achieve – capturing the many identities of the Connemara landscape through a form of cultural mediation. It is, however, with some trepidation that I attempt to draw parallels between two quite distinct forms of documentary production. The process of mapping and film-making are obviously different in many ways. What I want to suggest is that Collins and Robinson share a similar approach in their own forms of documenting Connemara to create a place-based art form that magnifies the intricacies of this important landscape, while it also reduces the primacy of the 'maker' in the process.

*Figure 12* Connemara topography (from Pat Collins's film *Tim Robinson: Connemara*, photo by Colm Hogan).

*Tim Robinson: Connemara* is not only a documentary about the cultural and geographical elements of Robinson's map-making and writing, but is also about Robinson's technique of capturing the essence of place, a method that comes back full circle to Collins's primary aim in the documentary. Collins's film unfolds another dimension of Robinson heretofore overlooked: the very act of map-making for Robinson is a documentary process. Throughout Collins's film there is a sense that Robinson, rather than functioning as a *subject* of the documentary, is the *agent* of his own documentary. To this effect, we can recognise Robinson as a documentary map-maker – or as he describes it, the process of 'translation of a geography into a graphic image' – who captures the Connemara landscape through the visual and written form of map-making, a form that bears resemblances to documentary film-making.[3]

At this stage it seems necessary to briefly recognise Collins's sustained documentary film-making practice, which continues to focus on subjects (predominantly people and places) in Ireland. Collins has been making perceptive documentaries in Ireland for over two decades. A strong case could be made that he is currently Ireland's most influential documentary film-maker because his work concentrates primarily on the relationship between Irish culture and landscapes. Indeed, various commentators have categorised his work as a form of cinematic landscape art; for example, in two of his more recent films – *Silence* (2012) and *Living in a Coded Land* (2014) – Collins takes viewers on psycho-geographical journeys through Ireland, traversing contested histories, geographies and cultures.[4] Collins has also explored themes of Ireland's past through landscape and culture in *Oileán Thoraí* (2001, a documentary about the lives and landscape of the islanders on Tory Island, nine miles north of the coast of Co. Donegal); *Beyond the Mountain*

(2005, a short documentary about the mining industry in Allihies, Co. Cork); and *Loch Dearg* (2008, a documentary about a St Patrick pilgrimage in Co. Donegal called Station Island). He remains most notable for his documentaries about Irish cultural figures, such as Frank O'Connor, John McGahern, Michael Hartnett, Gabriel Byrne and Nuala Ní Dhomhnaill.

Transitioning from Collins's previous cinematic work to documentary more generally, it is necessary to point out that documentary film-making follows a mode of production that characteristically uses non-actors, or 'real' people, in order to demonstrate a high level of realism.[5] In *Tim Robinson: Connemara*, just as the title suggests, two of the non-acting subjects propel the motivation of the film – both Robinson as writer/cartographer and Connemara as the landscape accompanying the overarching geographical imagination of the film. We might also ask what exactly is the film documenting? How do the subjects contribute to an overall purpose for the documentarian? How do the subjects make their own statements merely by functioning 'normally' during the filming? These are all important questions to pursue when analysing any documentary but they are particularly important for my purpose here because both Robinson (the map-maker) and the Connemara landscape (the mapped) are the key subjects for Collins (the film-maker).

Connemara represents one of the main foci of the film, but this landscape is more precisely viewed through the maps and writings of Robinson, who is the other subject in the film. As Jack Ellis and Betsy McLane point out in *A New History of Documentary Film*, documentaries typically concentrate on more than the general human condition, which is usually the purpose of narrative film-making. Documentaries do not organise the structure of the film chronologically, nor do they shape a narrative developed around characters. Documentaries centre on a subject, purpose and/or approach, all of which are defined and viewed differently depending upon the documentarian.[6] Bill Nichols argues that 'documentary is not a reproduction of reality, it is a *representation* of the world we already occupy. It stands for a particular view of the world, one we may never have encountered before even if the aspects of the world that is represented are familiar to us.'[7] There is a striking similarity if we compare Nichols's definition of documentary with one of Robinson's reflections on map-making: 'For if cartography is not necessarily more helpless than other modes of *representation* in the face of the world, it has its own characteristic failings, which the blanks on a map, essential to its legibility as they are, reveal with disconcerting candour.'[8]

Collins's documentary is primarily about three related and, in some ways, overlapping subjects: Connemara, Robinson as figure in Irish culture and Robinson's unique process of mapping/map-making. The first subject of Connemara is quite clear since it highlights one of the most recognised and symbolic terrains in Ireland. Robinson's status as a cultural figure in Ireland, as the second subject of the film, provokes a more compelling discussion. After all, Robinson is a transplant to the west of Ireland. Originally from Yorkshire, England, he later spent time studying mathematics at University of Cambridge and then lived in London and Vienna in the

1960s as a visual artist, a practice that informed his subsequent work as a map-maker. Collins's film focuses on Robinson's documentation process of Connemara, not Aran or the Burren, through his maps and writings. This occurs particularly through his critically acclaimed *Connemara* trilogy – *Listening to the Wind* (2006), *The Last Pool of Darkness* (2008) and *A Little Gaelic Kingdom* (2011) – which are non-fiction accompaniments, also considered to be 'topographical writing', to his map of Connemara.[9] But, like the successes and challenges of many documentarians, Robinson's outsider status provides another dimension when he articulates the topographical and cultural contours of the Connemara landscape. His mapping and map-making, like that of Nichols's definition of documentary, is a *representation* of a world forged out of the experience of the 'maker', resulting in a process of unfolding where both the familiar and unfamiliar merge to form the documentary product.

The most important element of Collins's film involves Robinson's process as a map-maker, which resembles in many ways the process of documentary film-making as a visual and cultural method of capturing the people and landscapes on a particular geography. Robinson produces another form of visual documentary through his process of map-making, which, as I have argued elsewhere, constitutes both technical topographical cartography and cartographic prose (writing that could otherwise be called 'literary cartography').[10] Robinson does not produce a prototypical documentary in the genre of early English language documentaries akin to Robert Flaherty or John Grierson, particularly because he focuses on another documentary form outside of film-making. Instead, Robinson underscores the unreliability of objectivity when reproducing another form of representation as a detailed view of Connemara in both topographic and literary maps.

There are three specific ways in which Robinson deploys his own form of documentary map-making:

> In the basic geographic act of mapping I find three conjunctions: that of the place mapped with the one who maps it; that of the mapper with the map itself; and finally that of the map with the mapped – this last confrontation that tests the worth of the first and second.[11]

These three 'conjunctions' of mapping appear in Collins's cinematic depiction of Robinson's process, and they will serve as the overarching guides for the remainder of this essay. One caveat that is important to acknowledge when analysing Robinson's mapping and map-making is that he believes this process, much like the making of a documentary film, to be an imprecise form of documentation. The subjective reality in what is often perceived in both documentary film-making and map-making as 'objective' dominates Robinson's thinking on the subject:

> I'm not very interested in maps from the technical point of view, so I will be brief on how I went about producing this one, and move on to the more interesting questions of what is it like to make a map – insofar as I can untangle my memories of the process – and why maps are, finally, so unsatisfactory.[12]

Drawing from these distinctions of map-making, the following three sections examine the questions and methods that go into making a map and mapping for Robinson, while also drawing parallels to this process as it unfolds in Collins's documentary.

## The map of place and the map-maker

The first conjunction underscores the importance of the mapped place with the person who maps it. For Robinson, this could constitute the 'ABC of earth wonders', as he calls them, of his maps of Aran, the Burren and Connemara.[13] All three examples demonstrate a definitive primacy of place in Ireland. Location becomes the principal focus for both the documentary map-maker and film-maker. As important as it is for the map-maker to document the significance of place, so too must the documentary film-maker capture the landscape as more than just a detached topography or an external location for the film. Documentary film perhaps more than any other film form captures a sense of place, which includes not only the topography but also the culture that creates such an emotional and personal connection to a specific place.

The American film-maker Robert Flaherty was an early champion of making documentaries as a way to capture a place. Possible film subjects would often materialise into focus for Flaherty after a specific place was selected, and then he would scrutinise the ways in which that place might beneficially present itself on film. In the context of Robinson and the Aran Islands, Flaherty's film *Man of Aran* (1934) remains a prime example. Incidentally, it was this film that initially brought Robinson and his partner Máiréad to the Aran Islands in 1972. A case could be made that the English-language documentary began with Flaherty's *Nanook of the North* (1922), a film about an Inuit family that lived in a remote region of northern Canada. The Arctic topographies in northern Canada are as much a subject in the film as are the Inuit family, just as the Aran Islands are as much a subject as are the indigenous islanders in *Man of Aran*.[14] While these two films in particular have deservedly received criticism for the ways in which they represent indigenous peoples, they are nevertheless examples of early forms of English-language documentary film-making that demonstrate the importance of location and place when determining the subject and purpose of the project.

The necessity of location entails filming in a specific place, or what is known as shooting on location. Filming on location can be technically difficult to achieve because of lighting (electricity, transportation, weather, etc.) and shooting (the process must be supported by handheld or waist camera movement in order to cross difficult terrains), and therefore complicates the process and increases the cost of the film. Regardless, the film-maker has few alternatives in order to capture the realistic elements that are required. Filming on location presents an important parallel to map-making: the sense of place must be recorded in person and not afterwards in abstract spaces of sound stages or studios. Place, as I use the term here, promotes the idea that landscape is infused with meaning and experience. According to Kent Ryden, 'place is created when experience charges landscape with meaning'.[15] The geographer Yi-Fu Tuan has similarly noted that space and place are contingent upon modes of experience that undergo alteration depending upon geographies in which they are situated. Spaces are more open and less undefined while places are often charged with meaning and emotional association.[16] Robinson describes his map-making of place in 'A Connemara Fractal', an essay that explains his mapping process:

> My business in making maps is not just the recording of all the oddities and singularities I pick up on my way along convoluted walks like this one through Connemara. I, too, want to create meaning, but meaning of a specifically geographical sort. So, I am particularly interested in or excited by the points at which the thread of my explorations crosses itself, as it were, from various directions, and can be knotted firmly, that is, memorably, in a way that elevates a mere location to the status of a place.[17]

Both Robinson and Collins demonstrate the importance of place in their own processes of filming and mapping by focusing on Connemara as a landscape that may appear to be an undefined space when in actuality it is a place resonating with meaning.

Documentarians often attempt to promote an understanding or incite interest in the subject (place or person) being filmed.[18] On the most obvious level, *Tim Robinson: Connemara* centres on the relationship between Robinson and Connemara. On another level, Collins spotlights a documentarian in Robinson, and for just under an hour during the running time of the film Collins demonstrates Robinson's approach and process as a topographical and imaginative map-maker. Connemara appears to function as the central subject in the film, particularly in the traditional way that we can view a subject in documentary film-making as a contrast to narrative film where the interrelationships between characters, or what is simply called drama, literally takes centre stage. Perhaps this is why Collins's entire film uses the writings of Robinson, with Robinson actually reading his own work as part of the overlain and also interwoven narrative of the film. The shift occurs when Robinson mediates his own experience with Connemara.

One example in the film that demonstrates Robinson's map-making in connection to the primacy of place is when he remarks in an interview, 'When I am walking the terrain, I am the point of the pen moving over the map. When I am drawing the map, the point of the pen is me moving over the terrain.'[19] He goes on to explain that this rubric he follows breaks through the layers of 'interference and forgetfulness between the experience of being in the place, and the experience of drawing the map'.[20] The map, he maintains, contains a 'sensation of being in that place', not only an accurate account of events.[21] By mapping Connemara as a documentary form, with its inherent attention to subject, approach and process, Robinson, as he comments elsewhere, 'was trying to preserve the texture of immediate experience'.[22] Collins then duplicates this process of 'immediate experience' as a film-maker by capturing Robinson's own procedure and attempts to convey it to the viewer through the documentary film form rather than through a map or book.

I am not claiming here that map-making and film-making are identical forms; indeed, they contain as many differences as they do similarities. There is a sense, however, that these two methods of capturing a sense of place do in fact have significant crossover, despite the relatively little amount of critical attention they have received as similar practices. The way in which maps and films are created, as well as how they relate to one another, has garnered some interest among film scholars. In *Cartographic Cinema*, one of the few books examining the relationship between films and maps, Tom Conley states that 'a film, like a topographic projection, can be understood as an image that locates and patterns the imagination of its spectator'.[23] And while Conley's argument emphasises the use of maps in film, rather than my intention to stress the similarity between map-making and film-making, he nevertheless identifies the core similarity about how these two forms document and articulate spaces. Conley goes on to argue,

> A map underlines what a film is and what it does, but it also opens a rift or brings into view a site where a critical and productively interpretive relation with the film can begin. A corollary is that films are maps insofar as each medium can be defined as a form of what cartographers call 'locational imaging'.[24]

The map represents Robinson's own film form where 'locational imaging' occurs and where both the map and film are visual representations of interior and exterior landscapes. The pen, or his body walking over the terrain, functions as the recording device similar to the camera. This is why the best way to understand Robinson's documentary form, in addition to watching Collins's illustration of this process unfold in the film, is to read/view his maps of Connemara, as well as his other maps of the Aran Islands and the Burren.

Similar to making a documentary film, which attempts to articulate a point without dramatic motivation, 'the making of a map', as Robinson admits, 'is many things as well as a work of art, and among others it is a political, or more exactly an ideological, act'.[25] Robinson recalls that in map-making the 'purpose of identification was to short circuit the polarities of objectivity and subjectivity, and help me keep faith with reality'.[26] As described here by Robinson, map-making presents an unexpected parallel to documentary film-making, where the outcome challenges the objective and subjective polarities in order to produce some form of coherent reality void of previous prescriptions. Collins's film demonstrates this balance by producing a significant number of frames that only detail the Connemara landscape as a place mapped, which are then contrasted with frames of Robinson walking in the landscape as the map-maker. The film presents a noticeable juxtaposition of both subjects – Robinson and Connemara – as separate while they are also interconnected modes of production in terms of process and articulation.

**Map and the map-maker**

The second conjunction that Robinson mentions is the relationship between the map and the map-maker. If we translate this into film, then this would entail the relationship between the film and the director. Both examples reveal the connection between the artist/creator and the art/object. In the documentary mode, distance and proximity ebb and flow to create what may initially appear to be only objective when in actuality it is simultaneously subjective. As much as the process of filming records the subject for the film-maker, the act of walking similarly records the landscape for the map-maker to make the map.

One example that demonstrates the relationship between the map and map-making is when Collins films Robinson walking through Roundstone Bog. Robinson reflects (in voice-over narration),

> and having selected this particular stretch of coast because its near unmappability perversely suggested the possibility of mapping it, I had felt the idea of walking its entire length impose itself like a duty, a ritual of deep if obscure significance through which I would be made adequate to the task of creating an image of the terrain.[27]

As Robinson reads this quote in the film, a tracking shot with an extreme close up horizontally moves along the actual map of the coast near Roundstone in Connemara. Collins often employs this technique throughout the film in order to increase intimacy and proximity for the viewer. Such a scene demonstrates that Collins subtly recognises that film-making shares similar qualities to map-making, which is to say they both capture the essence of place through observation, listening and a certain kind of visual language. They also offer an imprecise form of documentation that fails to *completely* understand the place or person being documented.

Walking has always been a valuable way for Robinson to document the landscape. Robinson usually appears on screen when he is walking somewhere in Connemara during the film. This 'live action' walking displays the importance of Robinson's particular method of mapping in connection to the map. Eamonn Wall identifies Robinson's use of walking as an activity and trope that frames his overall mapping and writing project.[28] For Wall, 'walking is the primary building block of Robinson's methodology and underlines his style'.[29] Although Wall's quote pertains mainly to Robinson's writing style, one could make the argument that he takes the same approach to both his non-fiction writing and map-making, and the two complement each other through the method of documenting place. In doing so, both literary and topographical cartography contribute to form

Robinson's complete mapping project of Connemara, but they are understood and then transformed through the act of walking. Robinson comments, 'I aspire to a compensating gift of walking, not in a way that overcomes the land but in one that commends every accident and essence of it to my bodily balance and my understanding.'[30]

Walking exemplifies not only Robinson's method of mapping, but also his personal relationship to place through the body. The map-maker, like the documentary film-maker, must embody the subject by immersing oneself in it. Collins once remarked in an interview that 'film-making is not like a normal nine-to-five job. To a certain extent you have to become consumed by what it is you are making.'[31] Wall equally maintains that for Robinson, 'He is intensely involved in his work while, at the same time, he maintains a level of detachment to ensure that the primary focus is on place rather than on its recorder.'[32] Documentarians, both as film-maker and map-maker, submerge themselves into the making of each project, as well as the place and subject supporting it. Walking creates an embodied understanding of place for the recorder that, according to Robin Jarvis in *Romantic Writing and Pedestrian Travel*, produces a 'multiplicity of appearances and the particularity of actual landscape'.[33]

As Collins films Robinson's own particular 'gift of walking', particularly as a way to understand the land, a remarkable correlation emerges between Robinson the map-maker and Robinson the performer. In this sense, then, Robinson's own cartographic experience contains a performative quality in relation to the landscape. Collins's cinematic depictions of Robinson's documentary map-making process provide an imbricated quality. A layered process materialises during the scenes of Robinson walking across the landscape as though it were a performance, not only as an actor in the film, but also as a routine that he performs every day over the past forty years during his method of map-making as the map-maker. In an essay in the collection *Crossroads: Performance Studies and Irish Culture*, J'aime Morrison recognises that Robinson has created his own 'politics of movement' where his movement guides his writing and the 'narrative wanders between cartographic observation and poetic reveries' in order to 'evoke the contours of the land'.[34] By acknowledging cartography as a form of performance, in so far as the embodiment of body/subject is integral to the discipline, we can illuminate Robinson's method of reconciling the relationship between map and map-maker in performative terms that result from the embodied movement of walking. An explicit example of this dynamic becomes evident when watching Robinson performing himself as a character in Collins's documentary. Robinson narrates through his own writing, while we see him on screen move through Connemara on foot, challenging discursive histories and memory.

If theatrical performance can be considered a text, as it has been described in performance studies, then landscape performance can also be seen as a text where agents act and move across it. As a mode of non-fiction film-making, documentary serves as an ideal medium to present the peripatetic landscape performer who circumnavigates the topographies of the land. Walking contributes to the function of map-making for Robinson, but it also creates a dynamic relationship between the body and the landscape. The audience witnesses how Robinson's movement negotiates between the places associated with the map and the map-maker in the visual documentary medium. Connemara, in this context, is used as a site of performance where walking constitutes one of many actions that draws performer and audience into the terrain that is simultaneously represented on screen and as part of the mapping process.

**Map and mapped**

One of the comparative elements that struck me between Collins's depiction of Robinson and Robinson's representation of Connemara is the amount of space that exists in the respective map and film. The so-called empty spaces develop in the film through both silence and the visual vastness of the long take technique coupled with the extreme long shots of Connemara (see Figure 12). Robinson claims in the film that 'the most revealing features of a map are its blank areas'.[35] Unpacking this admission is perhaps a provocative way to end this essay because it encapsulates the third conjunction of the map and mapped process. The map, as a material object that tenuously reflects the mapped, contains the codified system of the mapping. The perceived empty or blank space in map-making and documentary film-making magnifies the relationship between the map and mapped. This occurs through a process of unfolding, without definite answers to the probing inquiries of absolute truth, and overrides the drive to obtain a finished or end product. Christine Cusick has argued that 'Robinson's ambulation, much like his writing, is not exclusively a quest for destinations. It is more often a mode of discovery. By putting foot to sod, he is not merely finding answers but discovering the questions.'[36] Robinson's process of unfolding combines his peripatetic form of exploration of the mapped (Connemara in this case) with his writing and map-making to arrive at a map (material product).

What occur in these spaces of unfolding that promote processes of discovery are as much a part of place as are the dots or lines indicating mountains or lakes on the map, or, on the other hand, film edits as part of cinematic technique of the film map. In the film Robinson calls these methods of map-making 'ornamentation', which is 'the ground of the map'.[37] He maintains that for the general cartographer white spaces on the map represent the ignorance or preconceptions one has about a place, while the black lines and various dots peppering the map signify the real ground or topography of the map. What is important for the map-maker to recognise, he goes on to explain, is that ignorance about the place transcends any sense of ornamentation one can express on the map. What is referred to as 'blank' or 'empty' spaces are, of course, merely expressions for uncluttered space, rather than space without meaning. Blank also denotes the unknown or unexplainable aspects of mapping, an attempt to catalogue the ineffable qualities of a place. As Robinson once admitted, 'Although I have been making maps for a dozen years now, cartography, in the sense of a general desire and competence to make maps, remains alien to me.'[38] Therefore, the empty qualities of Connemara mirror the process of documentary map-making and film-making, where the procedure entails 'a mode of discovery' without a known conclusion or product. Such a method must be approached without preconceived ideas and the willingness to allow the process of unfolding to occur.

In the opening shot of the film Collins exemplifies the connection between map and mapped by using an establishing long-take shot of the rocky and boggy terrain over a flat section of Connemara. What immediately comes to mind when watching this shot is the immense silence and vast amount of space in Connemara. The relationship between the map and mapped ultimately is based upon communication, or the language of a place. How does a film speak? Or, for that matter, how does a mapped landscape speak? How does a map communicate? Do they contain their own language or does the film-maker or map-maker choose the language? In the film, Robinson admits the 'death rattle is nothing compared to the silencing of life. Or the silencing of language.'[39] Here, silence and language share a similar relationship with place. Just as language contains an infinitely complex array of meanings, silence communicates other levels of depth and intensity.

There is a whole sequence where Collins films point of view shots (from the viewer's perspective) that last a few seconds, and each shot captures different parts of Connemara. He begins with a shot of the valleys strewn with granite and stagnant pools of water, moves to a shot of the mountains covered in mist, then shows a high angle bird's eye view shot of the Atlantic shoreline squeezed in by wild grasses, and finally shows a thicket of verdant ferns amidst a grove of trees. During this entire sequence, there are only diegetic sounds of the external shots, which mostly consist of wind blowing through the leaves of the ash trees and waves crashing upon the rocky shore. Not only do silence and space guide the viewer throughout this succession, Collins also captures the subject of Connemara here much like Robinson's own non-fictive accompaniments to his Connemara map in his trilogy writings. This series of zoom shots end with an extreme close up of what appears to be lichen on an ash tree. The bark configuration resembles a map, with boundary lines and dots running their course (see Figure 13). This symbol seems to suggest that the actual map, with a decoded language that the reader/viewer can understand, may look very similar to the bark of a tree, which is indecipherable and presumably meaningless. The map may simply be the mapable version of the mapped place that the map-maker privileges. Throughout the film the map is Connemara, not the bark of a tree, but the metaphor suggests the subtle distinction the visual form takes in the process of documentary map-making. Such a process depends upon the quality of language and listening when constructing the map, while also subjectively deciphering the mapped as an agent in the discourse of the map.

*Figure 13* Bark of tree (close up resembling a map) (from Pat Collins's film *Tim Robinson: Connemara*, photo by Colm Hogan).

The qualities of listening and sound might be the primary relationship when thinking about the map and the mapped. When comparing the film's empty diegetic sound of the landscape exteriors to the white spaces on the map, there is a sense that the wind denotes the past sweeping through this landscape that both Collins and Robinson attempt to understand through their own forms of documentary. Good documentarians listen to and record what is being said, presented or articulated in the vastness of expression. Collins minimises the non-diegetic sounds in the film; in fact, the soundtrack we do hear layered over the film blends with the actual diegetic sounds of Connemara. This merging indicates that Collins, much like Robinson, wants to listen to the landscape (or the mapped), not only record it as a film (or the map). Robinson goes so far as to parallel sound as a way of recognising the history and memories of the mapped landscape when trying to produce a map. Robinson comments in the film that the 'sound of the past is an agonistic multiplicity'.[40] Collins spends a significant amount of time panning shots across the coastline of Robinson's Connemara map, with cross-cutting shots cut back and forth between the actual terrain of Connemara and the extreme close up shots of Robinson's map. During one of these sequences, Robinson reads from the preface of *Listening to the Wind*: 'The ear constructs another wholeness out of the reiterated fragmentation of pitches, and it can be terrible, this wide range of frequencies coalescing into something approaching the auditory chaos and incoherence that sound engineers call white noise.'[41]

Building on the idea of 'white noise' as the chaotic sounds of the empty form of silence penetrating both Collins's film and Robinson's map, I shall conclude by suggesting that these wide ranges of frequencies swirling through the Connemara landscape are a way of understanding the map and mapped of Robinson's documentary form of map-making.[42] According to Robinson, the intent of the *Connemara* trilogy is to explore three fundamental elements of the landscape: 'the sound of the past, the language we breathe, and our frontage onto the natural world'.⁴³ Sound, language and experience all contribute toward understanding the map and the mapped. They also constitute the principle guiding Collins's depiction of Robinson, on the one hand, and Robinson's description of Connemara, on the other hand. Documentaries rely on the unfolding process of the subject being filmed or mapped. The film/map becomes a product of this process but the viewer never completely understands or records the experience of the film-maker or map-maker. To this end, documentary map-making produces a living product from an equally living process and Collins's film reveals, through Robinson's work of maps and writings and methods to record them, that documentary can also be understood as a form of map-making.

## Notes

1. Tim Robinson, *Setting Foot on the Shores of Connemara and Other Writings* (Dublin: Lilliput Press, 1996), 77.
2. Tom Conley, *Cartographic Cinema* (Minneapolis: University of Minnesota Press, 2007), 5.
3. Cited in Eamonn Wall, *Writing the Irish West: Ecologies and Traditions* (Notre Dame, IN: University of Notre Dame Press, 2011), 29.
4. Derek O'Connor (Irish Film Institute) specifically used the term 'psycho-geographical journey' in his promotional review of *Silence*. See Pat Collins, *Silence* (Cork: Harvest Films, 2012), at the 15th Annual European Union Film Festival. Accessed 3 May 2012, www.eufilmfestival.com/night04_12.htm.
5. Jack C. Ellis and Betsy A. McLane, *A New History of Documentary Film* (New York: Continuum, 2005), 1.
6. Ellis and McLane, *Documentary Film*, 2.
7. Bill Nichols, *Introduction to Documentary* (Bloomington: University of Indiana Press, 2001), 20. Original emphasis.
8. Robinson, *Setting Foot*, 78. My emphasis.
9. Robinson refers to himself as a 'topographical writer', in addition to other themes that have associations with space in his work. See Tim Robinson, *My Time in Space* (Dublin: Lilliput, 2001), 10. Robinson's map of Connemara took many years to complete and was finally published in 1990.
10. See Derek Gladwin, 'The Literary Cartographic Impulse: Imagined Island Topographies in Ireland and Newfoundland', *Canadian Journal of Irish Studies* 38:1–2 (2014): 158–83. For a range of perspectives on literary cartography, see Robert T. Tally Jr (ed.), *Literary Cartographies: Spatiality, Representation, and Narrative* (New York: Palgrave Macmillan, 2014).
11. Robinson, *Setting Foot*, 15.
12. Robinson, *Setting Foot*, 78.
13. Robinson, *Setting Foot*, vi.
14. Ellis and McLane, *Documentary Film*, 3. It is worth noting that the indigenous islanders in Flaherty's film were ironically actors portraying lives of those who live on Aran, thereby defying the purpose of documentary film.
15. Kent C. Ryden, *Mapping the Invisible Landscape: Folklore, Writing, and the Sense of Place* (Iowa City: University of Iowa Press, 1993), 221.
16. Yi-Fu Tuan, *Space and Place: The Perspective of Experience* (Minneapolis: University of Minnesota Press, 1977), 3.
17. Robinson, *Setting Foot*, 100.

18　Ellis and McLane, *Documentary Film*, 2.
19　Pat Collins, *Tim Robinson: Connemara* (Cork: Harvest Films, 2011).
20　Collins, *Tim Robinson*.
21　Collins, *Tim Robinson*.
22　Robinson, *Setting Foot*, 77.
23　Conley, *Cartographic Cinema*, 1. For more on mapping and cinema, see Sam Rhodie, *Promised Lands: Cinema, Geography, Modernism* (London: British Film Institute, 2001). Another useful study, although more focused on visual culture and landscape, is Fred Trunger, *Filmic Mapping: Documentary Film and the Visual Culture of Landscape Architecture* (Zurich: Jovis, 2013).
24　Conley, *Cartographic Cinema*, 2.
25　Robinson, *Setting Foot*, 3.
26　Robinson, *Setting Foot*, 77.
27　Collins, *Tim Robinson*. See also Robinson, *Setting Foot*, 18.
28　Wall, *Writing*, 7.
29　Wall, *Writing*, 10.
30　Tim Robinson, *Connemara: Listening to the Wind* (London: Penguin, 2006), 5.
31　Jackie Keogh, 'Filmmaker Shows Confidence in the Small', *The Southern Star* (11 October 2012): n.p. Accessed 8 May 2013, www.southernstar.ie/Home/Filmmaker-shows-confidence-in-the-small-11102012.htm.
32　Wall, *Writing*, 14.
33　Cited in Wall, *Writing*, 14.
34　J'aime Morrison, '"Tapping Secrecies of Stone": Irish Roads as Performances of Movement, Measurement, and Memory', in Sara Brady and Fintan Walsh (eds), *Crossroads: Performance Studies and Irish Culture* (Basingstoke, UK: Palgrave Macmillan, 2009), 75.
35　Collins, *Tim Robinson*.
36　Christine Cusick, 'Mapping Placelore: Tim Robinson's Ambulation and Articulation of Connemara as Bioregion', in Tom Lynch and Cheryll Glotfelty (eds), *The Bioregional Imagination: Literature, Ecology, and Place* (Athens, GA: University of Georgia Press, 2012), 140.
37　Collins, *Tim Robinson*.
38　Robinson, *Setting Foot*, 75.
39　Collins, *Tim Robinson*.
40　Collins, *Tim Robinson*.
41　Collins, *Tim Robinson*. See also Robinson, *Listening*, 2.
42　For more on listening and sound in Robinson's work, see Gerry Smyth's essay in this collection titled '"About Nothing, About Everything": Listening in / to Tim Robinson' (Chapter 11).
43　Robinson, *Listening*, 3.

# II
# Topographic writing and narrative

# 5

# 'And now intellect, discovering its own effects': Tim Robinson as narrative scholar

## *Christine Cusick*

> Seeing comes before words. The child looks and recognizes before it can speak. But there is another sense in which seeing comes before words. It is seeing which establishes our place in the surrounding world; we explain that world with words but words can never undo the fact that we are surrounded by it. The relation between what we see and what we know is never settled.[1]
>
> – John Berger

## Introduction

There is a small cafe table in my campus office, and on either side of the table are mismatched chairs from the dusty storerooms of our campus tunnels. In these chairs, and by the light upon the table, I meet with students about their writing, their professional futures and sometimes their fears. I always keep a book on this table, and one day a curious student paused before he left and asked, looking at the overflowing bookshelves against the walls of my modest office, why I typically choose just one book to rest on that table. I told him that it is the book that I am thinking about at that time. The one that I have read, or that I am rereading and that I am trying to understand. The book changes, but when I meet with students I like to remind myself that I'm also still figuring things out, so I keep that book within reach. Politely listening, the student hesitated before responding, until a look of concern spread across his face and he said with respect and earnestness: 'But that same book has been there for a long time. Are you *still* figuring it out?' The book that the student was looking at is *Stones of Aran: Pilgrimage*, and he was correct: it has been on my mind, and my table, for a long time. And to an extent, I suppose that perhaps I never really want to let it go.

In Robinson's preface to *Pilgrimage*, he tells the now famous story of his and his partner's Máiréad's first day on Aran Island in the summer of 1972: 'On the day of our arrival we met an old man who explained the basic geography: "The ocean", he told us, "goes all around the island." We let that remark direct our rambles on that brief holiday, and found indeed that the ocean encircles Aran like the rim of a magnifying glass, focusing attention to the point of obsession.'[2] To this day, forty-three years later, having traversed literally every inch of this island terrain, Robinson still returns to this beginning, stating in a July 2014 interview with the French programme Étonnants Voyageurs that when he heard this description he felt that 'that seemed enough of a guide to the geography for me to start off'.[3]

As academics bound to the treaty of specialisation, we often spend years of our professional lives studying one cultural era, one author, one historical moment. It is not unusual for us to dedicate careers to these content areas, but there is typically one moment that marks the beginning, the moment that Virginia Woolf recalls when she asks 'you know the little tug – the sudden conglomeration of an idea at the end of one's line: and then the cautious hauling of it in, and the careful laying of it out?'[4] For Robinson, the ocean 'that goes all around the island' was enough of a tug to inspire him to respond to a postmistress's suggestion that he make use of his hours of foot exploration by creating a map of the rugged island terrain. Four decades later, it strikes me as peculiar that interviewers still express astonishment that Robinson has spent so much time studying a region as ecologically and culturally variegated as Aran and Connemara. I suspect humans would not question a literary scholar's lifelong dedication to the works of James Joyce, for example. Perhaps this simply speaks to the bias that humans embody, often without realising, toward the worthiness and complexity of human culture. And perhaps it is also because Robinson's treks are somehow understood as something other than a method of study. This essay is, in one sense, an exercise in speculation: what would happen if we upended these biases, if we saw a geographic region as worthy of such close study and if we reimagined the intellectual methodologies that we invoke to gain understanding?

**Narrative scholarship as approach**

In a 2010 interview I asked Robinson how he would respond to a description of his non-fiction as a form of 'narrative scholarship'. He responded with the following words: 'I try to present any scholarship there may be in my work as a personal experience, that of a learner. The smell and heft of a tome matter to me as much as the lift off of a seagull. All experience, bodily or mental, becomes the matter of a book.'[5] Robinson's choice of the word 'learner' embodies the ethos of his approach. Stepping gently, with curiosity, and in this movement recognising the layers of human experience, his perception and access to understanding is part intellect and part sensory, both of the same whole.

Within the context of academic discourse, Robinson's response articulates what some may identify as a fundamental paradigm shift of an academic discourse, a shift that creates space for personal narrative. Such scholarship requires that we examine our assumptions about the sources of knowledge and about our hermeneutic relationship with this knowledge.⁶ Such scholarship asks academics to present intellectual insight without an assumption of objectivity, authority or even correctness. This movement is not in and of itself a new phenomenon, nor is it contained within one discipline.

Steven L. Winter, a constitutional law professor, for example, explores the turn to narrative scholarship in legal studies, offering conditional agreement when 'supporters of narrative scholarship argued that storytelling captures modalities of meaning that cannot be expressed in more conventional legal analysis'.⁷ Moreover, recent studies, such as *Telling Stories: The Use of Personal Narratives in the Social Sciences and History*, examine both the opportunities and challenges of integrating personal stories into evidence-based research, while feminist scholars such as Rachel Blau DuPlessis and bell hooks have long articulated the necessity for autobiographical criticism as a methodology for reasserting that the personal is political, particularly in the context of feminist theory.⁸

This study, however, invokes the term 'narrative scholarship' from its specific application for environmental criticism. Ecocritic Scott Slovic offers the following rationale for narrative scholarship: 'We must not reduce our scholarship to an arid hyperintellectual game, devoid of smells and tastes, devoid of actual experience … [We must] analyze and explain literature through storytelling.'⁹ Within environmental criticism, what distinguishes this mode of enquiry is a conscious awareness of how the sensorial experience of an individual in biological context shapes the cerebral encounter.¹⁰ Ian Marshall also articulates this point: 'Narrative scholarship is a way of putting into practice the ecological principle of interconnectedness. To look at something from some objective distance implies that you are outside it, not part of it. To be aware of our role not just as observer but as participant.'¹¹ Implicit in this ecocritical sensibility is that narrative scholarship situates intellectual analysis within its material context: 'If we believe that writers are influenced by places as well as texts, it makes sense that a careful scholar, as a matter of credibility and authority, should check those sources, making use of what Simon Schama calls "the archive of the feet".'¹²

While Slovic is often recognised as a founding voice of formal narrative scholarship, Terry Gifford notes that 'narrative scholarship has, in a sense, been an assumption behind American nature writing since John Muir's first published essays'.¹³ At the same time, Gifford points out, 'This kind of writing is generally frowned upon in the United Kingdom with the suspicion that such personal narratives are probably too self-indulgent and uncritical.'¹⁴ There are, according to Gifford, noted exceptions to this critique, among them the work of writers such as Tim Robinson.

For Robinson, the narrative process is embedded and bound to his steady trek and grounded in mutuality. He manifests stylistic unity as both a cartographer and a prose writer. His mapping is the result of walking to learn, and so this ambling is nearly always the context for a participant in his composing processes: 'it occurs to me that there is at least a coherence between this style of drawing and a cluster of images that surface everywhere in my writing, centering on the human pace, the step taken'.[15] At the same time, his record is not merely of his own journey. Rather, his walking is a tool of investigation as he seeks and listens to the stories of both planned and unplanned encounters. Gifford argues that 'narrative scholarship derives its distinctive strength' from 'mutual meaning-making resulting from the integration of reading and being, of historical research and personal experience'.[16] Robinson's prose is constantly moving, across the page, and within itself. He brings research knowledge to his walks and his walks give him the questions to take back to his books, and he captures the simultaneity of this 'meaning making' through word to page. In 'Interim Reports from Folding Landscapes', Robinson reflects on his decisions about perspective and proportion: 'In all these choices I was trying to preserve the texture of immediate experience. I had a formula to guide me and whip me on through the thickets of difficulties I encountered: while walking this land, I am the pen on the paper; while drawing this map, my pen is myself walking the land.'[17] Robinson's purposeful attention to this immediacy is what allows his step, his pen, to dialogue with his subject in a gesture of authentic reciprocity.

This immediacy, however, is bound to both the part and the whole of Robinson's experience. He sees the path before him, but through his investigation, through his invitation to the local inhabitants to participate in the narrative process, he honours and seeks connection of time and encounter. In his essay 'Landscape and Narrative', North American nature writer Barry Lopez writes,

> Perhaps a black-throated sparrow lands in a paloverde bush – the resiliency of the twig under the bird, that precise shade of yellowish-green against the milk-blue sky, the fluttering whir of the arriving sparrow, are what I mean by 'the landscape' ... These are all elements of the land, and what makes the landscape comprehensible are the relationships between them. One learns a landscape finally not by knowing the name or identity of everything in it, but by perceiving the relationships in it – like that between the sparrow and the twig.[18]

For Robinson, the relational characteristic of the landscape includes details of avian flight as well as the elemental composition of the turf, but it also includes the ownership history of village homes and the kindness of a village woman who gives him chocolate biscuits with her stories. Robinson records these details not as mere anecdote but as necessary narrative threads that he darns together as he softly but knowingly begins to record his landscape

### Listening to the sea

Robinson often invokes memories of his now well-known treks across the Connemara landscape as subject for his writing, one part of a layered description of the bogland that he hikes, the tidelines that he maps and the human stories that he records. In an interview conversation with Robinson, I asked him about this narrative process, about how memory and experience coalesce in his writing. He responded with the following example:

> In *Stones of Aran: Pilgrimage* I describe a bay on the Atlantic cliff coast of Árainn called Blind Sound because from sea it looks like a sound between two islands but a vessel entering it would soon discover that it is closed or blind. Hundreds of pages later, in *Stones of Aran: Labyrinth* I come to a holy well a mile or so away on the other side of the island the legend of which concerns a blind man who was led to it by the sound of a voice from the well, and was cured there. The story suggested to me that this might be the right place in which to sit with my eyes closed and listen to the sounds of the island. Doing so, I heard the breakers crashing in Blind Sound. I wrote about how appositely the name Blind Sound came in – 'as if to make the point that to the making of a point every other point is apposite' – that is, that the world is total interconnectivity. In fact the connection between the Sound and the well is linguistic so it hardly emanates from place, except in the general sense that we all do, as does all we do. Language short-circuits reality, creating a moment in which one feels the world supersaturated with meaning, with self-reference.[19]

This explication of narrative process uncovers the care with which he listens to given stories. At the same time, it reveals the centrality of Robinson's island terrain as his interpretive lens. He first views and interprets the sound from the position of the sea, an interpretation that changes as he nears the land and more clearly sees its position against the terrain; his sensibility emanates from his physical inhabitation of both sea and land. Further to the point, as Robinson notes, his narrative negotiation of this sound resurfaces in his next volume, hundreds of pages later, when he learns the community's narrative interpretation of the Blind Sound. It is at the intersection of these two experiences that he sits against the aural resonance of the breaking waves that mark the meeting place of land and sea. The tideline is thus the subject of his narrative, but it is also a part of his epistemological lens through which he is able to perceive the linguistic and existential meaning of the single word 'blind', and then to communicate these layers by connecting reader to material experience.

Environmental philosopher Christopher Preston argues that 'mind does not pre-exist place and then utilise it as a resource in order to expand its already existing capabilities. The physical environment is not just an instrument for mind but a source of the very possibility of its existence in the first place'. According to Preston, it follows then that 'the physical environment supplies some of the very architecture of the mind and is integrated into some of its very processes. Mind thus depends on place rather than merely uses it.'[20] Set against Lopez's argument that narrative identifies and places into relationship the part and particle of the whole landscape, Preston's attunement to cognition underscores the impact of the material on human perception. Within this paradigm, human knowledge and meaning is juxtaposed with the physical environment rather than imposed upon it. And, as an enactment of narrative scholarship, such enquiry avoids the risk to the extent that is possible within the confines of language, of privileging a single human's interests.[21]

Robinson encounters his subject with an analytical approach, his interpretive lens, as well as the conditions of his human interactions, defined and shaped by the physical demands of the sea. His mapping treks are, to an extent, reliant on the human capacity to traverse and navigate sea boundaries. In *The Last Pool of Darkness*, for instance, he recalls his first visit to Oileán Mhic Dara:

> I was working my way up and down the boreens of An Aird Thoir in the course of making my map, in the autumn of 1981, when two lads came across their fields to see what I was up to. Joe and Máirtín proved so friendly and interested in my work that it occurred to me to ask if they would like to take me out to the island, and they gladly agreed to do so. They had already been out setting a tram net at seven that morning, and their wooden currach was on the shore below their house.[22]

What begins as a neighbourly gesture stirred by curiosity in Robinson's trek evolved, as do so many of his conversations with local inhabitants, into a shared experience of negotiating the intrepid clash of land and water, uncovering the particularities of this union and the changing ways in which humans discern and inhabit it:

> We first made a clockwise circuit of the island. The expanses of shelving granite outside the storm beach were delightful to walk on; the seaward space was immense, a dim suggestion of the Aran Islands' long profile hardly interrupting the measureless farness of the horizon. Spot found a half-grown seal pup on the shore and barked at it from a safe distance as it snorted and lithered to the sea, where a watchful adult seal was showing its head above the swells. A rock shelf thrusting out into the waves that the lads named for me as An Leic Dheirg, the red flagstone, was a purple-brown field of *creathnach*, a sort of dulse, the sweet edible seaweed the women of coastal Connemara used to gather to dry in the sun and hawk around the hill-farming areas in exchange for wool.[23]

A prose mapping that could easily revert to Robinson's internalisation of encountering this island for the first time seamlessly unfolds into a relational experience of navigation, the approach of the landing place maps moments of encounter: boatmen with ambling writer; granite with sea; canine with seal; rock shelf with waves; seaweed with wool; women with island economy. And embedded in each of these points of encounter is the crash of water against land, not merely as a literary effect, but as an element of perception that impacts the possibilities, and impossibilities, of human knowledge of this place.

Robinson offers the reader detailed narrative access to his island footpaths, his water crossings and his human conversations and encounters, but in the midst and at the end of each study he is decidedly a writer, and the sea's presence is also articulated in this space of recollection and record. He explains, 'My writing desk and computer are in a downstairs room that has the sea lapping along two walls of it at high water, and when a yacht rounding the head of the pier looms in a window behind me, two or three simulacra of it seem to manoeuvre and flit from vista to vista.'[24] Even within the walls of his harbour edge home, however, the sea does not recognise human-constructed boundaries:

> The landscape is folded into the building, not just visually but in respect of all the senses: the cracking and creaking of a hooker's mainsail being raised in the harbour calls us to the windows; the chill rush of a squall sends us hurrying around to position the rounded beach cobbles we use to stop doors slamming; there is a salty tang of damp (and an obscure, many-legged, skipping shore-fauna) in corners of the storeroom.[25]

Cognisant of the poetic sensibility of these sounds and scents, Robinson is not lost to them; he also acknowledges the practical and harsh edges of his home place: 'around the equinoxes, a spring tide coinciding with a south-easterly gale can set big waves on us, rushing across the bay, bursting over the sea wall into the courtyard and forcing their way through crevices of ancient stonework in our foundation to well up through floors'.[26] Robinson's writing space is at once part of and separate from the sea, built environment both resisting and surrendering to its force (see Figure 14). In the end, though, Robinson admits his tender embrace of the intrusion: 'Although I grumble at the having to empty the bottom shelves of the bookcases and stack boxes of map-covers on top of filing cabinets whenever one of the fishermen calls in to warn us of an exceptionally high tide, secretly I relish the sea's occasional visits.'[27]

The sea's proximity to Robinson's place of sleep and pen connects and distances his writing process from his subject. For while there is a billowing line between the domesticity and terrain on this edge of Connemara, there are also reminders of modernity, the glow of a screen, the wheel of a chair, for which Robinson is not apologetic:

> The world beyond the windows sustains the house on our average studious and solitary working days too. There are in particular the tidal doings of the seaweedy shore immediately under the window at my elbow as I write, which I can rest my eyes on by spinning my office chair a quarter turn to the left from the computer screen ... In winter there is often a heron, motionless, a monk worn grey and sinewy by fast and prayer, waiting to stab a fish out of the water as I wait to catch the word I need.[28]

Robinson is embedded in physical place not just as he studies his subject but also as he translates his perceptions into word and map, and from observation to creation he is looking to the physical realities beyond his eyes and beneath his chair, and awaiting his word as the heron anticipates his fish, both responding to the instinctive impulse toward sustenance, both turning to the sea.

*Figure 14* Robinson's library (photo by Nessa Cronin).

**Hesitation and meaning**

Despite Robinson's attunement to his own biological and philosophical embeddedness in physical place, there is always an undercurrent of caution in his conclusions. His human self is never fully integrated into this seascape, due in part to the philosophical and physical incongruities. Cultural geographer John Wylie's astute study of Robinson articulates this impasse, arguing that the 'insistent message' of Robinson's writing is that 'any such sought-after unification – of self and land, word and world – is *never* ultimately achieved, and is moreover recognized as unachievable'.[29] In the *Connemara* trilogy there seems to be an elegiac tone attached to this impossibility, not because Robinson seeks or even desires this unity, but because of the effect of this fracture on the earth:

> As to our own effects on the ground we stand on, our powers of creative destruction and destructive creativity are enmeshed inextricably ... A new species has arrived, carrying a dreadful weapon, the intellect. An arms race has begun, the axe evolves from stone to bronze to iron to steel. Great woods with all their sights and sounds go down into silence; the animals succumb ... And now intellect, discovering its own effects, acquires a guilty self-consciousness. At the last moment we try to conserve some shreds of nature, which are, in fact the waste products of our economy.[30]

And herein lies the paradox of human encounter, one that depends upon a degree of intrusion and disruption: 'Should we stand here discussing the origins of the bog, knowing that a footprint in sphagnum moss lasts a year or more, that the tuft of lichen we crush unseeingly has taken decades to grow?'[31] This moment of self-awareness captures the possibility of analytical and engaged narrative scholarship as a form of environmental criticism; the scholar of place is both committed to his subject and yet facing the consequences of this pursuit. There is no false pretense of objectivity that often stifles traditional scholarship, but at the same time, there is no false supposition of unadulterated subject: 'Sometimes when a snipe leaps from under my feet and goes panicking up the sky, I am appalled at my own presence in a place so old and slow and long suffering as Roundstone Bog.'[32] If a reader is to truly accompany Robinson on his path, she must accept the agony of self-consciousness as much as she embraces the beauty of the heron beyond the windowpane.

In the early pages of *Labyrinth*, Robinson describes the Aran topography of Na Craga, the crags, remarking, with awe, at the desolation of this field network of drystone walls:

> If labour reveals its presence at all it is by sound, and almost solely by the clank of stone on stone. One hears how, over centuries, this landscape was created by the placing of stone on stone, how it is nowadays barely maintained by replacement of the fallen stone, and how it will lapse into rubble, stone by stone.[33]

Even in their desolation, and the echo of their rubble, these stone walls imply a lived presence, the necessity of allowing the young to decipher their own manner of being on the fields: 'Araners learn this as part of learning to walk. I have seen an Aran father stand back, watchful but not interfering, as a toddler heads up a six-foot wall. If that child does not leave the island he or she will grow up able to cross walls with such fluency one cannot see how it is done.'[34] The question that compels much of Robinson's narrative investigation is what happens if the child does leave, or if, at the very least, this manner of climbing is never recorded for those who are left? Will that moment of a father's gaze be met with silence when the child fails to return over that same wall? Even in these early pages of his prose, Robinson meets these questions with an impending sense of loss: 'I suspect that the child will leave Na Craga to the old men, and the old men are already leaving it to its ghosts.'[35] And yet, Robinson lifts his boot, his pen, toward the next wall's layers of might and rubble, eagerly anticipating the climb.

It seems the recurring task for a reader of Robinson is to come to terms with, and perhaps even reconcile, his compulsion to record what he accepts is now a land that rests in the hands of its ghosts.[36] But is not this always the task of narrative? When humans tell stories, when we write close studies of history or literature, is not this always an act of piecework: the task of threading together the edges of the past to the present in the hope of breathing one more moment of existence into this moment or text, enough oxygen for one more reader? Robinson's approach to mapping Aran and Connemara has been

*Figure 15   Dry stone wall* (drawing by Angie Shanahan).

decidedly diachronic. His prose records the immediate moment, but inevitably places it within its proper evolving context. In writing about a deserted Famine village, Rosroe, Robnison reflects: 'I believe that right living in a place – as I try to life in Connemara – entails a neighbourly acquaintance with those who lived there in previous times.'[37] Having spent some time with Robinson's words, I hesitate to suggest that he would ever seek to prescribe a didactic script for 'right living', but perhaps the notion that the manner in which we exist in a place, whether it is as a neighbour, a scholar, a fisherman or a post-mistress, is somehow the source of our meaning as human animals traversing tiny pieces of the universe.

## The next step

Despite a recurring sense of impossibility and inadequacy in Robinson's work, there is a strand of hope that his treks, his words and ink marks on a page matter:

> For over twenty years now I have been living in and exploring with manic attention a rather limited patch of that globe ... And if the countless footsteps I have taken in these three terrains have not in some sense carried me beyond their horizons, if the work I have done there does not have wider relevance, then I have cast away a large proportion of my life.[38]

The international fascination with Robinson's 'footsteps' is striking testimony to the fact that what he has done in this patch of the globe indeed has 'wider relevance', a reality that this collection alone articulates. However, I would like to argue that his work has particular relevance for how scholars, including but not limited to those concerned with environmental criticism, approach our subjects of study. Christopher Preston argues that a 'non-anthropocentric ethic looks beyond simply social and cultural factors when trying to answer the question of why knowledge is shaped the way it is. Such an epistemology insists that the physical realities of the environments in which beliefs are formed are relevant to the ways people know.'[39] It is of course rare that scholars of a field such as Irish Studies would ever approach a text without considering identity factors such as colonial history or gender. But what if *how* we come to understand these texts also has something to do with our point of encounter? The imaginative design that we as scholars bring to it, and the hermeneutic framework that we invoke to understand it. And what if both this architecture and framework has something to do with our own material context as well as that of the text? We witness the tactility and care with which Robinson engages with these regions, but in his gesture he knows that the end result is about more than these regions. Perhaps what we can take from this influence is the desire for more expansive, more mindful scholarship, the sort that Gerry Smyth suggests embodies 'a willingness to attune itself to the inaudible music of that same text'.[40] I wonder if we could be more purposeful in making connections between the contexts for production and the histories embedded in the literary texts that we study, not as a matter of polemics, but rather as a means of uncovering how the various components of non-human nature are entrenched in the material culture that infuses our ways of knowing. Perhaps we must be willing to step beyond our scope of expertise, walk across our campuses and into our communities to invite stories that we may not, initially, understand. Robinson is asking us to reimagine our methodology, not in the sense that he is overtly suggesting we all put on our boots, though I would argue that that is a crucial part of it, but his work, and his process, does in fact model for us that if our goal is understanding then our ways of knowing the world must engage every means of interpretation that we possess, every sense of mind and body.

Cultural geographer Susan Stanford Friedman suggests that 'the story of a given space and time that "history" tells us is just that, a narrative whose ordering principles reflect the implicit or explicit theories of its narrator'.[41] As scholars, we are trained to identify these patterns in the work and process of our subject, but how often are we reflective of our own starting points? Or, do we instead allow our scholarship to speak for our own myopic interests rather than for the subject that we claim to honour? Narrative scholarship is certainly not the only mode of enquiry that reimagines our approach. However, perhaps what Robinson's work suggests to us is that there is value in inviting stories that have not yet been published, which may not initially fit within our academic discourse, in recording and valuing the stories that we hear along the way, and perhaps even being honest about our own.

In the final chapter of *A Little Gaelic Kingdom*, which Robinson reminds the reader in the Author's Note was last to be written despite its second place in the trilogy, there is a moment of convergence that seems to draw Robinson to his starting point while recalling the layers of record that he has given voice in the decades since his early treks: 'All the peninsulas and archipelagos of south Connemara seem to yearn towards Aran, which has an almost transcendental status in their consciousness.'[42] As often occurs in this trilogy, Robinson pauses his hike to study the vista against a horizon, wondering about his place against its magnitude:

> Thinking now of the luminous cleanliness and bell-like resonance of Aran's limestone rock sheets, their parallel fissures pointing one to the edge of clear-cut cliffs ... I realize what a difficult terrain is south Connemara: multidirectional from every point, so complex in form it verges on the formless, disputing every step with stony irregularities, leachlike softness of bog or bootlace-catching twiggy heath. Often when visitors ask me what they should see in this region I am at a loss. A curious hole in the ground? The memory of an old song about a drowning? Ultimately I have to tell them that this is a land without shortcuts.[43]

John Wylie compellingly argues that 'the paradox haunting Robinson's writing is that the more he says, the more the words accumulate, thousands of pages of them, the more Aran and Connemara withdraw from view'.[44] Perhaps this impossibility does linger upon and maybe even 'haunt' Robinson's work, but I think that this is as he would exactly have it, never desiring 'shortcuts' into these terrains, but actually relishing in all that he will never know.

In the 2014 interview on Étonnants Voyageurs, Robinson said, 'I have an endless amount to say about Connemara. But the trilogy is finished, and with it a project that has taken me about forty years has come to an end. And at the moment, I am trying to arrange my life so that I am facing a blank sheet of paper, once again – in physical reality, and in my mind.'[45] As Robinson permits this turn for himself, my sense is that the most pressing question he will ask of himself he wrote in *Listening to the Wind* just as he turned from an uphill climb north of Mám Éan:

> Looking back now, from the present moment of writing to that moment of retrospection, I see that, as it has worked out, I was looking at the ground trodden by this book and which I have come to regard as my neighbourhood ... Have I dealt fairly with the land and its people?[46]

If his words on the page are any indication, I think that the answer that he may hear as the gale wind lifts off of the Atlantic is a definitive, and perhaps, hushed, yes.

## Notes

1 John Berger, *Ways of Seeing* (London: Penguin Books, 1990), 7. 'And now intellect, discovering its own effects' in the chapter title is from Tim Robinson, *Connemara: Listening to the Wind* (Dublin: Penguin 2006), 56.
2 Tim Robinson, *Stones of Aran: Pilgrimage* (London: Penguin Books), 10.
3 Tim Robinson, 'William Fiennes, Tim Robinson: La beaute du monde', Étonnants Voyageurs: Festival International du Livre et du Film. 7–9 June 2014. Accessed 1 October 2014, http://vimeo.com/97608775.
4 Virginia Woolf, *A Room of One's Own* (New York: A Harvest, 1989), 5.
5 Christine Cusick, 'Mindful Paths: An Interview with Tim Robinson', in *Out of the Earth: Ecocritical Readings of Irish Texts*, ed. Christine Cusick (Cork: Cork University Press, 2010), 210.
6 I am reminded here of Glen Love's words in *Practical Ecocriticism* (Charlottesville: University of Virginia Press, 2003): 'research goes on within a biosphere, the part of the earth and its atmosphere in which life exists. In some of the literary texts that we study and discuss, this enveloping natural world is a part of the subject on the printed page before us. But even when it is not, it remains as a given, a part of the interpretive context, whether or not we choose to deal with it' (16).
7 Steven L. Winter, *A Clearing in the Forest: Law, Life and Mind* (Chicago: University of Chicago Press, 2005), 127. While conceding this value, Winter warns of shifting too far toward the story as knowledge source, particularly when listening for legal culpability: 'In telling a story, the narrator invites us to enter a constructed world. Because we share basic cultural assumptions about how that narrative world is constructed, we know automatically that we are being asked to view that constructed world from the point of view of the protagonist' (128).
8 Mary Jo Maynes, Jennifer L. Pierce and Barbara Laslett, *Telling Stories: The Use of Personal Narrative in the Social Sciences and History* (Ithaca: Cornell University Press, 2008). Rachel Blau DuPlessis, *The Pink Guitar: Writing as Feminist Practice* (Tuscaloosa: University of Alabama Press, 2006). bell hooks, *Bone Black: Memoirs of a Girlhood* (New York: Holt, 1997).
9 Scott Slovic, *Going Away to Think: Engagement, Retreat, and Ecocritical Responsibility*. (Reno: University of Nevada Press, 2008), 28.
10 Neil Evernden contends that because of recent scientific discoveries of cellular biology, humans more fully comprehend the extent to which 'there is no such thing as an individual, only an individual-in-context, individual ... defined by place' (103). 'Beyond Ecology: Self, Place and the Pathetic Fallacy', in Cheryl Glotfelty and Harold Fromm (eds), *Ecocriticism Reader* (Athens: University of Georgia Press, 1996), 92–104.
11 Ian Marshall, *Story Line: Exploring the Literature of the Appalachian Trail* (Charlottesville: University of Virginia Press, 1998), 8.
12 Marshall, *Story Line*, 8.
13 Terry Gifford, *Reconnecting with John Muir: Essays in Post-Pastoral Practice* (Athens: University of Georgia Press, 2006), 108.
14 Gifford, *Reconnecting*, 108.
15 Tim Robinson, *Setting Foot on the Shores of Connemara and Other Writings* (Dublin: Lilliput Press, 1996), 213.
16 Gifford, *Reconnecting*, 106.
17 Robinson, *Setting Foot*, 77.
18 Barry Lopez, *Vintage Lopez* (New York: Vintage Books, 2004), 6.
19 Cusick, 'Mindful Paths', 207.

20 Christopher Preston, *Grounding Knowledge: Environmental Philosophy, Epistemology and Place* (Athens: University of Georgia Press, 2003), 116.
21 Gifford, for example, reminds us that to a UK audience narrative scholarship has historically been viewed as 'self-indulgent' and 'uncritical' (108). In a similar voice of warning, Michael Cohen identifies some versions of narrative scholarship as a 'praise-song school' of criticism, which, he argues, is 'not sharply analytical but gracefully meditative' (21, 22). 'Blues in the Green: Ecocriticism under Critique', *Environmental History* 9:1 (January 2004), 9–36.
22 Tim Robinson, *Connemara: Last Pool of Darkness* (Dublin: Penguin, 2011), 142–3.
23 Robinson, *A Little Gaelic Kingdoml*, 144.
24 Robinson, *Listening*, 144.
25 Robinson, *Listening*, 144.
26 Robinson, *Listening*, 145.
27 Robinson, *Listening*, 145.
28 Robinson, *Listening*, 147.
29 John Wylie, 'Dwelling and Displacement: Tim Robinson and the Questions of Landscape', *Cultural Geographies* 19:3 (2012), 368. Original emphasis.
30 Robinson, *Listening*, 56.
31 Robinson, *Listening*, 57.
32 Robinson, *Listening*, 57.
33 Robinson, *Labyrinth*, 11.
34 Robinson, *Labyrinth*, 13.
35 Robinson, *Labyrinth*, 13.
36 Kelly Sullivan's astute analysis in this collection describes Robinson's relationship with audience as one of 'readerly responsibility', one that is made possible by the promise of 'not knowing'. See Chapter 6.
37 Robinson, *Connemara: A Little Gaelic Kingdom* (Dublin: Penguin, 2011), 301.
38 Robinson, *Setting Foot*, 210.
39 Preston, *Grounding Knowledge*, xi.
40 Gerry Smyth, 'About Nothing, About Everything': Listening in / to Tim Robinson', Chapter 11.
41 Susan Stanford Friedman, *Mappings: Feminism and the Cultural Geographies of Encounter* (Princeton: Princeton University Press, 1998), 195.
42 Robinson, *Gaelic Kingdom*, 370.
43 Robinson, *Gaelic Kingdom*, 370.
44 Wylie, 'Dwelling and Displacement', 379.
45 Robinson, 'William Fiennes, Tim Robinson'.
46 Robinson, *Listening*, 402.

# 6

# Not knowing as aesthetic imperative in Tim Robinson's *Stones of Aran*

## *Kelly Sullivan*

In Pat Collins's 2011 film, *Tim Robinson: Connemara*, Robinson describes his map-making in terms of what is unknown, explaining that the white space of the black-and-white maps represents 'the state of ignorance with which one starts'. As if to drive home the point that the cartographer or writer of a place can at best gesture toward an entirety, he goes on to say, 'your ignorance transcends any amount you write about the place or can express on one sheet'.[1] He makes a similar point, this time arguing from the perspective of the reader or viewer, in an essay about his map-making process: 'we could not use or even bear to look at a map that was not mostly blank'.[2] In describing the means by which he created maps of the Aran Islands, Connemara and the Burren, Robinson also describes the aesthetics of writing about place; his maps and his books both function as 'conceptual model[s] of the terrain projected onto paper ... representation[s] of spatial relationships in a symbolism that facilitates calculations'.[3] In the same essay he explains that even at its richest, consciousness holds only a fraction of the infinite number of geometric relations between an individual and the world around him; yet this 'rudimentary element of geometry, the relationship of topological inclusion, is the kernel of all the complexities of social and ecological belonging'.[4] And we but barely grasp it.

Consciousness, an infinity of relations, the individual and the world beyond: these are concepts that recur again and again in Robinson's books and essays. In *Stones of Aran: Pilgrimage*, he asks whether it is possible to forge all the contradictions of the human world – geology, biology, personal history, myth, politics – into 'a state of consciousness even fleetingly worthy of its ground'. He proposes that such a synthesis would not be just a work of art, but a step beyond: 'it would be like a reading of that work'. 'Impossible,' he says. And yet it is precisely this aesthetic and ecological experiment that drives both volumes of *Stones of Aran* (*Pilgrimage* and *Labyrinth*).[5]

Thus Robinson's masterpiece of ecological prose, the *Stones of Aran* diptych, suggests failure at the outset. The book will not succeed in narrating 'the good step', in summing up 'that unsummable totality of human perspectives upon [these islands] which is [his] real subject'.[6] Nonetheless, it is precisely this concept of a failure to know the world presented as an aesthetics of 'not knowing' that forms the driving imperative of *Stones of Aran*. It is a necessary and insurmountable ignorance in face of the infinity of perspectives and interactions with any given space with which Robinson grapples and ultimately embraces in the two volumes. In *Pilgrimage*, not knowing coupled with authorial self-doubt about representing place focuses the narrative as a quest for knowledge, a literal pilgrimage around the perimeter of the island of Árainn in an effort to circumscribe that which can be studied, that which 'holds out the delusion of a comprehensible totality'.[7] Karen Babine links Robinson's non-fiction writing to Montaigne's *essais*, quite literally 'attempts', in Chapter 8 of this section on topographical writing and narrative.[8] Robinson's longer prose works are also 'attempts' at grasping that which cannot ever be known. The interchange of attempt and re-attempt compels us forward through the text.

*Labyrinth*, published ten years after *Pilgrimage*, enacts a dramatic shift, illustrating the fundamental unknowability of both places and individuals by using this opacity as a shield. The reader's inability to know helps preserve a place and culture rapidly disappearing in the modernising and ecologically threatened world. Yet even as monument or elegy to a disappearing landscape and culture, Robinson shows that a book is no more comprehensible than the place it purports to represent. *Labyrinth* escapes from its author, showing us that study, scholarship, literature and art are at best tentative gestures toward a deeper knowledge and 'the good step'. Yet taken together, *Stones of Aran*, for all its obliquity and sense of failure, ultimately calls on its audience to develop ways of reading that are adequate not only to the essays and books, but also to a philosophy of inhabiting place that is 'fleetingly worthy of its ground'. An aesthetics of not knowing goes some way toward ameliorating the inherent problem (present in all writing) of representing that which cannot be represented. Robinson approaches this fundamental bind in his work, but, as Gerry Smyth elegantly argues in Chapter 11 of this collection, 'there is at the same time a question mark over the existence at this time in the Western academy of a reading technique equal to the challenge of Tim Robinson's writing'.[9] Much like finding the good step equal to its ground, striving for ways of reading worthy of the literature we enjoy is a fundamental question, even the basis of all literary criticism. With this essay, I hope to offer yet another tentative *essai* toward a reading of Robinson's work adequate to his words on the page and to the good step he proposes.

Robinson's *Stones of Aran* and his more recent three-volume *Connemara* defy easy categorisation. They are compendiums, something beyond nature writing or travel guide or history book; they might best be described as anatomies in their compilation of deep history, geography, culture, folklore, memory, philosophy and environmental studies into detailed word-maps. If they demand critique, the complaint levelled might be that they hold an overabundance of information and deter readers through sheer volume and monumentalism. And so it might seem wrongheaded to argue that Robinson's organising principle in *Stones of Aran* is an aesthetics of not knowing. How could such knowing volumes fail to know everything?

In order to make this claim, I want to briefly examine slightly skewed cognates for 'not knowing', tracing the persistence of this principle through Robinson's own word, ignorance – unknowing, unlearned and, etymologically, 'not knowing' – to a term usually employed in a derogatory sense: stupidity. At the root of stupidity is stupor, stupefied, stunned. Thus stupidity links to a physical state, a stunned dumbness, even muteness, manifestations appropriate to Robinson's physically grounded interaction with place. Giorgio Agamben connects 'studying' to 'stupefy', arguing that those who study the world are shocked, unable to grasp what they want to learn and yet simultaneously unable to let it go.[10] As Natalie Pollard glosses it, instead of a fault or shortcoming, this physical arrest in the face of an onslaught of knowledge means 'the scholar ... is always "stupid"'.[11] But, she continues, 'Agamben's lines might be willing to render stupid the hypothetical "scholar", but not their own act of scholarship.'[12] Stupidity in this formulation is a state of flux that leaves the scholar both awestruck and actively engaged; the desire to know coupled with the literally physical arrest in the face of infinite ways of knowing calls for a creative reaction, scholarship. In Robinson's case, studying and recording facts and impressions – analysing a particular and circumscribed space – simply stupefies anew, even as the writer creates an archive of work. The study will be a failure, and yet it is the process of the work that provides renewal and fascination; a sense of stupidity shocks him into study, but study does not lead back through ignorance to a state of comprehension. Instead, the scholar remains humble before the work.

In *Stupidity*, Avital Ronell touches on stupidity's relationship to knowledge, arguing that it 'does not allow itself to be opposed to knowledge in any simple way, nor is it the other of thought ... it consists, rather, in the absence of a relation to knowing'.[13] Robinson describes his own writing in terms of comprehension and understanding, claiming that 'I have written about many matters I do not understand (but if I restricted myself to what I do understand I would be wordless). Sometimes I have followed the sound of words, trusting them (as a writer does; it is the difference between a writer and an intellectual) to lead me into sense'.[14] Thus Robinson's own writing process occasionally requires he divest himself of a relation to knowing and proceed by 'unknown' paths – sound, subconscious, serendipity. Trusting and proceeding as if blind – unknowingly – also marks Robinson's visually determined work of map-making. Instead of drawing a solid line for 'the coast, a fiction, the high-water mark' on his first map of the Aran Islands, he recreated in the act of drawing 'the hourly give-and-take of land and sea' (see Figure 16).[15] This give and take is also the stupor and subsequent study

Figure 16  Map of Árainn by Robinson.

of the scholar faced with immensities; here it stands in for the constant flux and change from knowing to not knowing that we bring to any lived place, especially one as intimately walked, mapped and studied as Árainn. Even his retrospective 'knowledge' of the creative act is one Robinson recalculates and qualifies: 'all this is to formalise in retrospect a process that was tentative and instinctual, and indeed to fill up with ideals the blanks on the resultant map'.[16]

In *Connemara: The Last Pool of Darkness* Robinson turns to a familiar metaphor when explaining the work of logical philosopher Ludwig Wittgenstein: that of the island, a body of land bounded by an immensity of water. Robinson describes Wittgenstein's project in the *Tractatus Logico-Philosophicus* as an effort to 'draw a boundary in language' that transcribes all which can be said, and leaves outside that boundary everything else. But Robinson goes on to say that, for the philosopher, 'what mattered was exactly that which cannot be said'.[17] Robinson writes of Wittgenstein as a model of intellectuality for him, and so it is no stretch to believe that his regard for Wittgenstein's passions goes beyond mere intellectual history and cleaves close to what the latter author conceives of as his own creative and writerly project. In describing Wittgenstein's work, Robinson thus opens up his own: 'To delineate a coastline is to delimit the land and simultaneously to indicate the illimitable ocean beating upon it; to show the limits of what can be said is to acknowledge the immensity of what Wittgenstein calls *das Mystische*, the transcendental and ineffable.'[18] Wittgenstein's sixth principle very briefly treats of that which lies outside of the logical propositions of science, mathematics and logic; but in the brevity of words lies the immensity of meaning. In its English translation, he adduces 'there are, indeed, things that cannot be put into words. They make themselves manifest. They are what is mystical.'[19] In Wittgenstein's *Tractatus*, it is only what we know that can be said. But it is what we cannot put into words – that which takes active not knowing – that is most fascinating and, finally, most worthy of pursuit. *Stones of Aran* pursues a 'good step' made manifest through recognition of all we cannot know.

Robinson's foray into unsummable totalities takes as subject 'a little piece … cut out of the world, marked off in fact by its richness in significances'.[20] An island is 'already a little abstracted from reality, already half-concept' and so 'holds out the delusion of a comprehensible totality'.[21] Curiously, it is abstraction and concept that appear 'knowable' in this formulation, whereas reality itself remains forever ineffable: *das Mystische*. But Robinson's formulation here functions in parallel to his project of mapping and representing a place, a landscape that stands as the 'reality' beyond himself and beyond the reader. By aligning abstraction and concept with what can be known, he holds the literal landscape as just beyond knowledge, that which is mystical and not only unknown, but actually unknowable.

And yet, we must proceed with caution into our description of this unknown. For Robinson, 'mystical' might be the wrong word. If *Stones of Aran* ultimately seeks to illuminate and make manifest something that cannot be put into words, it does so through a carefully scientific, logical accumulation of data. The narrator of

*Pilgrimage* abjures quasi-spiritual mysticism even as he collects folklore and traces the connections between place names and geographical features of landscape. In one sense, then, the volumes constantly seek that which can and must be known, a logical basis of interpretation. When possible, Robinson demythologises folk wisdom with an objective ear. An episode in *Pilgrimage* recounts an afternoon walk in which he finds himself suddenly 'standing in a circle of ragged, blackish mushrooms'. He tells us that he 'respect[s] ancient forms' even if he does not believe in fairies, and so he left the circle at a place where rock interrupted the fungi.[22] Friends walking with him were no help. They refused to join him 'in this sinister circle, perhaps only because at that stage of the walk no two of the four of [them] were on speaking terms'.[23] Later an old man of the island tells him the circle means he is an angel and his prayers will be answered. Ever wary of received knowledge, Robinson teases out the historic derivation of this folk wisdom: while working in a convent, a young nun once asked the man to pray for her because she was going through trouble. Robinson surmises that, her troubles over, she ran to him in happiness, exclaiming that he was an angel and his prayers were answered. Received ideas, the book seems to say, are not knowledge but instead block rational understanding. And yet, knowing the derivation of belief both demythologises and remythologises. Although the narrator discovers 'the spore from which the whole magic circle of misinterpretation had grown', the passage ends without a biological explanation for the circle of black mushrooms, or an emotional explanation for the obliquely hinted personal history; instead it subtly emphasises not knowledge at all but a 'respect' for symbolic forms even as they are discovered through nature. The text that appears a logical explication is, in fact, just another way of not knowing.

Robinson's aesthetics of not knowing serves several purposes in *Stones of Aran*. Perhaps foremost, not knowing helps guard against the potential violence enacted in any representation of a real place. An anxiety about representation permeates *Pilgrimage*, as it does much place-based writing of the twentieth century, including the work of Edward Abbey, Annie Dillard, Barry Lopez, W. B. Sebald, and Robert Macfarlane. Representation is limitation, in this formula, and Robinson strives to create a work that becomes a living document, something itself subject to the ravages of time that reshape and remake Aran on a daily basis. Only in this way can he avoid the two-dimensionality and constriction of static, knowing description. In a late passage in *Pilgrimage*, the author explains why he will not include photographs in his 'composed book' in which 'the persistency, recurrency and interpenetration of images … are to be as modulable as those of themes in music'.[24] Photographs 'would preserve certain images from the time-flow of writing and reading – and nothing, living or dead, can be allowed that exemption within the covers of this book'.[25] He notably includes writing and reading in this time-flow, thereby acknowledging that a book, like any work of art or any step across geographical space, brings with it temporal specificity. The book we read today will not be the same one we read tomorrow because we will not be the same readers.

We do not know our future selves, just as the author no longer knows his past self or the past work that self authored. And so a book enters the living stream as a living document, something recreated with each reading, and something unknown at each moment of its creation.

This time-flow likewise prohibits any static representation of a place. Any singular description would limit the boundlessness of Aran's unique and fragile landscape, and risk misrepresentation. In an effort to avoid maligning places of spectacular meaning, Robinson takes multiple approaches, as he does with Dún Aonghasa, a prehistoric fort placed high on the cliffs of Inishmore. The fort garners a full twenty-two pages of *Pilgrimage*, but the reader quickly realises Robinson's repeated attempts at showing us his version of this place come in reaction to its frustratingly 'known' quality as a tourist site: 'the ancient and remote fort is a cog of a world-wide machine, hauling up a chain of expectations almost as predictably as a ski-lift'.[26] The more visitors to the dún and the more interpretive interventions and amenities available to them, the more the experience will be diminished; the reader cannot but understand Robinson's unease in adding yet another filter between the reader and the real space. In fact, Robinson's first physical description of the dún is the one that feels most comfortable: 'the old fort guards its privacy by a bland vacancy of expression; it gives one the impression not only that one has seen nothing but that there is nothing to be seen'.[27]

Despite this presentation of nothingness, Robinson elaborates on two approaches, but carefully writes that it is the book itself which commits him to this, and 'to the attempt to fuse them into some more adequate awareness of what it is like to be in Dún Aonghasa – but still I often look around the place, baffled and a little despondent, and feel that the citadel might after all fall more readily before the casual glance of a tourist'.[28] Robinson's knowledge of the fort, like so many interpretive placards, now stands in the way of experience. Truly inhabiting the hilltop dún would require that state of not knowing which enables stupefaction. After many pages of detailed geological and archeological description, history and mythology, Robinson leads us back out with 'a slight sense of exasperation'. If Dún Aonghasa is another stop on the pilgrimage toward the good step, it is one Robinson feels has yielded nothing. The good step requires manifestation beyond what can be known, and the narrator concludes 'once again I have failed adequately to be in this strange place'.[29] In a coda, 'Perdition's Edge', he tries a final time to inhabit the monumental dún, and discovers a metaphor in the cliff edge, the line that divides what can be stated and known from that unknown beyond. Where the cliffs reach their highest, the flora grows lowest to the ground under the spray and blast of stormy seas (see Figure 17). And here 'the scale of what is underfoot shrinks as the edge is approached, where all commentary on this world of fascinating minutiae is brought up short by the overwhelming generalization of the void'.[30]

The sublimation of the particular details *Stones of Aran* contains into the 'overwhelming generalization of the void' of all we do not know places the two

*Figure 17* Cliffs of Árainn (photo by Kelly Sullivan).

volumes on a scale of reality such that they could never adequately convey complete knowing. They are predestined to fail. Robinson tells us that he believes he will not even adequately communicate his own view of the place. Yet this self-doubt is in fact an artistic sleight of hand: the author humbly acknowledges the immensity of the unknown, and in that gesture gains the reader's trust. A brash confidence in his ability to represent Árainn would make us wary and defensive readers, but doubt and imminent failure render us complicit in the quest for the good step, or at least an adequate one. Thus his self-doubt instills trust, a paradoxical reworking of the relationship between reader and writer. John Ruskin, in *Stones of Venice*, the titular model for *Stones of Aran*, approaches his pursuit of a truly comprehensive analysis and history of Venetian gothic architecture with the self-confidence of an educated historian.[31] Although he also acknowledges the connection between instinctual knowledge and truth (what, for Robinson, we might call adequacy), Ruskin claims that even if his reasoning can be 'proved insufficient, the truth of its conclusions would remain the same'.[32] In the introduction to his book, Ruskin assures the reader 'he will find the certainty of every statement I permit myself to make, increase with its importance; and that, for the security of the final conclusions of the following essay, as well as for the resolute veracity of its account of whatever facts have come under my own immediate cognizance, I will pledge myself to the uttermost'.[33]

Ruskin and Robinson ultimately pursue the same goal: they each attest to their book's credibility. But in the case of *Stones of Aran*, readerly trust comes not through assurances, but rather because we understand the beck and call of not knowing, and we come to see the immensity of the place through Robinson's own humbleness in the face of it. In a section appropriately titled 'Fear of Falling, Fear of Failing', Robinson contrasts 'the good step' to An Troigh Mhairbh, 'the dead step'. Island lore says an English officer was urged on by his wife to step further out on a narrow ledge until, facing her challenge, he took 'the step that isn't there' and fell to his death. The writer might fear failure as the officer feared falling, but Robinson explains that we are fascinated with failure and its biological equivalent in mortality 'because of the inadequacy of our step to the earth that bears it'.[34] As readers, we trust Robinson's fascination – his own word – with failure, and his commitment to honour and enact the vibration between stupefaction and study. A fear of failing means renewed effort at finding 'the good step'. Impossibility in knowing is the guarantor of adequacy.

Robinson's self-doubt appears genuine in *Pilgrimage* and especially in *Labyrinth*, and yet we must also note his insistence that these volumes are composed: as the author himself argues, composition allows for modulation. Tone, doubt, uncertainty and not knowing are all formulations here, and carefully worked in prose that is stylistically confident. Doubt and non-knowledge take on the pose of a kind of calculated ignorance, one that we might read as ultimately self-protecting. Robinson presents us with his own limited perspective but simultaneously acknowledges the shackles of such limitation. The aesthetics of not knowing goes even further in *Pilgrimage*. Through gaining the reader's trust and complicity in the pursuit of the good step, Robinson creates a narrative drive through the otherwise overwhelming display of facts, history and nature. Pollard describes stupidity's 'rich temporality' in relation to contemporary poetry, and this richness is equally applicable to *Stones of Aran*: 'Stupidity is not the emptying out of history, a moment of stupefied ahistorical suspension; rather, it shows up as an excess of history; a dense jostling of accounts, happenings and narratives; a piling up of voices interrupting across the past and present, clamouring for our attention.'[35]

Robinson uses similar language in describing what he sees as his book's approximation to the ideal, all-knowing book he cannot compose. What he writes 'has a certain flirtatious but respectful relationship' with the image of a book about which he dreams, the book that is preliminary to the good step. Like our view of Dún Aonghasa, Robinson's Árainn is multiply filtered from the 'reality' of the landscape. The pages we hold might be 'a muddled draft of [that ideal book], or more usefully a demonstration of its impossibility; for the multitudinous, encyclopaedic inscription of all passing reality upon a yard of ground is ultimately self-effacing'.[36] Not knowing provides a narrative purpose, a way for us to wade through the excesses of history alongside Robinson, even as the shortcomings of any one individual's consciousness ultimately determine that we will fail to find our way.

Robinson's narrative route through *Stones of Aran* occasionally follows a continuum from self-doubt to self-effacement. To read all the signatures of a place, one would need to become Emerson's 'transparent eyeball'. Nature surely influences *Stones of Aran*, and Robinson's idea of self-effacement finds precedent in Emerson's sense that 'all mean egotism vanishes ... I am nothing; I see all; the currents of the Universal Being circulate through me'.[37] As the narrator leads us through the labyrinth of information and perspectives, he also catches himself out from too easily 'reading' meaning in landscape. Robinson alludes to Joyce when he describes the beach and bay of Port Mhuirbhig, a quiet cove with a fine sand shoreline. A still evening with a low sun and no one else around 'can tempt one into the error of thinking: "signatures of all things I am here to read"', but even as he attempts to inscribe seabirds, the marks of waves, the scrawl of lugworms, he recognises that his boots, which 'add the stamp of authenticity', only authenticate his 'witnessing' of the natural processes, not those processes themselves. This desire for self-effacement finds a parallel in not knowing; both concepts offer simultaneously positive and negatives answers to Robinson's abiding question: 'is there a way of inhabiting a moment, rather, of knowing all its bays and headlands intimately and lovingly, without that familiarity making one half-blind?'[38] An aesthetics of not knowing contains the answer; inhabiting a moment requires both ultimate consciousness, a superhuman awareness and knowledge, and a loss of consciousness through which there is simply no relation to knowing.

*Pilgrimage* concludes with a return to that familiar metaphor of the island of the explicit in an ocean of the illimitable. In order for his book to be an 'island out of the sea of the unwritten it must acknowledge its own bounds, and turn inward from them', Robinson asserts, and thereby delineates adequacy out of not knowing, a not quite success culled from certain failure.[39] This turn inward, to the labyrinth, also marks a stylistic shift in Robinson's *Stones of Aran*, with the second volume rangier, longer, more personal, less honed. The prose itself seems to signal that we are entering the brambles and rabbit warrens and no longer have that pen-sharp outline of an island as guide, however false such a shoreline may be. The notion of turning inward and acknowledging bounds indicates that *Pilgrimage* is a book about failure, and that this failure is itself the only accurate representation. If Robinson falls short of preparing for his reader 'the good step', he also lets us know that he has 'acted out to the best of [his] capacity the impossibility of interweaving more than two or three at a time of the millions of modes of relating to a place'.[40] Failure is, in fact, the point; the inability to know is what Robinson wants his reader to know. If something has been made manifest in the process of not knowing and standing stupefied before the immensity of place, the narrator most hopes it is this drive to understand what is incomprehensible, and remain humble before it.

Yet the book itself goes one step further and – in his formulation – acts on its own. Although Robinson clearly distinguishes the volume we hold in our hands from 'that imaginary work of art preparatory to the taking of [the good] step', the very logic of the paragraph indicates that the book, the text, becomes an entity separate from the author who conceives of it. In fact, it seems to function of its own volition, and undertakes 'the conceiving of what [he] knew to be inconceivable'.[41] *Stones of Aran* becomes unknown to its author, and Robinson's turn inward in the second volume indicates that it is not only landscape that requires our heightened consciousness in order that we more delicately place ourselves in the world; we must also be conscious of our individual immensities. Instead of condemning his project to the dustbin, Robinson shows us that not knowing as stupidity and not knowing as fascinated study are as proximate as mirror images. In order to conclude *Pilgrimage* and turn to the interior, he will 'walk on, out of the state in which nothing matters into its mirror image, more vivid like all such, in which everything matters'.[42] The text pivots on a key interpretive word in this conclusion to *Pilgrimage*, reshaping the 'adequate' of the initial step which would entail a consciousness 'even fleetingly worthy of its ground' for an adequacy that is not quite equal to but acceptable enough.[43] This adequacy will metaphorically end the quest for the good step and close the looping pilgrimage of the first volume so he can set the book free for reading.

Robinson's assent to a book of prose that does not encompass all he wanted, and yet might just do, ultimately feels not despondent but sonically celebratory. Internal rhyme and chiming end-sounds mark congruencies in his prose that illuminate landscape subconsciously or beyond the inadequate limits of what we 'know': 'This will do for now: the adequate step will be one light and sure enough to carry such explosive significances across tricky ground. This will do; this will get me across An Chois.'[44] The pairing of now/ground and do/Chois indicates that adequacy and knowledge are at their best subconscious and instinctual. It is the sound of words and not only their sense that enables a leap from a debilitating knowledge to a liberating habitation. The step across the water between Straw Island and Inishmore, An Chois, teaches us to trust instinctively what we cannot see if we look directly, and to understand that, if we 'read' properly, we experience far more than we can ever know.

If *Pilgrimage* is a volume dedicated to the pursuit of something its author knew to be inconceivable, then *Labyrinth* enacts for us the impossibility of conceiving of ourselves, our landscapes, or even our perceptions of ourselves within our landscapes, as knowable entireties. Robinson opens with what sounds like a riddle: 'unquestionable answer to unanswerable question' and tells us he must begin this second volume at the same place the former closed so they can be stored away, 'like two mirrors face to face'.[45] But if this image seems one of resolution, it is of a labyrinthine defeat. Two mirrors face to face must endlessly reflect each other, regardless of who may or may not witness this reflection, and thus, placed away, they enact an infinity of unseen images. As a kind of memento mori, Robinson's diptych must find a certain symmetry and cohesion so that it can reach an end, and yet the author acknowledges, as he begins again, that the period of time and geographical distance between the composition of the first book and the second seems insurmountable. In order to penetrate the 'huge amount of material' he has collected and gathered, he turns to a moment of not knowing that was previously disguised as knowledge.[46] He turns to a mistake. In an earlier essay, he had corrected Synge for referring to the maidenhair fern as dúchosach in his meditative *The Aran Islands*; now, ten years later, Robinson has heard Aran people use that name. Chastened by this earlier confidence – 'As if Synge, with his deep, intuitive eyes, cares whether or not I have more facts on Aran than he!' – Robinson retreats to the notes and data in order to grow humble and find intuitive approaches once again.[47]

But these reams of notes and even Robinson's careful correction of his earlier essay quickly raise the question of who it is who quests through this book. What drives us forward, as readers? In fact, if Robinson appears initially cowed by the immensity of information he has compiled at the start, his prose style is if anything more assured; the task of finding a way through this maze of information falls to us. And the bulk of data transmitted in *Labyrinth* – 450 pages of text before notes and indexes – risks immediately enacting for us that sense of stupor and awe in the face of monumental knowledge not yet our own. The impression that we are somehow on our own in *Labyrinth* is not accidental; Robinson's attitude about the big island has changed between the publication of the first book and the second, a period of time during which he and his partner moved from Inishmore to Roundstone in Connemara. But more importantly, the idealistic energy and sense of possibility that infiltrates *Pilgrimage* seems diluted or denigrated here to an almost hunted sense of bare survival. In nearly every chapter the narrative registers a weary distaste for the rapid change and destruction of what we now perceive to be a dynamic and sacred place. *Labyrinth* becomes a cat-and-mouse game with the reader; instead of fellow-travellers on a quest to discover 'the good step', we pursue Robinson through the labyrinthine interior of the island and of his memories of his own time spent there, ferreting out, where we can, spectacular moments of discovery and knowledge from a text that works hard to obscure intimate details.

And yet, *Labyrinth* is a book of contradictions. In his effort to write a work adequate to place, the narrator strives to present to us his own quest, thereby shielding the real place with the known limitations of the personal. He carries the reader along through mental and even physical anguish, ultimately making this sedimentary, rich book more intimate than *Pilgrimage*. At one late point he must reassure himself with '[p]atience, my hand. Patience, my mind. Patience, my heart. Your book will be finished yet.'[48] He includes a third-person history of his own time on the island, writing a cool, disinterested summary of his career when he reaches 'The Residence', the house in which he lived in the village of Eoghannacht. Immediately following this biography comes a nine-page interlude of tender reminiscences of his private time in that house, a section which includes excerpts from his diary. He ends the section with a moment that gestures toward the immensity of *Stones of Aran* and suggests that not knowing can be a way to protect and preserve both individual experience and a real place. He declares that making love to his partner has been a 'sustaining joy' and announces 'There's a certainty! And where else but in the secret heart of my book could I dare such simplicity? From where, proclaim it to so wide a world?'[49]

This moment of declaration hides inside a book full of stylistic and thematic conglomeration, a defensive act or a record of impossibility on the behalf of a real place, one to which Robinson promises to remain faithful. The impossibility of *Labyrinth* is not only of artistic self-doubt and the inability to represent a place. A feeling of profound temporal loss and mortality permeates the volume. Searching again for Poll Talún, a hole that opens to an undercut in the cliffs below about which he had written in *Pilgrimage*, Robinson fails to find it. He quickly discovers that a section of cliff and the blow hole had fallen into the ocean, and 'a page has been torn out of my book'.[50] *Labyrinth* feels saturated with the sense that we can never truly know the island, and someday, with the exception of the 'inadequate' descriptions in *Stones of Aran* and other accounts, we will no longer even know what we do not know; 'Aran is a dying moment.'[51] Like the old fort, the island guards its privacy, but to an obscure end. Here an aesthetics of not knowing is a dark philosophy at best.

If 'not knowing' is a particular narrative strategy in *Stones of Aran*, and if we read the two volumes as deeply concerned not only with explaining or describing or capturing the good step, but also, perhaps more emphatically, with instilling in the reader a sense of what this good step might entail, then the ethics of narrative in Robinson's work is one of readerly responsibility. Not knowing becomes both a shared literary challenge for reader and narrator, and also a means by which the author can insist his reader discover her own imaginative and ecological engagement with place. Thus ethical dilemma and narrative become one and the same driving need: that of finding a way to experience and dwell that is worthy of a real place. Because *Stones of Aran* enacts for the reader the very obtuseness of things unknown, the reader feels the same sense of not knowing that troubles and motivates the author. In *Stupidity*, Ronell, via Flaubert, describes the perturbing indelibleness of stupidity as something we cannot understand because it 'resembles a natural object – a stone or a mountain. One cannot understand a stone or a mountain.'[52] Flaubert links the seemingly stupid to monuments in stone, and finds the *bêtise sublime* in graffiti carved on Pompey's column in Egypt, a sight he witnessed on his travels. How stupid for someone named 'Thompson' to have carved his moniker across the monumental column, and yet Thompson's was an act that makes him – unknown as he is – a part of history.[53] This Victorian vandal literally leaves his signature for others to read, but it signifies nothing except that which we make of it. It is the act of reading that imparts meaning, for there is nothing there to know.

'This is not a travel guide but an elegy. A memorial. You're holding a tombstone in your hands. A bloody rock,' warns Edward Abbey in his 1968 *Desert Solitaire*.[54] Abbey rapidly diminishes the description of his prose narrative from something useful and knowledgeable to its most mute state of not knowing, that of obstinate rock in the Utah desert. And in a fundamental way, Robinson's *Stones of Aran* may also be reduced to intractable masses of limestone if we read them as sacrifices to 'the sacred rite of our times, the acquisition of fact' and do not accept what Robinson openly states: 'that my book can only achieve its end by relinquishing its all-inclusive aspirations'.[55] In his postscript to *Labyrinth*, Robinson echoes Abbey, telling us he has chosen 'the slow deposition of facts and observations, coalescing and fusing under their own weight into tablets of stone'.[56] This is his choice in the either/or predicament of spatial existence. The two options, as he sees them, are 'to be simply present and not to know and remember it, or to be reflectively aware, which implies the mediation of imagery, of mirroring': he can choose not knowing, or choose to present us with the data and accumulation and let us make of it what we will.

This postscript to *Labyrinth* evokes Hy Breasail, locally known as Beg-ara, the Lesser Aran. This is an imagined place, surviving in oral legend and folk memory, an island that 'drifts off the west coast of Ireland like flotsam from the wreck of Atlantis'.[57] In describing an imagined place as the final statement on a labyrinthine and monumental accumulation of fact, Robinson presents the aesthetics of not knowing as readerly responsibility, indeed, as creative opportunity. By asserting his Aran is the Lesser Aran, a mirage and therefore unknowable and impenetrable, he gestures to the strangeness both of the place and of literary prose projects. To 'know', to believe one knows everything or knows enough, would be a greater failure than a 'failed' book: it would be to shirk the duty to learn and discover. Instinctively feeling his way with sound, with words as sensory guides, Robinson ultimately offers a model for imaginative reading and imaginative dwelling. And further, *Stones of Aran* requests its readers follow the logic of not knowing to imaginative engagement, forging their own nodes of experience and making their own adequate steps.

Like the first circuit of *Pilgrimage*, this reading of Robinson's *Stones of Aran* must conclude where it began, with his own reflections on map-making, an artistic process set at an oblique enough angle that it seems to clarify that too well-known act of writing. Ignorance, Robinson said, is the state from which we start, but it is also the state in which we end. The map is primarily blank – indeed it must be so for us to bear looking at it. This is because we need emptiness 'to fill with our own imagined presence, for a map is the representation, simultaneously, of a range of possible spatial relations between the map-user and a part of the world'.[58] Not knowing allows for possibility and imaginative dwelling; we must end in ignorance because in order to continually press on as scholars and as readers, we need the presence of immensity before us. 'The whole layout of [the Burren] map', Robinson writes, 'breathes order, lucidity, certainty. But through its precise gridwork show, I suspect, many tiny darks.'[59] It is through not knowing that we find the 'tiny darks' that are serendipitous; these moments may not be 'the good step', since such a step must always be an unattainable ideal, but they are close approximations, and they are just enough.

## Notes

1. *Tim Robinson: Connemara*, directed by Pat Collins (Cork: Harvest Films, 2011).
2. Tim Robinson, *Setting Foot on the Shores of Connemara and Other Writings* (Dublin: Lilliput Press, 1996), 106.
3. Robinson, *Setting Foot*, 106.
4. Robinson, *Setting Foot*, 105.
5. Tim Robinson, *Stones of Aran: Pilgrimage* [1986] (London: Penguin, 1990), 3.
6. Robinson, *Pilgrimage*, 3.
7. Robinson, *Setting Foot*, 1. In an essay comparing mapping (in both a cartographic and a literary sense) in Robinson's work to that of Canadian John Steffler, Derek Gladwin highlights what he calls 'unknowing'. Gladwin places more emphasis on what he calls the '*unknowability* of our own interiors' but links this quest for self-knowledge to the process of mapping and reading a landscape (3). See Derek Gladwin, 'The Literary Cartographic Impulse: Imagined Topographies in Ireland and Newfoundland', *Canadian Journal of Irish Studies* 38:1–2 (2014).
8. See Karen Babine, 'Tim Robinson and Chris Arthur: In Defence of the Irish Essay' in this volume (Chapter 8).
9. See Smyth's '"About Nothing, About Everything": Listening in /to Tim Robinson' in this volume (Chapter 11), p. 174.
10. Giorgio Agamben, *The Idea of Prose*, trans. M. Sullivan and S. Whitsitt (Albany: State University of New York Press, 1995), 64, quoted in Natalie Pollard, 'The Fate of Stupidity', *Essays in Criticism*, 62:2 (April 2012): 1.
11. Pollard, 'Stupidity', 1.
12. Pollard, 'Stupidity', 1.
13. Avital Ronell, *Stupidity* (Chicago: University of Illinois Press, 2002), 5.
14. Tim Robinson, *My Time in Space* (Dublin: Lilliput, 2001), 9.
15. Robinson, *Setting Foot*, 15.
16. Robinson, *Setting Foot*, 15.
17. Tim Robinson, *Connemara: The Last Pool of Darkness* (London: Penguin, 2008), 32.
18. Robinson, *Last Pool*, 28 and 32.
19. Ludwig Wittgenstein, *Tractatus Logico-Philosophicus* [1922] (Project Gutenberg, 2009). Accessed 13 November 2013, www.gutenberg.org/ebooks/5740.
20. Robinson, *Setting Foot*, 1.
21. Robinson, *Setting Foot*, 1.
22. Robinson, *Pilgrimage*, 56.
23. Robinson, *Pilgrimage*, 56.
24. Robinson, *Pilgrimage*, 160.
25. Robinson, *Pilgrimage*, 160.
26. Robinson, *Pilgrimage*, 64.
27. Robinson, *Pilgrimage*, 65.
28. Robinson, *Pilgrimage*, 65.
29. Robinson, *Pilgrimage*, 83.
30. Robinson, *Pilgrimage*, 84.
31. For more on the relationship between Ruskin and Robinson, see Moya Cannon's essay in this volume, 'Thirteen Ways of Looking at a Landscape: The Poetic in the Work of Tim Robinson' (Chapter 7).
32. John Ruskin, 'Preface', *Stones of Venice*, Vol. 1 (1951). (Project Gutenberg, 2009), n.p. Accessed 13 November 2013, www.gutenberg.org/ebooks/30754.
33. Ruskin, 'Preface', n.p.
34. Robinson, *Pilgrimage*, 100.

35 Pollard, 'Stupidity', 133.
36 Robinson, *Pilgrimage*, 176.
37 Ralph Waldo Emerson, *Nature* [1836] (Project Gutenberg, 2009), n.p. Accessed 20 November 2014, www.gutenberg.org/ebooks/29433.
38 Robinson, *Pilgrimage*, 227.
39 Robinson, *Pilgrimage*, 282.
40 Robinson, *Pilgrimage*, 277.
41 Robinson, *Pilgrimage*, 277.
42 Robinson, *Pilgrimage*, 272.
43 Robinson, *Pilgrimage*, 12.
44 Robinson, *Pilgrimage*, 278.
45 Tim Robinson, *Stones of Aran: Labyrinth* (Dublin: Lilliput Press, 1995), 1.
46 Robinson, *Labyrinth*, 3.
47 Robinson, *Labyrinth*, 4.
48 Robinson, *Labyrinth*, 402.
49 Robinson, *Labyrinth*, 297.
50 Robinson, *Labyrinth*, 95.
51 Robinson, *Labyrinth*, 95.
52 Ronell, *Stupidity*, 11.
53 Ronell, *Stupidity*, 12.
54 Edward Abbey, *Desert Solitaire* [1968] (New York: Touchstone, 1990), xiv.
55 Robinson, *Labyrinth*, 400 and 438.
56 Robinson, *Labyrinth*, 455.
57 Robinson, *Labyrinth*, 451.
58 Robinson, *Setting Foot*, 106.
59 Robinson, *My Time*, 89.

7

# Thirteen ways of looking at a landscape: the poetic in the work of Tim Robinson

## Moya Cannon

I owe Tim Robinson an enormous debt. Through his maps and writing he has introduced me, as he has so many others, to intimate corners of the landscapes and sea edges of the Aran Islands, the Burren and Connemara. Anyone opening one of Tim's books or maps cannot but be immediately arrested by the quality of attention manifest there – by the combination of precision and resonance, the access to and obvious delight in a wide variety of academic disciplines and yet the ability to draw these disciplines together within a wonderfully poetic and frequently playful discourse. On an early visit to the pier-end house in Roundstone called Nimmo House (see Figure 21 in Chapter 9), which he shares with his partner Máiréad, I found him working on his major cartographical project, the map of Connemara. He was painstakingly taking small printouts of names of mountain ranges and fringing them with a scissors or blade so that they could be curved and could thereby follow the contours of the mountain ranges as he pasted them onto the master copy of the map. It is some equivalent practice in relation to language, his painstaking care in the harnessing of words and syntax to faithfully follow reality, which provides so much of the tension and energy of his work.

I first encountered Tim Robinson's work in his map of the Aran Islands. I was already well acquainted with Inis Oírr, the smallest of the Aran Islands, having taught there over several summers, and having come to know the sea-bitten limestone pavement, the remains of Bronze Age settlement, the tiny medieval churches, the wells, the labyrinth of stone walls and the tiny room-sized fields contained within them (see Figure 18). To have every promontory and cove of this small complex territory lovingly charted and named was a great gift, as was the subsequent Burren map and the later map of Connemara. These were Ordnance Survey maps brought to a most unusual point of refinement. In the way in which a musician's house might be charted among megalithic tombs and medieval church ruins, the maps seemed to operate not only on a plane of space but also of time, and to throw a trawl net out to map traces of human culture. There is a multi-dimensional quality

*Figure 18* Stone walls of Árainn (photo by Kelly Sullivan).

to the maps and about the entire body of written work which is reminiscent of the three-dimensional, computer-generated maps of the seabed to which we now have access. Mountain ranges and gorges are revealed where we had previously seen only flat water. Maps are meant to render the three-dimensional into two dimensions, by virtue of the sleight of hand of contour lines. Nothing, in one sense, could be more two-dimensional than the Folding Landscapes' maps, their austere black-and-white clarity a testimony to the drawer's training both as a visual artist and as a mathematician. Yet, in their fine attention to detail and in their ordering of that detail, in the naming, for instance, of particular rocks on the south coast of Inis Oírr, *Dún na Ní*, *Trácht Míl*, *Béal a' Chaladh*, we seem almost to hear, not only the rush and sough and suck of creamy foam over long flat tongues of limestone, but the whispering of pre-Famine seaweed gatherers in their graves in the high sand dunes of the medieval churchyard a little further north (see Figure 19).

The reading of the maps is, at this point, coloured and illuminated by the pleasures of the written work, the main body of the Robinson oeuvre. His first book, *Stones of Aran: Pilgrimage*, is transgressive of academic boundaries in the quietest of ways. It is an account of a walk around the periphery of Árainn, the largest of the Aran Islands, a fragment of limestone laid down near the South Pole some three hundred million years ago and isolated from what is now the mainland of Co. Clare by one of the many great inundations. In *Pilgrimage*, this scrap of the earth's crust, approximately eight and a half miles long by two and a half miles

*Figure 19  Sand dune graveyard, Aran Islands* (drawing by Angie Shanahan).

wide, is viewed through the lenses of oceanography, botany, ornithology, geology, history, sociology, natural history, folklore and several other disciplines. The narrator passes from one academic discipline to another as nimbly as an Aran Islander as Tim's description passes from one tiny field to another via an almost invisible stile in the limestone wall. In spite of the rigour of his research, there is an extraordinary lightness and poetic quality to the work. A few years after I had first read the volume, I was sitting in Tim and Máiréad's snug library-cum-sitting-room (see Figure 14 in Chapter 5), when, on the shelf behind my chair, I noticed a number of editions of John Ruskin's *Stones of Venice*. I should have made the connection earlier. The devotion to precision was the same, as was the elegant concision of the prose and the passion for the genius of a particular water-defined place.

The first time that I met Tim was by the Ballinahinch river, seven miles or so from Roundstone, in Connemara. I was visiting another extraordinary man, John Moriarty, whom I had met a short time previously. I do not think that they knew each other well at this point but, over the next twenty years, until John's death from cancer in 2007, a deep and unusual friendship developed between them. I have never known two people to differ so greatly in their perspectives on reality and yet to hold each other in such high regard. In their philosophies they appeared to agree about nothing – Tim being defiantly materialistic and John, whose knowledge of literature, mythology, philosophy and religions was as encyclopaedic as Tim's in other areas, being barely willing to admit that the material world existed except as spirit in hibernation. It was as if one existed and was planted in that magnificently lit and folded corner of south Connemara as a foil and an intellectual sparring partner for the other. They shared a deep love of, and understanding of, the landscape of Connemara and of ecology generally; a respect for and dedication to intellectual endeavour and a profound human decency. I do not know whether, over the years of their friendship, either succeeded in eroding the position of the other even a fraction of a millimetre. I do know that Tim and Mairéad were immensely supportive of John in his last years. Tim dedicated his fiction 'Orion the Hunter' to John, and it can be read as Tim's own engagement with and ultimate dismissal of a mythological interpretation of reality, a dismissal which is not without a tinge of regret.

The story is a discussion between a mythological and a scientific world view, through the device of focusing on the constellation of Orion. The narrator is reading in bed, researching the name of Chios, when his little terrier whines to be let out into the garden. He gets up to oblige, lets her out, follows her into the garden and is star-struck: 'Above, spanning the perfect blackness, hung a huge empty framework of stars. My breath caught in my throat, as always when I am confronted – Orion the Hunter, a mile high, a thousand miles wide.'[1] The narrator describes that most familiar but enduringly magnificent of our constellations, points out, with characteristic Robinsonian precision and lyricism, a slight upward curve in Orion's belt 'like the sequent notes of a scale, a hunting horn's bright echoing challenge'.[2] He focuses on a perfect Pythagorean triangle which his father had pointed out to him as a child, identifying it as Orion's bow. His father had also pointed out Sirius, the Dog Star, who kept faithfully near the heel of Orion. The narrator then returns indoors, picks out a number of reference books on his way back to bed and proceeds to research the various myths associated with Orion and Sirius. It transpires that Sirius does not belong to Orion at all, that the name 'Orion' means 'Dweller in the Mountains', that the myths associated with him are complex and violent and that his bow may, in fact, be a club. Casting about for some relief from this brutality, he remembers a painting once seen in the Metropolitan Museum of Art in New York – Poussin's *Landscape with Orion*, which presents a much more palatable version of the legend.

The story then shifts mode, as the door, which he had left open for his little dog to return, opens a little further and a figure slips inside:

> Every instinct in my body cried out that it was an animal – its scent immediately filled the room, as complicated as a thicket, with flowers and bitter berries and foxy dung

beneath – but I could see that it was human: a slight, ragged, dark-visaged male. My breath stopped in my throat, but he immediately showed he meant no harm by turning to leave his stick propped in a corner … he was old, he might have been newly delivered out of the ages like a corpse given up by a glacier … I could make nothing of him. Incomprehensibility was engrained in him like a darkness.[3]

The narrator recovers from his shock and proceeds to fling his thoughts at Orion, in a consideration of the absolute immensity of the space between us and the stars, a consideration less timorous but scarcely less awestruck than Pascal's often quoted statement: 'The eternal silence of these infinite spaces terrifies me.'[4] He quotes Whitman, 'I hold a leaf of grass to be the journeywork of the stars … and the mouse is a miracle to confound a quadrillion atheists',[5] and then he poses the question: 'What is the relationship of the leaf to the star?'[6] He asks Orion, 'Do you know anything of stars?', as he might interrogate a snowman about the nature of snow.[7] Focusing initially on Orion's eastern shoulder, the red giant Betelgeuse – which, though ten thousand times as luminous as our sun, is a mere dot to us, being 270 light years away – he then observes that this distance is as nothing compared to the size of the galaxy, 'a hundred billion light years across, built of a hundred billion stars'.[8] The galaxy, in turn, is tiny in the context of the universe. He then returns to the theme of the 'journeywork of the stars'.

Shifting from the bewilderingly macroscopic to the microscopic he asks Orion to consider the origin of life, the formation of molecules and amino acids in space, life which will fall with meteors onto a hospitable planet like ours 'and in no time at all, amoebae, birds of paradise, pyramids, computers'.[9] He moves on to discourteously deconstruct the constellation of Orion, telling him that the distances from Earth to Orion's several stars vary considerably, that his splendid outline is merely a trick of perspective, that

> viewed from the Pole Star, for instance, you do not exist. 'Constellation' is the name of an act, the quintessential human act of joining up the dots, leaping over the dark, stringing events into stories, stories into persons, persons into history. Before we came, stars unnamed bloomed and seeded and blew away like dandelion fluff.[10]

It was humanity who gave meaning to the universe, and, he then reflects, we may not always be here to keep up the pretence of meaning. If and when we destroy ourselves, which is very likely, there may or may not, he muses, be a few bewildered survivors. Should human consciousness cease to exist, he imagines the ever-expanding universe as 'the distracted clockwork bits of heaven ratcheting away, to no end, world without end'.[11] He asks Orion if he can provide any better vision of the future. Travelling perhaps at, perhaps beyond, the speed of light, has he any insight to share with us? Has he a privileged access to the future?

Under this bombardment of scientific information and questioning Orion has retreated into himself. All the wild visual, tangible, olfactory sensations which he had triggered had withdrawn into him 'and the room was left an empty geometry'.[12] He sat silently for a very long time, then, as the narrator's little dog barked sharply from the garden, he appeared to respond to it as to a signal, perhaps from the world of sense experience.

> The hunter stood, and stretched and yawned, took up his stick – it was a little bow ... with a knotted thong for a string – and stepped out into the glow of dawn. Very soon the room was emptied of strangeness, as if he were drawing after him long dim tatters, glittering streamers, billowy, starry banners.[13]

The mythological does not argue back, since that is not its mode. It merely absents itself. The story concludes with the little dog hopping into her basket and the narrator stretching out his hand to his sleeping companion, as domesticity reasserts itself after this strange starry incursion.

The storying of the heavens must have occurred very early in our own human story, sometime very shortly after that most mysterious of evolutionary leaps (if it was a leap), the birth of meaning. It is a storying which continues today, employing our most advanced and dizzylingly powerful technological skills and resources. As mentioned above, this story was dedicated to John Moriarty. In a footnote, it says that John 'used to come in at the door described and hold forth at the foot of our bed'.[14] According to Máiréad, John, on his way home from his weekly grocery shopping, used to waste no time on small talk but used to head straight into 'the abyss' of his own philosophical and metaphysical questioning. It is perhaps his ability to entertain a world view not only at variance with, but opposed to, his own scientific, rational prisms which gives Robinson's work its plenitude and its poetic quality. One of the characteristics of poetry, of any art, is the bringing together of disparate elements and of arranging them into a harmonious whole. This usually involves entertaining contradiction. As W. B. Yeats famously put it, 'We make out of the quarrel with others, rhetoric, but of the quarrel with ourselves, poetry.'[15] This is one of the reasons why poetry can so often calm us where nothing else can, reflecting, as it does, the way in which each human life entertains enormous contradictions.

Tim's engagement with folklore and mythology is manifest in the early map-making project. The place names of the Aran Islands, Connemara and the Burren are frequently literal and descriptive but many of them can be understood only in relation to the stories associated with them. Tim and Máiréad's early encounter with the Aran Islands was an encounter with a deeply storied landscape. I have said that *Stones of Aran* was a walk around the island viewed through several scientific prisms but also viewed through its myths and stories. Although the luminous prose of *Stones of Aran*, the *Connemara* trilogy and the collected essays and fictions reflect the author's belief in science and rationality, they are also characterised by a deep courtesy toward and hospitality to myth and folklore – world views at odds with Tim's own central vision. The Orion story expresses in microcosm what his writings on the Aran Islands and Connemara do on a larger scale.

I have said that I owe Tim Robinson a great debt. I think that Ireland also owes him a great debt. The Polish poet Czeslaw Milosz writes in his poem 'Bypassing Rue Descartes' of revisiting Paris as an older man and of remembering a youthful visit to that capital of world culture. The older man comes to the conclusion, 'There is no capital of the world, neither here nor anywhere else, / And the abolished customs are restored to their small fame / And now I know that the time of human generations is not like / the time of the earth.'[16] Robinson has restored to many customs and places in his chosen landscapes 'their small fame'. He has frequently done so in collaboration with Irish scholars, and with country people who have conserved the stories of their own places. He has given us the deepest of maps, the richest of 'readings' of landscapes and seascapes. He is one of the great restorers, or *re-storyers*, one of the quiet unravellers of imperium. His vision is conservationist rather than conservative. He and Máiréad have involved themselves again and again in environmental issues, most particularly in opposing the building of an airport on the environmentally sensitive Roundstone Bog and the inappropriate siting of windmills on the south of Inis Meáin (Inishmaan), in the Aran Islands.

His writings are in many ways reminiscent of the work of Barry Lopez in the Arctic, of John McPhee in his deep storying of the geology of the United States, of Robert Macfarlane in his tracing of old paths in England and Scotland, or of Nan Shepherd in her wonderfully poetic and rigorous account of the Cairngorms in the north of Scotland. However, his decision to base himself physically in his field of studies for half a lifetime, his background as a mathematician and as a visual artist, his scientific interests and his openness to folklore and myth give his work ever new and unexpected dimensions and perspectives, making it utterly unique. In the introduction to *Pilgrimage* he writes of the limestone karst of the Aran Islands – 'This bare, soluble limestone is a uniquely tender and memorious stone.'[17] The Aran Islands and Connemara are indeed fortunate to have such an eloquent, tender and memorious chronicler.

## Notes

1 Tim Robinson, 'Orion the Hunter', *Tales and Imaginings* (The Lilliput Press, Dublin, 2002), 145.
2 Robinson, 'Orion', 145.
3 Robinson, 'Orion', 145.
4 Blaise Pascal, *Pensées* (Project Gutenberg, 2009), 61. Accessed 3 March 2014, www.gutenberg.org/files/18269/18269-h/18269-h.htm.
5 Walt Whitman, 'Leaves of Grass', quoted in Robinson, 'Orion', 150.
6 Robinson, 'Orion', 150. Robinson's version of Whitman's quote is slightly different to the original, which reads, 'I believe a leaf of grass is no less than the journey work of the stars ... and a mouse is miracle enough to stagger sextillions of infidels.'
7 Robinson, 'Orion', 150.
8 Robinson, 'Orion', 151.
9 Robinson, 'Orion', 151–2.

10 Robinson, 'Orion', 152.
11 Robinson, 'Orion', 152.
12 Robinson, 'Orion', 152–3.
13 Robinson, 'Orion', 153.
14 Robinson, 'Orion', 188.
15 William Butler Yeats, 'Anima Hominis' Part V, in *Per Amica Silentia Lunae* [1918] (Project Gutenberg, 2009), 29. Accessed 5 March 2014, www.gutenberg.org/files/33338/33338-h/33338-h.htm.
16 Czeslaw Milosz, *The Collected Poems, 1931–87* (Penguin, London, 1988), 382–3.
17 Tim Robinson, *Stones of Aran: Pilgrimage* (London: Penguin, 1986), 4.

# 8

# Tim Robinson and Chris Arthur: in defence of the Irish essay

## Karen Babine

In what might be one of the most clever delivery systems of the primary argument of the non-fiction genre – that of truth in non-fiction – noted American non-fictionists Bill Roorbach and Dave Messer created a 'cartoon essay' to address the negotiation of truth and fact, between fact and fallible memory, between recollected dialogue and the subjective effects of experience. The cartoon narrator finally concludes: 'Try your best to be both accurate and artistic. Take it as a challenge to use the facts as your clay. And if you don't feel like you can do that, write fiction.'[1] While the cartoon aspect of the discussion might seem childish – a strange melding of forms – the truth/fact debate is an incredibly old argument, one that never comes to any conclusion, and one that often seems reductive to those who have participated in it before and come to their own answer. The post comes at the same time John D'Agata's book *Lifespan of a Fact* was released, a book that created yet another uproar among non-fictionists and reignited the age-old question about the negotiation between fact, fiction and the page, an argument that is not solely confined to non-fiction (and American non-fiction).[2]

Irish literature, perhaps more than other national literatures (including American), seems to be preoccupied with negotiating the zone between fact and fiction and the concept provokes an interesting interpretation of this question. In *The History of the Irish Novel*, Derek Hand observes,

> The story of the self, of its creation and its persistence, is the only important one and, yet, the truth of that, the reality of it, can only be approached in the guise of fiction. Again and again in Irish prose – from James Joyce to John McGahern – this line between fiction and reality, between biography and fantasy, is negotiated.[3]

When looking at shelves of other genres, as written by Irish authors, this same deep-seated negotiation is similarly applicable and is not confined to fiction, as Hand suggests. Historically, Irish drama, narrative poetry, the short story, as well as the novel, all move against and between and through the truth and reality found in the story of self. Non-fiction, however, represents a different set of priorities and complications.

In the 1960s and 1970s, as fictional techniques in non-fiction became more popular in the United States (with the rise of New Journalism and the non-fiction novel), the use of dialogue, characterisation, description and such in non-fiction became more standard. As the memoir form itself became more popular in the United States in the 1980s, the emphasis on telling the stories of others shifted to include telling one's own personal stories. Australian non-fictionist Mark Tredinnick considers the functional differences between genres and provides an excellent counterpoint to Hand, as a transitional moment to consider the connections between fiction and non-fiction, between narrative and lyric work, and to stretch the definition, to apply these ideas to the difference between memoir and essay:

> I am emphasizing the fictive, fanciful, inventive element of novels not to suggest that novels do not have regard to the real world, but to make a case for the qualitatively different *relationship* that essay writing forges between a writer and the world and between the writer and the reader. The work that fiction does differs fundamentally from the work that nonfiction does. And within nonfiction, the essay is a particular case, because it is an artistic form that does essentially lyric work: it is uttered in the first person, it engages with vernacular moments, places, and times, and tries to elaborate their nature; its success depends not upon the tale so much as the telling. The essayist imagines and tries to render what is real – deeply, structurally, poetically, eternally real – in a moment, in a place, in a life.[4]

And so while the memoir has staked its place in the Irish canon, alongside the traditional storytelling forms, what is largely missing from the Irish spectrum of non-fiction, however, is the essay, the Montaignian, personal or familiar essay, specifically.[5] There are, of course, various Irish exceptions (Hubert Butler (1900–91), for example) and others writing historical or satirical essays.[6] Of these practitioners, I wish to concentrate on the contemporary essays of Tim Robinson and Chris Arthur, who both operate almost exclusively in the non-fiction genre (though each has published small exceptions in fiction and poetry respectively).[7]

Why Irish literature lacks a consistent non-fiction community may lie in the ideas of 'literature' itself, especially in a culture with such a specific literary foundation. Literary critic G. Douglas Atkins, in *Reading Essays*, provides a common reasoning for this lack of literary attention to the fourth genre:

> The essay is unique as a literary form. Like Roland Barthes, I do not think it a genre (rather, he said, it is a-generic), nor do I consider it quite literature. It is 'almost literature' and 'almost philosophy,' a little of both, although not quite either – not a thoroughgoing thing. More: the essay hangs between – between literature and philosophy, creation and fact, fiction and nonfiction, process and product – born of tension, then, and with tension as perhaps its essential characteristic.[8]

This genre has long been disconnected from literature, shelved in rhetoric (especially classical rhetoric) and used as the genre-to-write-about-other-genres. This idea that non-fiction is not literature in and of itself is absurd: essayists use their language as a medium of art, they control the pacing of their prose through their punctuation and sentences, and something new can be gleaned upon each encounter and rereading. The difference is attention to art, not function. The Montaignian essay has no other purpose than to explore. It does not argue; it does not advocate; it does not inform; it does not push. Perhaps the disconnect is inherent in the distance between Montaigne's more 'frivolous' essays and Francis Bacon's more 'serious' essays.[9] An essayist does not have to rely on dramatic events for the subject. What drives an essay is what the writer's mind makes of it and, as a result, every subject is a possibility for an essay.

**A caveat**

The multi-layered nature of the essay as a form necessitates a multi-layered approach to studying it; that being the case, a bit of clarification is necessary. First, I realise that what I am about to discuss may sound like a manifesto for the existence and persistence of the Irish essay, but I am truly approaching this project in the exploratory, conversational, tentative spirit that has long been the defining characteristic of the essay form itself. Most of the scholarship on the essay itself has been done by Americans on American essays or considering the British periodical essayists. Many of the articles I reference in this chapter come from American literary journals, particularly *River Teeth* and *Fourth Genre*, two of the three main literary journals that solely publish non-fiction, and I am favouring the perspectives of practising creative non-fictionists who are also either academics or editors. In terms of audience, Chris Arthur is frequently published in American literary journals, earning frequent reprints and notable mentions in *Best American Essays*; his latest book, *On the Shoreline of Knowledge: Irish Wanderings*, was published by the University of Iowa Press in 2012, though his earlier works were published by the Colorado-based Davies Group. Part of this American attention to Arthur's essays is due to the culture of literary journals and university presses publishing creative work that has grown up around the rise of American creative writing graduate programmes. Robinson, on the other hand, rarely publishes in literary journals – yet when he does, the work often receives non-fiction-specific accolades. He has also been reprinted in *Best American Essays*, as well as listed in the notables.

Additionally, the majority of scholarship on the essay form – even on Montaigne's work – has not been done by creative writers. While Arthur has been referenced by writers Robert Atwan, Patrick Madden and other American non-fictionists as a shining example of the contemporary Irish essay in American creative writing circles, Robinson's work is primarily examined by ecocritics. Furthermore, most of the scholarship on the essay belongs to the realm of composition and rhetoric, not creative writing or literature. As a result, not only is there great opportunity for Irish essayists to write in this form, there is great opportunity for academics to study what they write. All perspectives contribute to the vitality of the creative non-fiction genre.

## Place: grounding the quotidian

What is important about the Irish incarnation of the essay is its deliberate emphasis on place: ideas, stories, histories, complexities and perplexities are inextricably rooted in the physical landscape, whether that is the natural environment or the built environment. Perhaps this should not be surprising, given the role that the land has played in Irish history, its wars and spiritualities, its feasts and famines. As a result, the everyday environment becomes the source of the ideas that the Irish essayist explores. Montaigne's *essai* becomes another non-fiction form to wrestle with the friction of the world, through the many levels consciousness and sentience occupy in the pursuit of meaning. This is where, as Robert Atwan observes, 'the writer's reflections on a topic become as compelling as the topic itself, when he or she searches for the larger theme behind an isolated issue or event, or when the craft and handling of material reveal a keen sense of a subject's true complexity, then I believe "essay" is the most accurate designation'.[10] Phillip Lopate argues in the introduction to *The Art of the Personal Essay*,

> The essayist attempts to surround a something – a subject, a mood, a problematic irritation – by coming at it from all angles, wheeling and diving like a hawk, each seemingly digressive spiral actually taking us closer to the heart of the matter. In a well-wrought essay, while the search appears to be widening, even losing its way, it is actually eliminating false hypotheses, narrowing its emotional target and zeroing in on it.[11]

Thus the work of Robinson and Arthur can serve as models to address Montaigne's essential question of 'What do I know?' or, as Sarah Bakewell observes in her biography of Montaigne, his essential questions of 'How to live?' and how to explore the 'perplexities of existence'.[12]

Robinson's choice not to write about his own life and dramas, but to use an essayist's attention to explore what is below the surface of these places and ideas, provides a distinctly unique way of observing and participating in the world that surrounds him. In 'The Fineness of Things', reprinted in the 2008 *Best American Essays*, Robinson considers an encounter with two mating moths, a moment of absolute ordinariness and in which 'the transparent space above our lawn that day was seething with messages'.[13] From there, he spins a mental journey that gives him room to contemplate how we look at ordinary things, daily events, creatures of little consequence, at what distance, and how multiplying and dividing our perspective by powers of ten yields a moment of blinding insight for both the writer

*Figure 20* Robinson writing at his desk in Roundstone, Co. Galway (from Pat Collins's film *Tim Robinson: Connemara*, photo by Colm Hogan).

and the reader: *How is the mating and death of a moth representative of a greater truth that cannot be accessed any other way? How many other ordinary events in life do we overlook because we do not consider them to be important?*

Likewise, Arthur's conscious, writerly recognition that what must be explored is not his own life emphasises that he is merely the agent of examination, not the subject, and his own personal history is only a part of how he comes to knowledge. He writes in 'Level Crossing' of the gyrfalcon – a move, that like Robinson, finds inspiration in the natural world – yet the essay is not just about birds of prey, but how 'my perception of these falcons is only an example; it embodies and points beyond itself to a far wider phenomenon. What interests me is the way we're taken from the familiar to the strange, from the known to the unknown, from the bounded, labelled, and limited into less easily contained expanses – and how we then come back again.'[14] The essay moves from the scientific aspect of the bird to his relationship with his former teacher Arnold Benington, who first introduced Arthur to falcons, to the idea of perspective illustrated by Sam Pickering's essays, various definitions and etymologies of 'level'. The essay becomes a place where all these disparate elements can come together to illuminate what happens when a place and people become fixated on one level of thinking, that 'surely [it is] a key role of education, of religion, of art, of science, of the varied components of human culture – however differently they tackle this – to foster an awareness of life's different levels and teach us how to cross between them'.[15] As a result, the essay form itself offers the space to explore the ideas behind events, underneath feelings, through wonderings.

Robinson and Arthur both particularise the local as the source of meaning. The idea of place as a repository for stories and ideas resonates more firmly in Irish writing (across all genres) than in other national literatures. As the American essayist Paul Gruchow writes,

> All history is ultimately local and personal. To tell what we remember and to keep on telling it is to keep the past alive in the present. Should we not do so, we could not know, in the deepest sense, how to inhabit a place. To inhabit a place means literally to have made it a habit, to have made it the custom and ordinary practice of our lives, to have learned how to wear a place like a familiar garment, like the garments of sanctity that nuns once wore. The word habit, in its now dim original form, meant *to own*. We own places not because we possess the deeds to them, but because they have entered the continuum of our lives.[16]

Landscape is often the first text we read, even before we have the language for it. Landscape has always been constructed, manipulated or otherwise filtered through human consciousness and thus a built landscape is just as natural as the uncultivated. With the building frenzy brought on by the Celtic Tiger and the effect caused when the housing bubble burst in Ireland, new considerations of place and identity are taking shape, most notably in the recent rise of Irish crime literature, particularly noir, which is intricately linked to urban place. But such considerations are also important to non-fiction.

Part of this attention to the local is because both writers find themselves 'out of place' – Robinson as an Englishman who has come to make the West of Ireland his home and Arthur as an Ulsterman who has come to make Wales and Scotland his home. Thus the essential qualities of the essay – and the elevation of the quotidian to the larger world of writing – becomes, as Atkins argues, 'the essay's inherent value and importance', which 'lie [in] attention and dedication to the particular, the local, and the rooted – in short, the familiar – [and] constitutes but half the story. Equally important is the essay's own rooting, at the same time, in meaning, the timeless, the universal'.[17] Robert Root writes that 'place isn't only a destination; it's also a threshold through which to enter time and text'. He continues, arguing for the use of place in several different contexts:

> Where we dwell influences who we are as individuals – it's the impulse to memoir and personal essay; where we venture reveals who we are as observers – it's the impulse to travel and nature narrative and reportage; where we go in the paths of others tells us who we are as readers – it's the impulse to cultural criticism.[18]

Therefore, there are several ways to consider essays of place: one is to adhere to the nature/place/travel essay, grounded in the natural or built environment, and taking from there one's essay subject, as Robinson often does; the second way is to find an essay's subject in an object that offers a different definition of 'grounded', and in the case of a majority of Arthur's oeuvre, he takes a quotidian object and examines it not only for how it places him in the scheme of his own life, but how it places him in his family, in his childhood memories, in the place he used to call home. Eamonn Wall writes: 'for Robinson, Inishmore will not only be surveyed as a place deserving of scrutiny in its own right, but it will also serve as "the exemplary terrain" that informs our knowledge of the world as a whole, the little informing the large, the micro informing the macro, and vice versa'.[19] For Robinson this means the natural world of the West of Ireland (Aran, Connemara and the Burren), and for Arthur this means (often) the built/domestic world of his childhood in Northern Ireland.

In 'Who Owns the Land?', from *Last Pool of Darkness*, Robinson takes this idea of land ownership in one specific corner of rural Connemara and uses his literal meanderings to echo the historical meanderings through the various 'owners' of this place to explore the idea that the harder one tries to claim a piece of land the more elusive that ownership becomes. He begins with characteristic attention to the primacy of language, placing himself and his geologist friend on top of 'Doon Hill, the 213-foot height of which was enough to let us see how the promontory of Errismore gesticulates in lesser promontories and throws out archipelagos as if to spend itself in splendid complexities before dwindling into the western ocean like a rocket into the sky'.[20] As he and his friend are accosted by the land's owner, Sailor MacDonagh, who had not remembered granting permission to Robinson to be on his land, Robinson realises that he occupies a precarious place, as one who is English by birth, but who has deliberately made this place his home. Robinson's physical exploration forms the spine of this essay, but his own experience is not the subject of the essay. In 'The Essential Prose of Things', Tredinnick writes:

> The essayist, though – at least the good one – speaks not *about* personal experience, but *from* it. The essayist relates engagement with the actual world ... The essayist does not write about himself, nor does he is making a creation that is independent of the ground from which it rose. He is witnessing a small part of reality. And for witness one must be both present, alert for what else is present, and detached – having the experience, involved in the figure, and reflecting upon it; one must be capable of encountering phenomena and noting one's encounter simultaneously.[21]

While the landscape is the repository for the stories of humanity and the larger question of who owns the land – and who can own such a land – it is in the exploration of those small details that will lead Robinson to the clearest answer, if such an answer is to be found. In his introduction to *The Art of the Personal Essay*, Lopate writes, 'there is something heroic in the essayist's gesture of striking out toward the unknown, not only without a map but without certainty that there is anything worthy to be found'.[22] This is something Robinson has both literally and metaphorically embraced across his work and in this essay in particular. What seems on its surface to be a fairly straightforward history of a place, moving forward in history until he deliberately moves it backward, at least as far as the people associated with it are concerned, actually reveals itself to be much more complicated. Robinson's mental sifting through the names and dates that have accumulated themselves onto Doon Hill find a moment of exposition, where Robinson takes a step back on the page and what all he has seen and thought might mean.

But larger still, at least on the level of the essayist's craft, is Robinson's attention to the space of the page, the rhythm and movement of the prose: how does one begin to inhabit a sentence, a text? His attention to the repetition of the hardness of the *p* (see italicised below) moves into amelioration with the repeated use of soft *h*, *n* and *l* sounds as the passage concludes, raising in pitch to a question:

> Now the relics of *p*ast times, which are *p*rincipally those of *p*ast owners, their *p*owers, *p*ieties and sorrows, are reduced to cattle-*p*ens on the farmer's land. *P*roperty rights and *p*ossessive attitudes cut us off from even these disparaged stones, leaving us as flimsy as *p*aper figures cut out of our backgrounds in history and nature. Where do we turn, to find a way of looking at the land, inhabiting it, loving it, other than that of ownership?[23]

This is a question that constitutes at least one layer of the majority of Robinson's work, from the very beginning of *Stones of Aran*, but he never comes to a final answer; each essay provides an answer that is accurate for that place and time.

Only when the actual turns are examined does the movement of the essay itself reveal how the quotidian and the place intersect: from the complications of human history Robinson turns into a space where humans and fairy coexist and that turns into a moment. He observes,

> To imagine the consciousness of other beings is to see through their eyes, and in the case of fairies this is revelatory. The fairies dwell in the hill, not on it, nor do they look down on it from tower house or bomber. They see the land from within, as do those who briefly visit them, the village seers and shamans, and it is of dazzling splendor.[24]

This idea of the brightness to be found from dwelling within leads Robinson to consider the geologic history of Doon Hill, which is of an extinct volcano. The interior world of the essayist, in this case, matches the brightness of the interior world of Doon Hill. Such a moment echoes non-fictionist Cynthia Ozick's observation that 'an essay is reflection and insight'[25] and 'an essay's heat is interior'.[26] Such a moment leads to considering the light and heat created by the essayist's interior world and the reflection of that light onto other surfaces and ideas.[27]

Arthur's 'Bookmarks' provides an interesting point of comparison to other ways that stories and meanings are layered in a place. 'Bookmarks' is, on its surface, about an everyday object, but this object allows Arthur to discover that there is as much value in his own act of seeing and thinking about these bookmarks as there was in his father's original stories represented by 'news cuttings, or old train or tram tickets, receipts, an occasional photograph or postcard, some pressed flowers, a leaf, small rectangles of cartridge paper with poems written out neatly in my father's hand. Each of these items is like finding a fingerprint of his life.'[28] That the same type of mental and emotional friction can be generated by the loss of a parent, as well as Arthur packing his father's books after his death, represents the Montaignian premise that ideas can be found anywhere and Arthur's consideration here leads the reader to wonder what a stranger might discern looking at one's own books or other similar objects.

This segmented essay – divided by numbered sections, as Arthur is wont to do – starts with an archaeological excavation in Nepal of an early Buddhist site and his consideration of this place he has never seen with this observation:

> Though the earth may hold no more echo of what took place upon it than a well-used library book retains traces of the multiple readings it has hosted, there remains a kind of spectral inheritance – invisible, undetectable but strongly felt – etched into the lineaments of the things around us and the places we inhabit.[29]

This moment gives Arthur the turn he needs to enter the next section, in which he transitions from the Buddha into his father's death twenty years before. The remnant of his life that he left behind, the task of packing up his father's things to take to a charity shop, these realities lead Arthur to think about the ways that 'archaeology via books' question whether what we discern about a person from their books may or may not be accurate.[30] Arthur pulls the lens back in the seventh section, bringing place back into the essaying:

> If you compare landscape and bookscape – the places my father walked and pages where the invisible footsteps of his gaze fell in the spectral journeying that reading affords – it seems obvious that the former would be more durable than the latter … Although he lived there for so long, few geographical locations in Northern Ireland retain any sense of his presence. Places have proved unexpectedly mutable … In contrast, his bookscapes are exactly as they were when he last walked in them.[31]

That a library and a home might be considered a landscape is important to Arthur's continued pursuit of self-discovery and encourages readers to redefine their personal landscapes. In classic Montaignian fashion, in the sense of valuing uncertainties over conclusions, Arthur ends the essay with a series of questions: 'How are we to make sense of the intricate contiguity of the extraordinary and the everyday? How can we maintain our balance, poised as we are between existence and extinction, the muddy and the miraculous?'[32] The answer to that question is found in attention to the quotidian itself, finding the specific in the universal and the universal in the specific that is the two-sided singular goal of the Montaignian essay. To consider what the Montaignian essay might look like in an Irish context, the link between place and idea must be an essential element.

## Space: uncertainty and coming to knowledge

The essay, as a form, requires uncertainty; Montaigne himself filled his essays with phrases of uncertainty – 'I think', 'it seems to me', 'perhaps' and so on – and as Sarah Bakewell writes in her biography of Montaigne, if nothing can be certain, what is the point? The reality – and opportunity – of the essay, then, becomes that uncertainty itself and 'makes everything more complicated and more interesting: the world becomes a vast multidimensional landscape in which every point of view must be taken into account'.[33] Essayists use the distance between memory and the present and the uncertainty that distance represents to discern meaning and to explore the possibilities offered by various juxtapositions of ideas, images and events. The space itself between past and present, between what we think should be and what is, between the experience of self and the experience of others, is what Arthur and Robinson bring to the page, why the essay itself as a form offers much to Irish non-fictionists. Robert Root writes that 'in nonfiction the writer ought to be pursuing truth. What's the point of recording or reporting or reflecting if it doesn't get the writer closer to a better understanding of the way real events, real experiences, real lives – including the writer's – work?'[34] In that process, the failings of memory will become apparent and yet, not only is this space *not* a failing, it is in that space where essays find the most fertile ground.

In Robert Root's book *The Nonfictionist's Guide*, the chapter titled 'This Is What the Spaces Say' uses the spaces he is writing about to make his point: this chapter is a complete meta-essay in itself.[35] He writes that 'space has become a fundamental element of design and expression in nonfiction' and 'knowing what the spaces say is vital for understanding the nonfictionist's craft and also for appreciating the possibilities of this contemporary genre; it also helps us to better understand the nature of truth in the segmented essay'.[36] Even as Arthur and Robinson are working toward other interpretations of the page and how they can affect how a reader literally reads the page, there is much room for Irish writers to stake their own claim on this space and experimental/lyric non-fiction. Though Robinson himself plays with the space on the page in more visible ways (the addition of mathematical principles and representations of fractals, as visual representations that break up the continuity of the text on the page), both he and Arthur add to what Atkins argues about the texture of the essay that 'made of layers, essays function laterally more than vertically, eschewing depth in favor of relations'.[37] As a result, the multiple ways of coming to knowledge and representing the pursuit of that knowledge on the page add to the layered depth of the essayist's deliberative process and the ultimate pursuit of truths and insights. The cracks in a writer's certainties, to where the friction creates light and heat, are the clearest opportunities toward the pursuit of truth and meaning.

Arthur, who prefers to meditate and explore the tension caused by everyday objects and family spaces, still finds in those subjects whole worlds to explore.

His essay '(En)trance' was originally published by *The Literary Review* and collected in *Irish Elegies*, and reprinted in the *2008 Best American Essays*. The essay starts with Arthur's knee-jerk tendency toward referencing himself as a writer, but the move to connect his own explorations (and failings – an important aspect of uncertainty) as an essayist with the physical objects of the pillars at Shandon is an important one because it foregrounds the uncertainties that lay ahead. The pillars themselves marked the way down the lane toward the farm in Co. Antrim where his mother and sisters called home, but the point of Arthur's work here is that the pillars no longer exist: a Baconian (certainty) essayist, even a writer of another genre, would have looked at the pillars (or the lack of them) as a moment of scenery, of backdrop, not inherently important in themselves. But it is a moment of craft that, as Arthur writes a few sentences later about the thickets behind the pillars, 'it straddled the space between boundary and territory'.[38] Arthur is free to move wherever his consciousness takes him, from Shandon and his own personal and familial history to the stone dogs that guard the entrances to Japanese shrines to Greek mythology and the power of dreams. But the purpose of the exploration, which considers the gates to dreams in Greek mythology (true dreams through the Gate of Horn, false dreams though the Gate of Ivory), Arthur wonders if a third gate is necessary:

> This Gate of Laurel is the entrance through which comes a crushing counterweight of fullness, a corrective for all the simplifications and superficiality with which we customarily clad things. It is a gate where the trance of mystery might be joined to those entrances whose thresholds we've grown so used to crossing as we make our way into our variously impoverished visions of the world.[39]

While the essay ends with a consideration of the eighteenth-century Zen teacher Ekai's text *Wu-Men Kuan/The Gateless Gate*, Arthur concludes the essay without concluding, preferring to dwell in a space of conscious liminality, but the larger question of uncertainty here is directly related to the essayist's refusal to come to a conclusion, to prefer to use the pillars at Shandon as an example of the question *what holds memory?*, rather than a specific moment for Arthur to detail his own history and experience in this particular place. The result, then, is space for the reader to substitute his or her own pillars into the essay and thus take part in the essaying of the idea. We come to understand by the end of the essay and the mention of the gateless gate that the pillars at Shandon, which no longer exist, are exactly this kind of gateless gate to the mystery of his family's farm.

Like Arthur, Robinson will follow his consciousness from Irish grammar, geological research, oral history and folk songs to personal experience and any number of other methods of gaining knowledge that still leave room for other possibilities and interpretations, but Robinson is much more subtle about it. If Robinson's purpose as he began *Pilgrimage* was to write a comprehensive map – and book(s) about the island – he quickly understood this would be impossible. By the end of *Pilgrimage*, Robinson has come to understand more completely the twisted nature of not only his subject matter, but also the form he is choosing to write in. He writes,

> It suggests the possibility of going round the island once again, looking at everything in more detail, or in others of the infinite of ways of looking. Perhaps a second circuit would be more rewarding now that my pace has been chastened by so many miles, my breath deepened by so many years. But for a book to stand like an island out of the sea of the unwritten it must acknowledge its own bounds, and turn inward from them, and look in the labyrinth.[40]

The use of high exposition, that which non-fictionist Bill Roorbach describes as 'having the quality of grand truth, of universal applicability', is essential here, not to the spectrum of certainty, but toward using the high exposition to allow for others to use their individual methods of gathering knowledge and attempt, themselves, to apply Robinson's ideas to see if they track.[41] While high exposition appears on its surface to aim toward certainty, the moment is still an attempt to come to meaning, to discern a possible way of understanding, at least for this moment, this place, this space of time.

The essay 'feasts on doubt', to quote Lopate in a recent column in the *New York Times*, but to do so is a learned behaviour.[42] Robinson's supposition that a second time around the island might yield further insights fits into the necessary quality of the essay to reconsider what has already been considered – and thus to create a distance, both temporal and page-based, between experience and what the experience means. Finally, the third element he identifies, turning inward, is essential to make one person's experience relevant to those who will never encounter it for themselves. He has a similar moment at the end of *Labyrinth* that echoes the same essayist ideals: 'A writing ... may incorporate spontaneities, but it is not the work of a moment and does not issue from a single mental state; like the step, a sentence holds open the possibility of returning, changed, to that point, to approaching it from the west rather than the east.'[43] For Robinson, space is as physical and literal as it is a function of an essay's energy.

Naturally, he ended *Pilgrimage* knowing where *Labyrinth* would begin, but the larger issue has to do with Montaigne's absolute commitment to avoiding absolutes. But the first lines of *Labyrinth* set a completely different tone and approach than they do at the beginning of *Pilgrimage*: 'Unquestionable answer to unanswerable question, this volume must close what the first opened, so that I can store them away safely, like two mirrors face to face.'[44] Yet he ends *Labyrinth* even more disillusioned with absolutes than where the book began, as he has now spent two full volumes trying to map the intricacies of this one, small island. His frustration – and the opportunity it brings, if one is feeling optimistic – is palpable, refracted in the mirror imagery he both begins and closes the book with. He writes in the final chapter, 'Lesser Aran',

> The either/or is this: to be simply present and not to know and remember it, or to be reflectively aware, which implies the mediation of imagery, of mirroring – and reflection multiplies mirrors as fast as mirrors multiply reflections. Writing is my way out of this labyrinth. But I am no abstract, deep-sea philosopher; if I raise a metaphor as a sail to catch the winds of thought, I am soon overturned by shoals, or fly to the horizon and lie becalmed there. Therefore I choose this Aran-building method, the slow deposition of facts and observations, coalescing and fusing under their own weight into tablets of stone.[45]

Here, Robinson essays a landscape of uncertainty. At one point during his Aran explorations, he mentions the difficulty of mapping specific places on the islands, because of the fragile nature of the limestone: he might return and find that the cliff has collapsed and the place is gone forever. Seen from a different light, the physical landscape of the Aran Islands, as well as Connemara, limestone and granite respectively, offer insight into a truth that may exist for a while, but may erode away, perhaps a distinctly Irish version of Heraclitus's dictum that 'no man may step into the same river twice'.

In 'A Connemara Fractal' Robinson takes the quotidian landscape of the place, in the form of the shoreline, and attempts to use a medium he believes will give him a definite answer – mathematics – only to find that his certainty gives him no answers. The movement of dimensions is important, from the map itself to Robinson himself walking through it in the rain, but the maps and Robinson's experiences preclude a way to express context, feelings and impressions. Examining the complexity of a shoreline through the lens of mathematics – which Robinson has a degree in from Cambridge University – is distinctly essayistic, as he attempts to come to meaning through an unexpected door. Certainty can be attempted, but never achieved, at least not in this context. This provokes the question, of course, that if Robinson could have come to certainty about the Aran Islands or Connemara, would he have been so compelled to investigate them to the degree he has? Likely not. He concludes the essay with the idea that

> discourse, too, is fractal: every remark suggests amplification and amendments, every thesis, critiques and refutations; this fractal nature of intertextuality is the clue to the way the text itself sits into the fractals of non-textual reality. To be true to the nature of fractals, I should go on indefinitely, but in this world of practicalities, I must at this point abandon the tangled tale.[46]

## Conclusions and possibilities

In the true spirit of Montaigne, I come to no concrete conclusions. Instead I see possibilities for writers of the Irish essay and opportunities for those who would study it. I see the Irish essay taking advantage of the current climate of non-fiction in the United States, to the mutual benefit of both, and in the case of Tredinnick, also in Australia. Melding the literary elements already present in Ireland – and the particularities of life in Ireland and its relationship to the global community – offers a specific creative opportunity. Atwan writes of his own criteria in putting together the *Best American* series:

> Though I found plenty of writers back then who wrote essays that were undeniably essays … I also came across a growing number of emerging writers who grafted elements of fiction, poetry, reportage, criticism, and especially personal memoir onto the familiar essay, creating some wonderful works of mixed or impure forms to reshape the essay rather than the other way around.[47]

Currently, various literary forms are already adopting what Atwan observes, especially considering the movement between memoir and fiction.

A series of important anthologies (edited by American creative writers) aims to situate the essay in various historical contexts and lay a path for what the essay's future might look like.[48] But another collection of creative writing/critical texts endeavours to advance the conversation of non-fiction by teaching the how-to of craft.[49] These two groups of texts, then, have very recently generated a third kind of text, a type that melds the critical and the creative in very important ways.[50] Additionally, movement in the American academy in the last twenty years might signal an increased opportunity for studying the Irish essay, a melding of creative–academic work that I hope will become the future of non-fiction studies, anchored in both the craft work of the creative essayist and the critical eye of the academic.

The opportunity is ripe for advancing the conversation of creative non-fiction in the academy, not just via American non-fiction, but other national non-fictions, as Tredinnick is doing with Australian non-fiction. As a result, several conferences dedicated solely to the discussion of non-fiction are also participating in the increase in global non-fiction studies.[51] As the essay itself has translated easily from Montaigne's sixteenth-century France to other places and times, perhaps the creative and scholarly work on non-fiction being done in these international spheres can encourage both creative writing and scholarship on Irish non-fiction in the future.

## Notes

1  Bill Roorbach and Dave Messer, 'Everything You Ever Wanted to Know about Truth in Nonfiction But Were Afraid to Ask: A Bad Advice Cartoon Essay', *Bill and Dave's Cocktail Hour* (blog). Accessed 28 March 2012, http://billanddavescocktailhour.com/everything-you-ever-anted-to-know-about-truth-in-nonfiction-but-were-afraid-to-ask-a-bad-advice-cartoon-essay/.

2 See Jennifer B. McDonald, review of *Lifespan of a Fact* by John D'Agata and Jim Fingal, *The New York Times*, 21 February 2012, as a good starting point to enter this particular conversation.
3 Derek Hand, *The History of the Irish Novel* (Cambridge: Cambridge University Press, 2011), 42–3.
4 Mark Tredinnick, 'The Essential Prose of Things', *The Land's Wild Music* (San Antonio, TX: Trinity University Press, 2005), 34. Original emphasis.
5 'Lyric' and 'experimental' non-fiction of the type that Americans John D'Agata, Jenny Boully, Lia Purpura, David Foster Wallace and others are known for is virtually non-existent in Irish non-fiction.
6 Shawn Gillen argues, 'Travel writing has long been a staple of Irish writing; the essay, too, has its own tradition, with Shaw, Wilde, Lady Gregory, and Yeats among the modern canonical figures associated with the essay and autobiography – but their work bears little resemblance to the creative nonfiction of our era.' But Gillen's observation underscores the work of writers whose primary genre is not non-fiction, writing occasionally in non-fiction, and these owe more to Francis Bacon than Montaigne. Shawn Gillen, 'Synge's *The Aran Islands* and Irish Creative Nonfiction', *New Hibernia Review* 11: 4 (2007): 130.
7 Chris Arthur, born in 1955 in Belfast and raised in Co. Antrim, was educated in Scotland and spent a significant portion of his professional career lecturing at various universities in Scotland and Wales, and now lives in Scotland with his family. He is the author of six essay collections.
8 G. Douglas Atkins, *Reading Essays* (Athens: University of Georgia Press, 2008), xii.
9 Christy Wampole offers a clear differentiation between Montaigne's *Essais* and Francis Bacon's essays:

> The word Michel de Montaigne chose to describe his prose ruminations published in 1580 was 'Essais,' which, at the time, meant merely 'Attempts,' as no such genre had yet been codified … Later on, at the end of the 16th century, Francis Bacon imported the French term into English as a title for his more boxy and solemn prose. The deal was thus sealed: essays they were and essays they would stay. There was just one problem: the discrepancy in style and substance between the texts of Michel and Francis was, like the English Channel that separated them, deep enough to drown in.

10 See Christy Wampole, 'The Essayification of Everything', *The New York Times*, 26 May 2013. Robert Atwan, 'Notes Towards the Definition of an Essay', *River Teeth* 14:1 (Fall 2012): 116.
11 Phillip Lopate, 'Introduction', in Phillip Lopate (ed.), *The Art of the Personal Essay* (New York: Anchor, 1995), xxxviii.
12 Sarah Bakewell, *How to Live, or A Life of Montaigne in One Question and Twenty Attempts at an Answer* (New York: Other Press, 2011), 4.
13 Tim Robinson, 'The Fineness of Things', in Kathleen Norris (ed.), *Best American Essays 2001* (New York: Houghton Mifflin, 2002), 236.
14 Chris Arthur, 'Level Crossings,' in *On the Shoreline of Knowledge* (Iowa City, IA: University of Iowa Press, 2012), 83.
15 Arthur, 'Level Crossings', 93.
16 Paul Gruchow, 'Home is a Place in Time', in *Grass Roots: The Universe of Home* (Minneapolis, MN: Milkweed Editions, 1995), 6. I deliberately mention Gruchow here because in Arthur's 'Essay on the Esse' he writes of the 1989 *Best American Essays*

as his entry into the genre; it is also the only year that Gruchow was reprinted in the main pages, a spectacular essay titled 'Bones' that concludes his book *Grass Roots*, from which 'Home is a Place in Time' is the first piece.

17  Atkins, *Reading*, xii.
18  Robert Root, *The Nonfictionist's Guide: On Reading and Writing Creative Nonfiction* (Lanham, MD: Rowman and Littlefield, 2008), 155.
19  Eamonn Wall, 'Walking: Tim Robinson's *Stones of Aran*', *New Hibernia Review* 12:3 (2008): 71.
20  Tim Robinson, *Connemara: Last Pool of Darkness* (New York: Penguin, 2008), 288.
21  Tredinnick, 'The Essential Prose', 35. Original italics.
22  Lopate, *Personal Essay*, xlii.
23  Robinson, *Last Pool*, 307. My italics.
24  Robinson, *Last Pool*, 310–11.
25  Cynthia Ozick, 'Introduction', in Cynthia Ozick (ed.), *Best American Essays 1998* (New York: Houghton Mifflin, 1999), xv.
26  Ozick, 'Introduction', xv.
27  It is worth mentioning that Ozick's ideas come from her introduction to the 1998 *Best American Essays*, which she edited, and for which she selected and reprinted Robinson's 'Orion the Hunter'.
28  Chris Arthur, 'Bookmarks', in *Irish Elegies* (New York: Palgrave Macmillan, 2009), 65.
29  Arthur, 'Bookmarks', 61.
30  Arthur, 'Bookmarks', 63.
31  Arthur, 'Bookmarks', 68–9.
32  Arthur, 'Bookmarks', 78.
33  Bakewell, *How to Live*, 130.
34  Root, *Guide*, 185.
35  Root's chapter was originally presented at the 2001 meeting of the Conference of College Composition and Communication, not to a creative writing audience, a moment that illustrates the multi-layered complications (and opportunities) of studying the essay.
36  Root, *Guide*, 85.
37  Atkins, *Reading*, 8.
38  Chris Arthur, '(En)trance', in *On the Shoreline of Knowledge* (Iowa City, IA: University of Iowa Press, 2012), 26.
39  Arthur, '(En)trance', 35.
40  Tim Robinson, *Stones of Aran: Pilgrimage* (London: Penguin, 1986), 282.
41  Bill Roorbach, *Writing Life Stories: How to Make Memories into Memoirs, Ideas into Essays, and Life into Literature* (Cincinnati, OH: Writer's Digest Books, 2006), 77.
42  Phillip Lopate, 'The Essay, an Exercise in Doubt', *The New York Times* (16 February 2013). Accessed 25 February 2013, http://opinionator.blogs.nytimes.com/2013/02/16/the-essay-an-exercise-in-doubt/?_r=0.
43  Tim Robinson, *Stones of Aran: Labyrinth* (Dublin: Lilliput Press, 1995), 455.
44  Robinson, *Labyrinth*, 3.
45  Robinson, *Labyrinth*, 455.
46  Tim Robinson, *Setting Foot on the Shores of Connemara* (Dublin: Lilliput Press, 1997), 102.
47  Atwan, 'Definition of an Essay', 115.
48  Phillip Lopate's *The Art of the Personal Essay*, Lydia Fakudiny's *The Art of the Essay*, Robert Root and Michael Steinberg's *The Fourth Genre: Contemporary Writers of/on Creative Nonfiction*, and John D'Agata's *The Next American Essay*, as well as others.

49 Bill Roorbach's *Writing Life Stories*, Dinty W. Moore's *The Truth of the Matter: The Art and Craft of Nonfiction*, Robert Root's *The Nonfictionist's Guide*, and others.
50 Specifically, *Essayists on the Essay*, edited by Carl H. Klaus and Ned Stuckey-French, and *The American Essay in the American Century*, by Ned Stuckey-French, and Mark Tredinnick's *The Land's Wild Music*.
51 The biennial NonfictioNow Conference (at the University of Iowa, which held its 2012 conference in Australia), *River Teeth's* annual Nonfiction Conference and the biennial In Praise of the Essay: Practice and Form symposium.

# III
# Place and the Irish cultural imagination

# 9

# 'But his study is out of doors':
# Tim Robinson's place in Irish Studies

## *Eamonn Wall*

We cannot make the argument that Tim Robinson is a traditional Irish Studies scholar; he is not an academic working and researching at a university; rather, he is a writer, as he says of himself, living with his partner Máiréad in the harbour master's house they have restored in Roundstone, Co. Galway (see Figure 21). Instead of having a library close at hand, Robinson has the ocean and a garden. Roundstone and its surrounding areas are defined by fishing and farming primarily, both of which are as ancient as they are important. Traffic passes through the neat village while its harbour and roads, in and out, connect it to the wider world. When people come to visit Roundstone, they do not come to study – it is not a university town – but to bathe in its beauty and enjoy its seafood. A majority of visitors will not know that Robinson, one of Ireland's most important and renowned writers, lives here and that his home is in plain sight.

Given Robinson's reticence, independence and the important roles played by solitude and anonymity in his calling as a writer, it is likely that Robinson prefers it this way. In those West of Ireland places (the Burren, Connemara and Árainn), or spaces to use the term he favours, Robinson is nowhere visible though everywhere present, as Flaubert might put it. At the same time, Roundstone is in the process of becoming a centre of learning, the harbour master's home having been deeded to the National University Ireland-Galway by its current occupants. Robinson, who has always worked at a remove from the academy, is bringing the university to Roundstone. Instead of the scholar making his way to the campus, the university, or a part of it, must relocate to the scholar's home, and this is a sure indication of the significance of Robinson's work as writer, cartographer and publisher, not to mention his generosity. Not only are the Robinsons making a present of their home and archive to NUI-Galway, and indeed to Ireland also it must be added, but they are also allowing for the continuance of their work. Their cultural enterprise is being gifted to the next generation in the same manner as a tune is passed on from an old fiddler to a young player. Both the

*Figure 21* The Robinsons' Nimmo House in Roundstone (photo by Nessa Cronin).

Robinsons and the old traditional musicians remind us, by making these gestures, of the responsibility all share for keeping traditions – whether of inquiry or performance – alive.

To designate Robinson as an independent scholar, given the suspicion that the term can somewhat unfairly arouse, would seem impertinent in light of his many achievements. As we have learned from David Lodge's explorations of the academic life in such novels as *Changing Places*, *Small World* and *Nice Work*, contemporary academic scholars are wont to be found hopping from classroom to library to airport terminal to scholarly conference.[1] After reading Lodge, one will surely conclude that the scholar's true place is the check-in desk at the airport nearest to campus, not the lecture theatre or laboratory. While attending conferences, scholars deliver papers while simultaneously developing new networks, keeping up to date with fresh developments and theories in the field, getting academic kicks and building more weighty résumés. In writing his Aran and Connemara volumes, Robinson has travelled great distances without ever venturing far from home: he resembles the Yaghans Bruce Chatwin describes in the book *In Patagonia* who 'were born wanderers though they rarely wandered far'.[2] Robinson's endeavours have about them a seriousness and depth that is wholly absent from Lodge's scholarly jet-setters. Robinson's sacrifices have been great though so too have been his rewards. Almost universally, his work is admired and even venerated. He is an independent scholar in the same way that Coleridge and Thoreau were – all three being transgressive authors too restless, physically and psychically, to be tied to desk and quad.

The Irish Studies field is no different from others in the academic world, with attendance at conferences both a necessity and a lifeline. If a scholar decides to remain at home, he/she will be left outside of discussions, will fall behind with regard to scholarly trends and become as dated as bell-bottom trousers and puffed-up shoulder pads. In addition, particularly in North America, scholars need to attend conferences in order to lessen the sense of isolation that is often the burden that the Irish Studies scholars must bear. On many campuses, the Irish Studies scholar is a lone wolf who works without peers. Frequently, Irish Studies scholars have had to fight hard to be allowed to teach and publish in the discipline. In fact, the founding of the American Conference for Irish Studies (ACIS) in 1963 was spurred by resistance. Historians and literary scholars, primarily, felt that professional organisations such as the American Historical Association and Modern Language Association did not consider Irish history or Irish literature worthy of specialised study. To found ACIS was a necessity and the new organisation, with its interdisciplinary outlook and desire to undermine ingrained attitudes, embraced the zeitgeist of the times. Academic organisations gather individuals into groups though the types of people who join such organisations are almost by nature solitary. In many respects, scholars are at their happiest, and some would argue at their best, while working alone in libraries, studies or offices. Whereas the conference is the place where research is presented to an audience of peers, it is a truth that neither the research nor the conference can be possible without the scholar's solitary work. Presentation is a reward for work while publication is confirmation of its originality and worth.

At a time – during the 1960s and 1970s – when Irish Studies scholarship was becoming increasingly professionalised, an important aspect of which was the gathering of like-minded scholars and teachers into institutes, departments and organisations, Robinson was stepping away from the metropolitan life of London, where the organisation had always flourished, to live on Inishmore. There, he would research and write his two seminal books – *Stones of Aran: Pilgrimage* and *Stones of Aran: Labyrinth* – before moving on to Roundstone where he continues to write his books and, with Máiréad, run his publishing venture, Folding Landscapes.[3] Though Robinson's scholarly work is of the same stature as that of such products of the academy as Declan Kiberd and Roy Foster, for example, he is hardly cut from the same cloth, having trained for his task first as a mathematician and later on as a visual artist. In fact, whereas Kiberd and Foster came to be an academic literary critic and historian, respectively, by choice, Robinson came to be the chronicler of Inishmore by accident. Though, like Kiberd and Foster, he is very much engaged with colonisation and its aftermath. Furthermore, Robinson's work is as rigorous as any produced by his contemporaries in the academy.

Robinson refers to himself as a writer and it is among a select group of contemporary writers, generally working independently of institutions, that he belongs: Gary Snyder, Rebecca Solnit, Terry Tempest Williams, Bruce Chatwin and Colin Thubron. Though this group is learned and distinguished, each has, like Robinson, done much of his/her work out of doors and at a remove from urban centres of learning. In this regard, they follow the botanist and marine biologist rather than the literary critic or historian. All of these writers, fiercely independent and original, share a distrust of large institutions ruled over by cadres of organisation men and women. Among senior Irish Studies scholars working within the academy, Kevin Whelan's and William J. Smyth's work come closest to Robinson's. In such volumes as the *Atlas of the Irish Rural Landscape*, which Whelan co-edited, and Smyth's *Map-making, Landscapes and Memory: A Geography of Colonial and Early Modern Ireland c.1530–1750*, we find gathered together and synthesised a wide variety of sources – ancient and modern – such as songs, poems, written histories, aerial photographs and so on, that reveal both scholars working comfortably across academic boundaries in the same transgressive manner that is a feature of Robinson's work. In Irish Studies, Geography, more than any of the other disciplines, epitomises the interdisciplinary approach that is supposed to underline scholarship in the field.[4]

The academy, for all its wealth and importance and for the quality of its writers and researchers, has not achieved the same influence as such writers as Solnit, Robinson, Snyder, Williams, Chatwin and Thubron, many of whom are household names in their own countries and internationally and have sold countless thousands of books. In the humanities, the university's reach is often quite narrow, a result of the frequently arcane and mind-numbing theoretical approaches paraded by scholars that alienate readers. It can happen that in an academic's hands a vibrant text can be reduced to concrete. In their focus on what is local, even when the location is exotic, Thubron and Chatwin, for example, have always been able to make what is being described a living part of our own world. Instead of alienating readers, their work draws them in. These writers, often using a bioregional focus, are concerned with showing off the world – wonders, warts and all – and not just their own learning. In no way is the integrity of their work compromised by the fact that they are writing for a larger rather than a specialised readership.

In great part, we are concerned with issues of genre and style as much as substance. Robinson and Solnit, for example, favouring the literary essay as their genre of choice, and working in Joycean plain style, have been able to connect widely with literate audiences. Though their work is deeply researched, it is, as poems, novels, short stories and plays are, primary work that the scholars working at universities must come to grips with. Rarely in Irish Studies, or in other disciplines, is the scholarly paper presented in the form of a literary essay, and this is a reminder of how different academic writing is from the work of professional literary artists such as Robinson, Chatwin and Solnit. In environmental writing and writing about Western American literature, there has been a marked movement away from the traditional academic paper in favour of more hybrid forms that stitch together elements of the scholarly and literary essay. In writing, style is personality and it is the presence of such in Robinson's and Solnit's work that make their books so engaging and which separates them for more depersonalised and less readable academic writing. As it continues to develop into its second half-century, Irish Studies, and its academic gurus tasked with the responsibility for training the next generation of scholars, might look toward the literary essay, and not the scholarly article, as the best mode with which to deliver research to audiences. Robinson's work will serve as the ideal guide.

Robinson's arrival on Inishmore resulted from his meeting Máiréad and having been exposed to Robert Flaherty's film *Man of Aran*. He provides the following account of how he began his mapping and writing about Inishmore, recounting how at the suggestion of the postmistress of Cill Mhuirbhigh, who had noticed his 'hand for the drawing, an ear for placenames and legs for the boreens', he drew his first map.[5] Robinson's skills as a walker, listener and artist were put to a practical use, one that would be of service to both islanders and visitors and which would shift his own life in a new direction. The postmistress's request and Robinson's resultant map signified the meeting of the visual and the scholarly, with the former more prominent at this point, and, most important, it was the impetus that served to push Robinson into deeper levels of description. It was a request – perhaps a commission or even a summons of sorts – that got Robinson moving in this new direction: it was one that would grow and swerve into his life's work.

It is also worth noting that by asking Robinson to make a map of Inishmore, the postmistress, a local person, was giving him, an outsider and foreigner, permission to record something of the topography and *dinnseanchas* of Árainn. Clearly, she had regard for Robinson as a person, writer and artist. Academics can carry with them a certain transparent arrogance, fed, ironically, by the acquisition of higher knowledge, so, in this instance, not being an academic but a man who had lived for a time among the islanders, whom they can come to know and trust, worked to Robinson's advantage. In isolated Irish communities, the post office was a centre of rural life: everyone must appear within its four walls to be recognised and judged with the postmistress serving as best judge of character and worth. Though neither he nor the postmistress would have realised it at the time, Robinson was being approved for authorship of the great book of Árainn. As Breandán Ó hEithir has pointed out, Aran Islanders have grown used to the figure of the *strainséir* – there have been so many of them – and have long practice at separating the real thing from the fake article. If needed, strangers/visitors have confirmed for islanders the importance of the places they call home:

> 'Did you ever ask yourself why so many people find those three limestone rocks in the Atlantic so terribly interesting?'
> 
> The short answer is that Aran islanders now take that interest for granted and find their own interest stimulated by the attention of others. To be an Aran islander is to be someone special, part of a long and many-faceted tradition, growing up in a bilingual community with an intense interest in its own history, just that trifle removed from mainland life and, perhaps, as one recent writer on island life commented acidly, imagining oneself to be just a shade better than most.[6]

That the postmistress's instincts were true is given confirmation by Ó hEithir, himself a native of Árainn, when he notes of Robinson that he is 'a Yorkshireman who came to look and stayed to learn Irish and restore the place-names to their original forms, thus putting the islands definitively on his magically detailed map'.[7]

Robinson was not an academic or what we might label as a professional scholar. On the contrary, he has always given the impression, voiced through his writings, of being the student rather than the teacher, the searcher rather than the finder, the man more in love with process than result. The postmistress's request was Robinson's letter of admission into his private university programme. It would be a mistake to underestimate the civilisation and society to which Robinson had been granted admittance. Whereas many universities, even the wealthiest and most famous of them, have been in existence for three centuries or more, Inishmore has been a centre of civilisation and learning, primarily oral rather than written to some degree, through pagan and Christian times. Its reach has been a long and enduring one: Robinson intuited this and set out to record something of its essence. Like all great scholars, within the quadrangle or walking independently on the outside, Robinson knew that he had chanced on something vital. He set out to learn some more and to report on what he had found.

Later on, in 1996, looking back at his time and work on Inishmore, Robinson was struck both by the continuity he noticed between his London and Inishmore periods and, conversely and simultaneously, by how much he was forced to reinvent himself on arrival on the island:

> Whereas I used to be dismayed by the breakage and loss caused by that sudden change in habit and habitat, nowadays it is the unchipped good order in which my little store of imagery accompanied me on the jolting journey from city to island that makes me wonder if it is ever possible to step beyond oneself. The question became sharp for me recently as I approached completion of the body of texts and maps I would have claimed had been inspired by my encounter with the West of Ireland, because in trying to foresee what I might do next I mentally revisited that earlier time of change, unwrapped some artworks stored away from my last year in London before the transition date of November 1972 – and discovered in them a concentrated abstract of the suite of images that had controlled my subsequent writing and is implicit in my cartography ... If so drastic a step as abandoning a career and a home, each of them close to a sort of cultural centrality, for an unknown language, an untried art and 'a wet rock in the Atlantic' is not sufficient to shake up one's deepest vocabulary, then where is there a possibility of self-transcendence.[8]

The artist, at a deep level, must allow him/herself to be guided by instinct and be given strength by chance. Furthermore, it is important for the artist to be flexible and pliable because the great new subject that he/she is presented with will require of him/her a trading in of an old vocabulary for a new one. At the same time, as Robinson discovered, the new vocabulary will borrow what it needs from the old, and then the project will move forward. It was as if, Robinson noted in hindsight, his whole life up to that point had been a preparation, or a series of preparations, for writing *Stones of Aran*.

Even in areas of academic pursuit like Irish Studies, the accidental and pragmatic play significant roles. Though the study of Ireland is by nature interdisciplinary, given that everything in the Irish world is part of an organic whole, Irish Studies organisations, drawing in scholars across academic pursuits, have occurred for practical reasons: it is likely that no single discipline, apart perhaps from literary studies, would ever have enough members to be viable as a separate organisation. Pragmatism has its rewards because the discipline is enriched by the meeting of scholars across areas of inquiry. At the same time, we should not assume that the interdisciplinary is a breeding ground for chaos where historians become experts on poetry and literary critics provide the last word on agrarian conflict. For the most part, the disciplines remain discreetly apart while accommodating each other's presence and expertise particularly at some of the crossroads points in Irish Studies – the 1916 Rising, for example. More than anything else, Irish Studies organisations function as umbrella unions under which many figures are gathered.

At the outset of his career as writer and cartographer, as he was taking on the research that would result in *Pilgrimage* and *Labyrinth*, Robinson seemed ill-prepared for the task at hand. He set out to practise three arts on Inishmore: map-making, mapping and writing, all three informed by his training as a visual artist. For all of his enthusiasm, Robinson faced steep hurdles: he did not know Irish, and was not conversant in the various disciplines the task required, being neither anthropologist, historian, folklorist, geologist nor natural scientist. Also, how could an outsider, an Englishman no less, find a way to penetrate Inishmore's oral culture, where its maps are written deepest? It seemed unlikely that one man could successfully duplicate the work of many and an act to madness to even try. T. S. Eliot had already persuaded us that the Renaissance man, who possessed a well-rounded store of knowledge, had disappeared in the early part of the seventeenth century. It was no longer possible, given specialisation and the emergence of the various sciences, for one person to know everything required to comprehensively describe a place and a people, and all that connected both to the history of the place, however small or large that place happened to be. It looked possible for Robinson to reinvent himself as a skilful map-maker; however, it seemed highly unlikely, despite the postmistress's encouragement, that he would be able to pull off his mapping project.

From the perspective of Irish Studies, Robinson's project was one that had to be completed outside of the university because it would likely have been considered too broad and transgressive to be accommodated within any ivory tower. In spending more time outdoors walking than inside writing, Robinson was following the path that had been cleared by William and Dorothy Wordsworth, Hazlitt and Thoreau, who wrote of William Wordsworth:

> Moreover, you must walk like a camel, which is said to be the only beast which ruminates when walking. When a traveler asked Wordsworth's servant to show him her master's study, she answered, 'Here is his library, but his study is out of doors.'[9]

Though they were not associated with institutions, the English Romantics and the American Transcendentalists, none more so than Samuel Taylor Coleridge, were serious scholars. Coleridge, in fact, is considered the founder of modern literary criticism. Another century and a half would pass before the academy would open its doors to writers. In universities, study is departmentalised so the writer working outside of its ivied walls has the advantage of being able to tread across boundaries. Thoreau puts this another way, 'a man's ignorance sometimes is not only useful, but beautiful, – while his knowledge, so called, is oftentimes worse than useless, besides being ugly'.[10] Many of the pithy asides that Thoreau provides in 'Walking' – an essay that is as much of an extended meditation on the West and the nature of the wild as it is an exploration of its given subject – can be applied to Robinson's presence in and writing about the West of Ireland:

> The West of which I speak is but another name for the Wild; and what I have been preparing to say is, that in Wildness is the preservation of the world ... I believe in the forest, and in the meadow, and in the night in which the corn grows ... Life consists with wildness. The most alive is the wildest ... In the desert, pure air and solitude compensate for want of moisture and fertility.[11]

Like Coleridge, Wordsworth and Thoreau, Robinson was attracted to wild and remote places that he, following their footsteps, recorded in formal, literary structures. The first three, all rebels in various ways, invented the West from their English and New England outposts. Robinson, on the other hand, recorded its temperature, gauged its history. At the beginning of *Pilgrimage*, he provides this statement of purpose:

> The detailed history of that sea, its slow changes in depth, temperature and turbidity, together with that of the life forms it nurtured, is preserved in the variations of the rock-layers themselves, and through its influence on the land forms carved out of those rocks, with which human developments have had to come to terms, impresses a characteristic series of textures – the ground of this book – on one's experience of the islands today.[12]

Though there is a wildness of place and independence of purpose present here, the author's intended progress is guided by such terms as 'detailed', 'slow', 'variations', 'textures' and 'ground' to indicate the extent to which each is considered and nuanced. Despite its grand purpose and range, Robinson's scholarship and writing are defined by modesty:

> Cosmologists now say that Time began ten or fifteen thousand million years ago, and that the horizon of the visible universe is therefore the same number of light-years distant from us. Let this number stand as the context, the ultimate context, of my writing and your reading of these words, that arise like an inwardly directed signpost at one particular little crossroads of reality, the intersection of my life with a spell of Aran's existence.[13]

The completion of a book is but a small mark made on the rock of time by an author, lucky and gifted enough, to have, for a moment, intersected with an island in the Atlantic. Others who have arrived in the West of Ireland, the Revivalists in particular, have often favoured broad-stroke description, larger than life events and the instant rather than the long view. Though much of this work is wonderful, it can only take its reader so far into the *dinnseanchas* (stories of places) of Árainn because it lacks the detached engagement that grounds Robinson's literary attitude. What Robinson shares with writers from Wordsworth to Joyce is a high-quality literary style, the antidote to the learned ugliness Thoreau decries.

Even though Robinson has always eschewed such labels as environmental when aligned to his own work, preferring to see what he produces simply as writing, there is clearly an environmental consciousness present in his many books. We can make the claim, as Christine Cusick has argued, that Robinson is both an environmental and bioregional writer:

> The power of Robinson's bioregional sensibility is that while it may begin with his own predilections and human biases, he permits it to evolve in concurrence with the logic of the nonhuman systems that surround him. At the same time, he understands these systems as embedded within the social history and present era of this region called Connemara.[14]

Of course, in labelling his work in this way we are assenting to the fact that his books are primary rather than secondary texts that belong on bookshelves beside Gary Snyder's *Turtle Island* and William Least Heat-Moon's *PrairyErth* and at a distance from more traditional scholarly works.[15] It is likely that Robinson would never have been able to complete his wide-ranging work had he been burdened by restrictive labelling: to be an environmentalist, or scientist, or linguist would have tied him down. Better and more useful to be a writer able to move wherever, and whenever, he liked. Doubtless, though, Robinson would agree with Thoreau's dismissal of Pope:

> The poet says the proper study
> of mankind is man –
> I say study to forget all that –
> take wider views of the universe.[16]

The human world is central in *Stones of Aran*; however, it is not privileged over the non-human. In this respect, Robinson's view echoes that of E. Estyn Evans, an anthropologist who has theorised on how Irish Studies research should be conducted. Evans believes habitat to be 'the total physical environment, and ... history the written record of the past. I would define heritage in broad terms as the unwritten segment of human history, comprising man's physical, mental, social and cultural inheritances from a prehistoric past, his oral traditions, beliefs, languages, arts and crafts.'[17] Evans also wrote that 'more would be lost than gained if academics were to be drawn away from their specialist fields, but they should be aware of what is going on beyond the fence: it is at the fences, along the borders, that discoveries are likely to be made'.[18] Evans was writing in 1973, just after Robinson's arrival on Inishmore. While it is a fact, as I have pointed out in relation to scholarly writing on the environment and the American West, that writer/scholars are more likely than they used to be to look across disciplines and utilise more reader-friendly genres, it is equally true that Irish Studies scholarship remains too rigidly focused and narrowly confined by department and discipline. Irish Studies scholars in universities have been slow to embrace Evans's view; on the other hand, writers and scholars working outside of academe have had little trouble embracing it.

In addition to bringing many realms of scholarship into play, as Evans advises, the writer and scholar should bring all of his/her own accrued experience into the task at hand. As Robinson found out when he retrieved his old artwork, the effort made in creating the various pieces was not wasted; instead, it prepared him for the challenge that mapping and writing about Inishmore would present. Almost exclusively in his work, the author's voice is objective and somewhat reserved though pulsing underneath is great passion. Robinson's style and personality result from his ability to incorporate and utilise aspects of his learning and experience, a combination that lends fluidity to what he knows, what he has experienced and how he writes. Unlike the professional academic scholar whose focus and voice can be narrowed by how they have been trained, Robinson, like his peers, allows himself to bring greater breadth to the subject at hand. This process is explained in terms of revelation by N. Scott Momaday, the great Kiowa writer: 'the imaginative experience and the historical express equally the traditions of man's reality. Finally, then, the journey recalled is among other things the revelation of one way in which these traditions are conceived, developed, and interfused in the human mind.'[19] Robinson's imaginative experience, as he learned in hindsight, has complemented his field studies and reading, and enhanced and sharpened his abilities as a writer and cartographer.

Inveterate conference-goers and socialisers, myself included, know that Irish Studies is thriving internationally. It is a delight to find the ancient and the modern, not always defined by age I should add, gathered in earnest debate and engaged conversation through all four seasons and on great arrays of subject matter. Increasingly, Tim Robinson's work is being discussed, mentioned and cited as academics register his importance as a scholar and writer. It is my hope going forward that something of Robinson's independence of spirit, and the great tradition he has emerged from, dating back to the British Romantics and American Transcendentalists, will allow scholars working in the academy, following his example, to spread a little wider their interdisciplinary wings.

## Notes

1 David Lodge, *Changing Places* (New York: Penguin, 1975); *Nice Work* (New York: Penguin, 1990); *Small World* (New York: Penguin, 1995).
2 Bruce Chatwin, *In Patagonia* (New York: Penguin, 1978), 137.
3 Tim Robinson, *Stones of Aran: Pilgrimage* (London: Penguin, 1990); *Stones of Aran: Labyrinth* (London: Penguin, 1997).
4 F. H. A. Aalen, Kevin Whelan and Matthew Stout (eds), *Atlas of the Irish Rural Landscape* 1st edn. (Toronto: University of Toronto Press, 1997). William J. Smyth, *Map-Making, Landscapes and Memory: A Geography of Colonial and Early Modern Ireland c.1530–1750* (South Bend, IN: University of Notre Dame Press, 2006).
5 Robinson, *Pilgrimage*, 11.
6 Breandán Ó hEithir and Ruairí Ó hEithir (eds), *An Aran Reader* (Dublin: The Lilliput Press, 1991), 3.
7 B. and R. Ó hEithir, *An Aran Reader*, 1.
8 Tim Robinson, *The View from the Horizon: Constructions by Timothy Drever 1972* (London: Coracle, 1997), 9–16.
9 Henry David Thoreau, 'Walking', in Wendell Glick (ed.), *Great Short Works of Henry David Thoreau* (New York: Harper and Row, 1982), 298.
10 Thoreau, 'Walking', 320.
11 Thoreau, 'Walking', 306–12.
12 Robinson, *Pilgrimage*, 2–3.
13 Robinson, *Pilgrimage*, 23.
14 Christine Cusick, 'Mapping Placelore: Tim Robinson's Ambulation and Articulation of Connemara as Bioregion', in Tom Lynch, Cheryll Glotfelty and Karla Armbruster (eds), *The Bioregional Imagination: Literature, Ecology, and Place* (Athens and London: University of Georgia Press, 2012), 135–49.
15 William Least Heat-Moon, *PrairyErth* (Boston: Houghton Mifflin, 1991). Gary Snyder, *Turtle Island* (New York: New Directions, 1974).
16 Robert Kuhn McGregor, *A Wider View of the Universe: Henry Thoreau's Study of Nature* (Urbana and Chicago: University of Illinois Press, 1997), 1.
17 E. Estyn Evans, *The Personality of Ireland: Habitat, Heritage and History* (Cambridge: Cambridge University Press, 1973), 3.
18 Evans, *The Personality*, 2.
19 Quoted in Least Heat-Moon, *PrairyErth*, 606.

## 10

# Maps, movements and migrants: reading Tim Robinson through Gluaiseacht Chearta Sibhialta na Gaeltachta

## *Jerry White*

In many ways, this basis of this essay is absurd. I am motivated here by a sense that Tim Robinson was somehow part of a movement during the late 1960s and early 1970s that has come to be known as Gluaiseacht Chearta Sibhialta na Gaeltachta, or the Gaeltacht Civil Rights Movement. When I contacted him last year to ask about this, he responded with a characteristic combination of generosity and uncompromising seriousness. He told me that 'although I have the highest respect for them and their work, an account of my relationship with the Gluaiseacht folk would make an extremely short chapter!' He told me, basically, that his connections to the Irish language were to be found elsewhere; he pointed to his early attempts to learn the language,[1] translation projects he's presently involved in, and so on.[2] He acknowledged that he knew the key organisers – Seosamh Ó Cuaig, Bob Quinn, Donncha Ó hÉallaithe – but in a way that was more friendly than political. 'I'm no good at committees and could not have added a feather to the movement's wingspan in Aran' is the way that he summed it up for me. Right then, case closed, yes? Ask the man himself, he says he's not involved, and that's it.

But I remain, foolishly I am sure, undeterred. The case I want to make here is not so much that Robinson was a part of a movement that he was clearly not. Rather, to put it in a slightly pompous literary theory kind of way, I want to read Robinson's work through Gluaiseacht Chearta Sibhialta na Gaeltachta. There are, I believe, defensible reasons for wanting to do that. The first has to do with the broad historical overlap here. Robinson came to Aran, came to the Irish-speaking world, in 1972, a period when Gluaiseacht had gained some momentum and had started to win some battles. It would have been difficult to be unaware of it entirely. Moreover, the movement's specific sensibilities, especially its intense commitment to localism, were completely consistent with the path that Robinson was forging for himself. What Gluaiseacht was trying to do in the late 1960s and early 1970s was to recentre the language movement in the communities where the language was still alive, still spoken as a matter of the everyday. That is, of course, fully consistent with Robinson's approach to Irish's importance, even if he was a bit more literary about the matter. In his 1976 essay 'Islands and Images', he writes how colonists saw Irish as a kind 'subversive muttering behind the landlord's back'; he goes on to say that 'this historical insult stings the sharper in Aran because Irish is its first language, and although with each generation some of the placenames are forgotten or become incomprehensible, thousands of them still bring their poetry into everyday life'.[3] The way that Robinson combines an awareness of how economic underdevelopment and language have been historically intertwined with an awareness of the everyday life *of a specific community* is, for lack of a better way of putting it, very Gluaiseacht.

Another compelling reason for me to try to read Robinson though Gluaiseacht is because, despite its historical significance, I expect that very few people indeed would come to this essay and understand a flip statement like 'it's very Gluaiseacht'. My general experience is that the movement has been entirely left out of most general histories of post-independence Ireland. To take but one illustrative example, *The Oxford Companion to Irish History* does not mention it at all; Nicholas Williams's very short entry on 'Gaeltacht' (the word for the areas of the Republic of Ireland that have been identified as being at least 80 per cent Irish-speaking and are subsidised with the aim of keeping them that way) manages to tell us that 'native governments have given the Gaeltacht preferential treatment', but the words 'Gluaiseacht', 'movement' or 'civil rights' do not appear. That same tome has an entry for 'civil rights', but Paul Ferguson discusses only the North and treats the term as synonymous with that region in the 1960s and 1970s.[4] One of the few exceptions I know of is Gearóid Ó Tuathaigh's contribution to Joe Lee's 1979 book *Ireland 1945–1970*. He summarises the period leading up to the movement this way:

> In the Gaeltacht, as elsewhere in rural Ireland, after the nadir of despair in the 1950s, the 1960s saw the first serious challenge offered to the defeatism and fatalism of a century. A group of articulate young radicals suddenly found its voice and began demanding policies to arrest the dissolution of and disappearance of its own community. These Gaeltacht radicals were generally well-educated, and like similar groups in Northern Ireland, were part of the global dynamics of youth politics and civil rights movements of the late 1960s. The new movement brought results.[5]

It is not so much that Ó Tuathaigh has Robinson himself in mind here; that is clearly not the case. But I do think that 'articulate young radical' is a good way of thinking of what he was becoming in the 1970s. I also think that the maps he has created and the greater knowledge of place names and the linguistic problems surrounding them qualifies as 'bringing results' (see Figure 22). Robinson arrived on Aran and set about making himself into a certain kind of artist and intellectual. Those kinds of artists and intellectuals in that part of the world gravitated heavily to Gluaiseacht Chearta Sibhialta na Gaeltachta.

*Figure 22* Robinson in a boat going to Deer Island (Árainn) (March 1984).

What kind of intellectual am I talking about, exactly? To a great extent I am referring to people who expressed the idealism of the late 1960s by leaving behind big cities and moving to marginal areas as part of a widespread rejection of materialism and careerism. Desmond Fennell, a political journalist who served as a kind of 'staff intellectual' for Gluaiseacht, wrote about the phenomenon in his book *Beyond Nationalism*, as he recalled his own move to the Connemara Gaeltacht in the late 1960s:

> Years later we discovered that in 1968, in West European countries and the USA, a trickle of families and individuals began to move voluntarily from the power centres to the peripheries. In France, 1968 is looked back to as the year when a resurgence of depressed cultures on the French periphery – Breton, Occitan, Alsatian and so on – began. In Paris there was a revival of interest in these cultures, and a drift of young intellectuals and artists from the capital to the regions in question. All of this, it seems, was connected with the 'May Revolution' of that year which rattled the French state to its foundations. In France, as elsewhere in the Western world, there was a sharp loss of faith in the 'centre' and a corresponding shift of faith to the periphery. It was around this time, too, that Welsh, Scottish, Cornish and Breton nationalism began a period of resurgence. Unknown to ourselves, therefore, when we moved west we were being moved by the 'sprit of 1968'. The Gaeltacht itself was being moved by the same spirit: it began to assert itself in the spring of 1969, here in south Connemara, and I participated in that assertion.[6]

This is not exactly the story of Robinson's arrival in Ireland, into the world of the Irish language, but it is not so far off. He writes early on in *Stones of*

*Aran: Pilgrimage* of that arrival to Aran, and of the reasons that motivated it. He recalls that in the early 1970s he had visited the place with his wife for a holiday, in some ways inspired by Robert Flaherty's famous film *Man of Aran*, and then writes how 'a few months later we determined to leave London and the career in the visual arts I was pursuing there, and act on my belief in the virtue of an occasional brusque and even arbitrary change in mode of life'.[7] Just a bit later, he recalls how as he first began his work on a map of Aran, he was struck by how different the process was from what he had known as a visual artist in the metropolitan centre: 'maps of a very generalized and metaphorical sort had been latent in the abstract paintings and environmental constructions I had shown in London, in that previous existence that already seemed so long ago, but I had not engaged myself to such a detailed relationship with an actual place before'.[8] That departure from a bustling metropolis, and from the intellectual practices that accompanied it, in favour of a new kind of practice that demanded *a detailed relationship with an actual place* is basically how Fennell defines the 'spirit of '68'.

The notion that this spirit is embodied primarily by a renewed sense of regionalism may seem highly inconsistent with the way in which it is generally remembered. Indeed, I'm not sure I agree with Fennell about the degree to which 1968 made French culture more region-aware. In the context of these islands, though, I think there is something to it. Indeed, another roughly contemporary figure for understanding Robinson and his connection to the ideology of regionalism is the English novelist and art critic John Berger. At just about the same time that Robinson was leaving a successful career as a London artist for the Gaeltacht, Berger was leaving his successful career as a London art critic and novelist to go in the other direction, toward Europe. He lived for a time in Geneva before moving in 1974 to a small village in the Savoie region of the French Alps, where he has lived ever since. Before moving to the Alps his best-known works were *Ways of Seeing*, a BBC television series and accompanying book that laid out a quasi-Marxist analysis of art history, and *G*, a modernist novel that wanders all over Europe and which brought comparisons with *Ulysses*. But after that, his best-known work was his trilogy titled *Into Their Labours*, three novels published between 1979 and 1990, which present the daily life of peasants in the French Alps in incredible detail.[9] In terms of subject matter, they are quite close to the concerns of Robinson's Aran books or his *Connemara* trilogy. What is different for Berger is the lack of attention to the politics of language. Indeed, while I am a genuine admirer of Berger's work, it seems to me clear that one shortcoming of the work he has done on the peasant Alps is the absence of any sense of patois, any realisation of those regions' linguistic difference (something that is front and centre in Raymond Depardon's quite comparable film series *Profils Paysans*, 2001–8).[10] One way to think of Robinson's work, then, is that he is halfway between someone like Fennell and someone like Berger: probably more literary than the former, and probably more political than the latter.

That said, there is a very interesting political correspondence between these three figures of Robinson, Fennell and Berger, one that illuminates a good deal about the politics of Gluaiseacht and indeed of Ireland as a whole. It is something of a cliché to refer to Irish society as particularly conservative, just as it is a cliché to note that the country basically has a choice of being governed by two different centre-right parties, Fianna Fáil and Fine Gael. What this slightly banal understanding of Irish politics conceals is the fact that within both parties there are substantial constituencies that most outsiders would identify as left. This is particularly the case in rural areas, where the co-op movement tends to be quite strong and there is serious, vigorous debate about the state's role in economic and cultural development. None of this is consistent with the pro-corporation, pro-laissez-faire economic policies that American Republicans have convinced the English-speaking world are synonymous with the word 'conservative'. Fennell has advised Fianna Fáil and has a reputation in Ireland as a conservative commentator; that is understandable inasmuch as his more polemical writings are peppered with constant attacks on the Anglo-American liberalism that he believes the Dublin-based elite has adopted as its new religion.[11] But his politics have always been anti-centralising and radically communitarian. Berger, a deeply committed Marxist who first came to prominence by skewering the British art world along leftist-materialist lines, has been just as acidic a critic of liberalism and its tendency to view individualist materialism as a universal value, an inherent social good. He writes very elegantly in his 'historical afterword' to *Pig Earth* about his notion of 'peasant conservatism'. He is very quick to point out that this 'has nothing in common with the conservatism of a privileged ruling class or the conservatism of a sycophantic petty-bourgeoisie'.[12] He sees peasant society as literally conservative inasmuch as its culture acts as repository (granary is the image Berger uses) of traditions that will allow life in highly marginalised communities to continue. He also writes,

> When a peasant resists the introduction of a new technique or method of working, it is not because he cannot see its possible advantages – his conservatism is neither blind nor lazy – but because he believes these advantages cannot, by the nature of things, be guaranteed, and that, should they fail, he will be cut off alone and isolated from the routines of survival.[13]

For Berger, this is mostly a matter of preserving traditional forms of agriculture and resisting the integration of larger scale or quasi-industrial modes of farming. He characterises peasant society as a 'culture of survival', which he sees as inherently opposed to a 'culture of progress' and thus in essence conservative.

Robinson is not as polemical as either Fennell or Berger on these kinds of issues, but he does speak about language and its connection to territory in similar terms. Consider this passage from early in *Pilgrimage*:

> Irish, the irreplaceable distillate of over two thousand years' experience of this country, which was poured down the drains in the rest of Ireland but which was carried unspoilt

even through the famine century in those few little cups, the western Gaeltachtaí of Aran, Connemara and parts of Donegal and Kerry, is now evaporating even here (as if a word or two disappears every day, the name of a field becomes unintelligible overnight, an old saying decides that its wisdom or foolishness is henceforth inexpressible), while what remains is splashed with the torrents of English. Many in Aran, as elsewhere, stake heavily on the future of Irish (and it is an awesome choice for parents to entrust their children's mental development, or a writer of a life's work, to an endangered language), but the cruel twists of history have put the survival of Irish in the hands of English; at least as essential as the dedication of Irish speakers would be a tolerance, indeed a positive welcoming, among English speakers, of cultural diversity, an awakening to the sanity of differences – and such wisdom is contrary to the stupefying mainstreams of our time.[14]

The images here are different – cups, not granaries – and Robinson is speaking of the future in a slightly more optimistic way than Berger does (with his invocation of the awesome choice faced by parents who do indeed stake their children's future on Irish), but there is a shared anxiety between the two about cultural survival and a more or less common analysis of that problem. At no point in his work that I am aware of does Robinson posit the future of Irish in strategies such as neologisms, spelling standardisation or grammar reform, all of which have plenty of history and current momentum at the governmental level. This would be as inconsistent with his ideas as would a suggestion that the peasant communities where Berger makes his home and tries to be a voice for would be economically stabilised by more modern manners of farming, something he explicitly rejects throughout his work. This sense of cultural survival can only be called conservative. It sees the most logical way for fragile communities to move forward as an awareness of what Robinson called in *Pilgrimage* 'the sanity of differences', the key difference for him, as for Berger, being a difference from de facto liberal views of modernity.

That focus on the continuity of spoken Irish in a community framework was absolutely central for Gluaiseacht. This was not, it is crucially important to remember, Gluaiseacht Chearta Sibhialta na *Gaeilge*. That would be the Irish Language Civil Rights Movement, and that's not the movement I am talking about here as being an important connection to Robinson's work. Indeed, no such movement has ever really existed. The Dublin-based Conradh na Gaeilge (the Gaelic League) has long been active in trying to secure rights for Irish speakers wherever they may live, and the 1970s saw an upturn in more militant activism around language rights, especially under the leadership of Maolsheachlainn Ó Caollaí. But historically, Irish-language political militancy has tended to focus specifically on Gaeltacht communities. One of the earliest examples of this is Muintir na Gaeltachta, founded in the 1930s to represent the people of the Gaeltacht (that is what its name means). Their biggest victory was the creation of a new Gaeltacht area in Ráth Cairn, Co. Meath, in 1935, something that they had lobbied and demonstrated for quite vigorously. That Gaeltacht, like the organisation itself, was populated mostly by Connemara people, and was meant to relieve certain pressures around congested farming land that was specific to Connemara.

The point of Muintir na Gaeltachta, then, was to create a viable community where the Irish language, by virtue of the people who actually lived in that community (that is to say, native speakers or people who had made a conscious choice to speak only in Irish), would be secure. This was exactly the model of Gluaiseacht Chearta Sibhialta na Gaeltachta. Their efforts were entirely focused on Gaeltacht communities, and their key demands focused less on the language itself than the control of those communities. This was spelled out in November 1970, when they released their plan for a Gaeltacht Authority that would govern all of these regions separately from Dublin. An article in the Irish-language weekly *Inniú* spelled this out in some detail, pointing out how the movement was calling not only for a governing body that would deal specifically with the Gaeltacht but also would be broken into local bodies, what they called 'Coistí Áitiúla', or local committees, one for each of the large Gaeltacht regions (Ulster, Munster and Connemara; they proposed that the Connemara and Ráth Cairn committees be combined).[15] This focus on Gaeltacht communities meant that Gluaiseacht was sometimes in a low-level conflict with more traditional language groups like Conradh na Gaeilge, whose mandate was always national; Fennell recalls in *Beyond Nationalism* that 'we called publically on the Gaelic language organisations to remove their headquarters from Dublin to Iarchonnacht [West Connacht, i.e. the Connemara Gaeltacht]'.[16] That sense that the Irish language will ever be a national vernacular is utterly absent from Robinson's work; he is passionate about the language but only insomuch as it is connected to communities and landscapes. Fennell also writes about the founding ethic of Gluaiseacht in ways that sound, to coin a phrase, positively Robinsonian: 'To their emphasis on *teanga* (language) we opposed our emphasis on *pobal* (people or community), maintaining that the language would look after itself if the communities which actually spoke it were stabilised through self-government.'[17]

There are echoes of this sensibility throughout Robinson's books, of course. *Pilgrimage* opens with a slightly rambling meditation on the notion of the 'good step', with Robinson writing of 'our craggy, boggy, overgrown and overbuilt terrain, on which every step carries us across geologies, biologies, myths, histories, politics, etcetera, and trips us with the trailing Rosa spinosissima of personal associations. To forget these dimensions of the step is to forgo our honour as human beings.'[18] In a 1990 essay titled 'Listening to the Landscape', he cautions that 'when talking about the land or the landscape speaking, do not forget that this is only a metaphor, suggestive in some contexts and baleful in others, and that in fact, the speaking is made up of the speech acts of countless individuals, each one in its unique historical and social setting'.[19] In *Connemara: A Little Gaelic Kingdom*, he dismisses some of the folklore collected by Seán Mac Giollarnáth by writing that

> I find these rambles rather tedious, however ancient their lineage and however rich they may be in the motifs folklorists have arduously catalogued and numbered. I value the book more for its introductory essays on the four old storytellers – including the

old Fenian Seán Mac Con Raoi – and its dim photographs of these almost fabulous beings in their great beards and thick waistcoats and heavy boots.[20]

He valued the book more for the *people* contained within.

One might think that Robinson's mapping work is influenced as much by the idealism of caint na ndaoine as is by Gluaiseacht. Caint na ndaoine (speech of the people) was a late nineteenth-century approach to literary Irish laid out by Canon Peadar Ua Laoghaire (Peter O'Leary), especially in his collections of occasional pieces published in 1922 as *Papers on Irish Idiom*. Fr O'Leary was a member of the Gaelic League and definitely inspired many of its more zealous Irish learners, but the basic idea of caint na ndaoine is quite close to Robinson's sensibilities.[21] In a late nineteenth-century essay titled 'The Irish of Keating's Time' O'Leary writes, '*For a living language, the books and the speech of the people should go hand in hand. What is printed in the books should be exact representations of what comes out of people's mouths.*'[22] What Robinson is adding to that approach, and what ironically brings him closer to the localist ideology of Gluaiseacht, is his non-neurotic approach to the *English* language. Robinson's goal has never been, along the lines of the Gaelic League's first president, Douglas Hyde, de-anglicising Connemara.[23] This is visible early in the Connemara mapping work.

In his 1985 book *Mapping South Connemara* (which collects the short articles he wrote for the *Connacht Tribune* as he worked on his map), we find this under the entry for 'Leitheanach, Oileán Gorm': 'The height north of the road is *Leitheanach Mór Hill* (we are close to the linguistic frontier here between the *fíor-Ghaeltacht* of Carna to the south and the anglicising influence of Cashel and its big hotels, and I record these mixed Anglo-Gaelic names as I find them.'[24] If people speak English in the areas that Robinson was mapping, then it's English that he mapped. His current home of Roundstone is very much part of Connemara but it is an English-speaking community. The gazetteer for the Connemara map duly notes 'ROUNDSTONE / CLOCH NA RÓN / the stone of the seals; it used often to be called Cloch Róinte, with the same meaning', but the map itself gives only the name 'Roundstone'.[25] In this strategy I can hear echoes of Donncha Ó hÉallaithe, a Connemara-based teacher (he lectures in Maths at Galway-Mayo Institute of Technology). Ó hÉallaithe had organised pirate television broadcasts in 1987 called Teilifís na Gaeltachta. Gluaiseacht won its first big victory in 1972 when it organised the pirate radio station Saor Raidió Chonamara and had basically embarrassed the government into setting up Raidió na Gaeltchta; Ó hÉallaithe had hoped to do something similar with television.[26] Ó hÉallaithe is a bit younger than Gluaiseacht leaders like Seosamh Ó Cuaig or Fennell and he has taken their place as the lead agitator for that neck of the woods. In a 2004 essay Ó hÉallaithe castigates national revivalist projects, which he sees both as a distraction from the real and demanding work of language stabilisation in the Gaeltacht and faintly oppressive in distinctly Robinsonian way: 'To change the language of Kilkerrin in East Galway to Irish would have done as much violence to that community's cultural life as changing the language of Cill Chiaráin in Conamara from Irish to English.'[27] Robinson is not a revivalist, and his lack of enthusiasm for that project is very much of a piece with Gaeltacht activism of many decades, indeed, right up into the present day. That activism has stressed the moral obligation to allow communities to control their own fate; that obligation of course extends to English-speaking communities as well. Anyone who looks closely at Robinson's Connemara map can see that.

But the place where you can really see pobal taking priority over teanga is in the gazetteers of the two 'big maps': *Connemara: A One-Inch Map* (1990) and *Oileáin Árann: A Map of the Aran Islands, Co. Galway* (1996). Although Robinson's work is seasoned with tales of encountering informants who he for one reason or another deems unreliable, he is quite adamant that his approach to documenting place names is led by the people who inhabit those places and not by externally imposed grammatical or syntactical standards, led by pobal and not by teanga. Readers of *Connemara: Listening to the Wind* may recall him trying to figure out the proper place name for Lough Doolagh: 'I deduce that the proper (i.e. original Irish) name of the lake is Loch Dúlach, or if it is more fussy about its own grammar than many a placename, Loch an Dúlaigh.'[28] Characterising as 'fussy' matters like the persistence of the nominative case where genitive is more correct is entirely consistent with the path he had followed in his mapping projects. The gazetteer to his big Connemara map of 1990 states, 'Some of the names on this map are grammatically defective (for instance a nominative is often used where a genitive would be more correct), but I have thought it better to write down the names as I hear them spoken, rather than touch them up in the name of pedantry.'[29] He also writes in that gazetteer:

> For townlands in the Gaeltacht, I have used the Irish names recommended by the Placenames Department of the OS [Ordnance Survey] … with only two or three exceptions (I do not see why a comfortable old name like Cill Bhriocáin should suddenly become Cill Bhreacáin, for example; such standardisation is still a cultural oppression, if a lesser one than anglicisation).[30]

The sort of pedantry and oppression that Robinson is alluding to is invariably Dublin-centric, led by government agencies or nationally oriented language groups who have a clear interest in normalising Irish's grammatical and orthographic conventions.

I myself would not argue that there is anything necessarily wrong with that kind of standardisation, especially in terms of orthography, but I can also see how it represented the first step in a journey whose end is the norms of urban dwellers becoming the norms for everyone.[31] That kind of centralisation certainly has a linguistic heritage, as France shows all too well. It is not hard to see why Irish speakers would not want to repeat in miniature the process by which modern French emerged by a post-revolutionary state wiping out the patois especially of the mountain regions in favour of a 'national' language actually based on the Parisian variant, all in the name of republican idealism about brotherhood. That sort of hostility toward the French model of centralisation, that sense of it as a form of 'cultural oppression', to use Robinson's phraseology, was very consistent indeed with Gluaiseacht's sensibilities. Fennell was a committed Europhile, but one who almost never offered France as an analogy. Instead, it was her bordering frère-ennemi that appealed to him. In a 1969 lecture he suggested, 'Switzerland's thinking has been done not in some remote capital city which saw only its mountains – and not even in a Swiss capital city – but in the 22 [sic] cantons of Switzerland, each with its own parliament and government, its own laws and budget, all linked together in a federal republic.'[32] Switzerland's most characterising aspect is the looseness of its federal arrangement, and this looseness is what leads Benjamin Barber's formulation that 'diversity is Switzerland's essence, drawing our interest, yet defeating our inquiries'.[33] What characterises Switzerland, what made it appealing to Gluaiseacht's intellectuals, brings us right back to the language of *Pilgrimage*: the fact that it is defined by cultural diversity, by an awakening to the 'sanity of differences'.

Fennell's 'teanga over pobal' formulation is a good way of understanding the Gluaiseacht idea in retrospective, but at the time there was a much more vivid image that actually explains Robinson's place in the community more precisely. On 28 February 1969 Fennell published a manifesto in *Inniú* titled 'Iosrael in Iarchonnachta' (Israel in West Connacht). It was a call for people to come to the Connemara Gaeltacht in the manner of Jewish people coming to Israel. This followed on an essay he had published in the *Irish Times* on 29 January 1969 called 'Language Revival: Is It Already a Lost Cause?', about half of which dealt with the Israel analogy. The 'Language Revival' piece is quite a long and slightly rambling essay; the *Inniú* article is a shorter and much more concise statement of the meaning of the Israel analogy. In Irish he summarised the plan this way: 'Irish learners who have trades and skills will come to Connemara and will make a "New Israel" there, as well as giving to the government a development plan that is run through the authority of the West Connacht Board.'[34] Although the plan never really moved forward, it set the terms of debate in the Gaeltacht for some while, and it is those terms that one can recognise in Robinson's description of his early days in Connemara.[35]

In *Stones of Aran: Labyrinth* he recalls his friendship with Breandán Ó hEithir, who he describes as 'the writer, broadcaster, and despairing but indomitable battler for the Irish language'.[36] He goes into some detail about a course that Ó hEithir was teaching which Robinson and his wife were part of:

> M and I first met Breandán at a course he was running for Bord na Gaeilge in An Cheathrú Rua in the autumn of 1978. The purpose of the course was to familiarise journalists with the economic life and institutions of the Gaeltachts ... It was

> clear why we had been so readily accredited; only four journalists had been tempted by the grant away from their metropolitan perspectives to spend a month visiting fish-processing plants and plastic-components production units in rain-sodden Connemara, and our participation made the course a little less of a numerical flop … It soon transpired that none of us had enough Irish to benefit from the projected programme, and we bowed our heads to a *crash course* in the twelve irregular verbs.[37]

I have inserted the italics there for a reason. In the *Irish Times* article that summarised some of the Israel analogy in English, Fennell goes into a lot of detail about the kinds of industry and infrastructure he hopes will emerge in Connemara (he is especially big on co-ops). He writes there (again with my own italics): 'Preference in new jobs and in regard to residence permits is given strictly to Irish-speakers. But special schools, as in Israel, give *crash courses* to those desirable immigrants who have unusual qualifications.'[38] Writer, map-maker and former land-artist would certainly count as 'unusual qualifications', and that sort of thing was more important to this kind of Gaeltacht scheming than the manifesto 'Iosrael in Iarchonnachta' might seem to indicate. In *Beyond Nationalism* Fennell recalls the happy arrival of the film-maker Bob Quinn, who had quit his job at RTÉ and come to Connemara, quickly gaining fluency in Irish and starting to make films in the language.

> By making films of his own in Gaelic, he and his wife Helen (who became an expert in casting and locations) drew other independent filmmakers to the district. Thus, within a space of five or six years, magazine-publishing, radio work, video and film-acting had become part of the life of Gaelic-speaking Connemara, and to these the Cois Fharraige co-op added printing and book-publishing.[39]

This is all to say that Fennell's 'Iosrael in Iarchonnachta' plan had several components: outsiders coming to Connemara and learning Irish; those outsiders having special skills not generally available in Gaeltacht areas; those skills very much including cultural expertise and artistic abilities; those specially skilled outsiders being given the support to learn Irish intensively by way of aiding their integration into this revitalised Gaeltacht. It is very easy indeed to see all of those components in the early days of Robinson's emigration to the Gaeltacht, as he drifted from Aran to Connemara.

So far I have been somewhat coy about Robinson's explicit engagement with Gluaiseacht Chearta Sibhialta na Gaeltachta, which is found in *Connemara: A Little Gaelic Kingdom*.[40] The title of one chapter refers to a march in 1974 sponsored by Gluaiseacht, a march 'Ó Charna go Bearna', from Carna to Bearna, the informal borders of Connemara. Robinson's account in *A Little Gaelic Kingdom* hits all of the notes that I have been playing here: the movement's hostility toward centralisation, the entwined quality of culture and economics, the presence of artists like Quinn, the importance of Fennell as a thinker, Seosamh Ó Cuaig's work as a journalist and politician, Donncha Ó hÉallaithe's sense of being a successor. Indeed, anyone looking for a very clear, concise and yet thorough summary of

Gluaiseacht overall could not possibly do better than this chapter. But Robinson is a realist, and the sense of his summary overall is that the movement's idealism has gone largely unrealised. On the 'Iosrael in Iarchonnachta' proposal, he writes that 'while the filmmaker Bob Quinn moved to An Cheathrú Rua in 1969 to set up his studio and cinema, few others heeded the call'.[41] He recalls watching footage of the march from Carna, and writes that 'on viewing the old film, it is the very smallness of the group, the evaporation of their oratory into the uncaring breezy spaces of Connemara, that movingly convinces one of their earnestness and the romantic appeal of their cause'.[42] I certainly know something about that romantic appeal, and that is, as it has probably become clear by now, a fairly significant part of what is leading me to try to link Robinson's work with this movement in which he has been clear that he had no part.

What I hope I have made equally clear is that you don't need to have taken a role in Gluaiseacht Chearta Sibhialta na Gaeltachta to have been strongly influenced by it. Toward the end of *A Little Gaelic Kingdom*, he explains his sense of distance from the movement in the same language he used with me in our email correspondence:

> I am unqualified to contribute to the sociolinguistic studies, the pressure groups, the educational inventiveness, 'the long march through the institutions', necessary to reverse the almost terminal decline of Irish. I can only offer an act of faith in its continued life, and hope that my own access to Irish, though limited, will allow me to show that language and landscape are the two wings, however bedraggled they be, on which south Connemara flies.[43]

For someone who is 'no good at committees', Gluaiseacht would have been not so much a long march as a long slog. But what has become clear to me is that Tim Robinson lives in, is nourished by and is a leading citizen of the Gaeltacht that Gluaiseacht has made, or at least the Gaeltacht they tried to make. We can see fairly clear shades of caint na ndaoine chez Robinson, but anyone who knows that movement knows that as an influence, it is not quite right: too invested in clerical idealism about uncorrupted rural folk. That's the thing about Robinson's work: he's clearly committed to his art and to his place, but you wouldn't really call him an idealist. He's too committed to the impurities not only of the landscape but of the people who live in the landscape to really subscribe to caint na ndaoine style dreaminess. 'It's *only through people* – through a continually shared existence, work and thought – that we'll give life or at least greater life to the language': that's Fennell writing in 'Iosrael in Iarchonnachta'.[44] That notion of the language's only future being 'trí chomhshaol cónaithe', through continually shared existence, is the essence of the 'pobal not teanga' approach of Gluaiseacht, and it's the essence of Robinson's artistic practice as well. He came to Connemara not as a metropolitan visitor but not as a native son either; that sense of being a committed outsider makes him seem very close indeed to someone like Berger. But many of the leaders of Gluaiseacht were those kinds of outsiders as well: Fennell, Quinn,

Ó hÉallaithe, none of them are from the Gaeltacht originally. Like Fennell they came west seeking a new kind of life, and their decision to stay entailed a decision to share in the work and the thought of the place. The spirit of '68, perhaps, but there is something to this that far outlives the memory of that failed Parisian strike, far outlives even the memory of the Prague Spring and its destruction. The Connemara that Gluaiseacht struggled for, the Connemara that Robinson now struggles to inhabit, is a testament to the belief that modernity can take many forms, not all of them English-speaking, not all of them urban, not all of them blissfully free of commitment to a specific place. What Gluaiseacht struggled for was, quite simply, an Ireland that would finally awake to the sanity of differences. What could be more Robinsonian?

## Notes

1 See especially Tim Robinson, *Stones of Aran: Labyrinth* (London: Penguin, 1995), 172–3 or 'Listening to the Landscape', in *Setting Foot on the Shores of Connemara* (Dublin: Lilliput, 1996), 151–64.
2 He pointed specifically to Seán Mac Giollarnáth's *Annála Beaga ó Iorrus Aithneach* (which forms the backbone of *Connemara: A Little Gaelic Kingdom* and is also a strong influence on his short 1985 book *Mapping South Connemara*) and Máirtín Ó Cadhain's *Cré na Cille* (which he and Liam Mac Con Iomaire are translating for Cló Iar-Chonnachta and Yale University Press, 2015).
3 Robinson, *Setting Foot*, 3.
4 See S. J. Connolly, *The Oxford Companion to Irish History* (Oxford: Oxford University Press, 1998), 216 and 94.
5 Gearóid Ó Tuathaigh, 'Literature, Language and Culture in Ireland since the War', in J. J. Lee (ed.), *Ireland 1945–1970: The Thomas Davis Lectures* (Dublin: Gill and MacMillan, 1979), 113.
6 Desmond Fennell, *Beyond Nationalism: The Struggle against Provinciality in the Modern World* (Dublin: Ward River Press, 1985), 130.
7 Tim Robinson, *Stones of Aran: Pilgrimage* (London: Penguin, 1986), 10.
8 Robinson, *Pilgrimage*, 11.
9 The novels are *Pig Earth* (1979), *Once in Europa* (1987) and *Lilac and Flag* (1990).
10 The films are *L'Approche* (2001), *Le Quotidien* (2005) and *La Vie moderne* (2008).
11 To take one among literally countless examples, he writes the following in *Beyond Nationalism* (the book that has the most to say about Gluaiseacht): 'The ideology which was rising to power in Dublin was anti-nationalist (or rather, anti-*Irish*-nationalist) and crassly materialist. It was made up of two ideological currents coming increasingly into fusion: on the one hand, the consumerist liberalism which was then, as for some years previously, the ideology of international capitalism; on the other, a revivified Irish provincialism. London, as the local capitalist power-centre, was the local relaying centre for consumerist liberalism, and it was, therefore, in its London-fashioned form – which had considerable American content – that this ideology impinged on Dublin' (35).
12 John Berger, *Pig Earth* (New York: Pantheon, 1980), 208.
13 Berger, *Pig Earth*, 208.
14 Robinson, *Pilgrimage*, 7–8.
15 See also 'Plean don Údaras Gaeltachta Ó Ghluaiseacht Chearta Sibhialta na Gaeltachta', *Inniú* (13 November 1970): 10. The article is unsigned, although it is probably by Ó Cuaig, who was a regular contributor to *Inniú* and one of Gluaiseacht's leaders.

16  Fennell, *Beyond Nationalism*, 141.
17  Fennell, *Beyond Nationalism*, 141.
18  Robinson, *Pilgrimage*, 12.
19  Robinson, *Setting Foot*, 155.
20  Robinson, *Gaelic Kingdom*, 179.
21  This was famously satirised in Flann O'Brien's 1941 novel *An Béal Bocht* (which he published under the name Myles na gCopaleen, a phonetically spelled version of 'Myles of the Ponies' and the name under which he signed his celebrated *Irish Times* columns). When one of the slightly psychotic Irish learners venturing into the countryside worries about the decline of people learning the language, he says, 'I don't think Father Peter has the word decline in any of his works, said the Gaeligore courteously.' In the original Irish version of the novel he says this in a phonetically spelled and quasi-unintelligible learner's patois that I know all too well: 'Nee doy lum go will un fukal sin "meath" eg un Ahur Padur, arsa an Gaeilgeoir go cneasta.' Flann O'Brien, *The Poor Mouth*, trans. Patrick Power (London: Picador, 1973), 49 / Myles na gCopaleen, *An Béal Bocht* (Dublin: Mercier, 1999), 42.
22  Canon Peter O'Leary, *Papers on Irish Idiom*, ed. T. F. O'Rahilly (Dublin: Browne and Nolan, 1922), 86. The italics are his.
23  I refer here to Hyde's famous 1892 essay 'The Necessity for De-Anglicising Ireland'. It has been widely reprinted; see, for example, Maureen O'Rourke Murphy and James MacKillop (eds), *Irish Literature: A Reader* (Syracuse: Syracuse University Press, 1987), 137–47.
24  Tim Robinson, *Mapping South Connemara* (Roundstone: Folding Landscapes, 1985), 15–16. The published map from 1990 gives these names as 'Leitheanach Theas' and 'An tO. Gorm (Blue Island)'. The latter is an interesting transformation inasmuch as the bilingualism that I am discussing here remains (few of the place names on the Connemara map are given in *both* Irish and English) but the grammar has been sharpened a bit (this would be an abbreviation for 'An tOileán Gorm', which is in the nominative case). 'Theas' means 'south'.
25  Tim Robinson, *Connemara Part 1: Introduction and Gazetteer* (Roundstone: Folding Landscapes, 1990), 2.
26  For details, see Risteárd Ó Glaisne, *Raidió na Gaeltachta* (Indreabhán: Cló Chois Fharraige, 1982), 11–32. He also discusses the debates around television in a kind of afterword called 'Agus an Teilifís?', 457–68.
27  Donncha Ó hÉallaithe, 'From Language Revival to Language Survival', in Ciarán Mac Murchaidh (ed.), *'Who Needs Irish?': Reflections on the Importance of the Irish Language Today* (Dublin: Veritas, 2004), 182.
28  Tim Robinson, *Connemara: Listening to the Wind* (London: Penguin, 2006), 33. Going to the website www.logainm.ie – maintained by Fiontar at Dublin City University, based on the work of the Ordnance Survey of Ireland and now the standard tool for these matters – does not clarify much. There is no entry there for Lough Doolagh or Loch Dúlach. They do have a Loch an Dúlaigh (whose English name they give as Ballindooly Lough), but that is just outside of Galway City and near the airport, and thus fairly far from where Robinson is mapping here.
29  Robinson, *Connemara Part 1*, 2.
30  Robinson, *Connemara Part 1*, 1–2.
31  I am strongly motivated here (quite selfishly, I admit) by Michael Cronin's recollection of the pre-standardisation days of Irish learning. He writes, 'The extraordinary difficulties faced by those wishing to learn Irish or write in the language prior to the emergence of broadly accepted orthographical and grammatical standards are too

easily forgotten.' Michael Cronin, *Translating Ireland: Translation, Languages, Cultures* (Cork: Cork University Press, 1996), 5.
32  Desmond Fennell, 'Connacht's View of Itself', in *Iarchonnacht Began* [1969] (Cill Chiaráin: Iarchonnachta 1985), 37. The pamphlet notes that this was a 'Lecture delivered in Community Centre, Castlebar, 28 May'.
33  Benjamin Baber, *The Death of Communal Liberty: A History of Freedom in a Swiss Mountain Canton* (Princeton: Princeton University Press, 1974), 12. This is actually a case study of the canton of Graubünden, which is the only canton where Romansh is an official language (the canton is trilingual, with German and Italian also being official languages). The Swiss constitution recognises Romansch as a 'national language' but not as an 'official language'. It is spoken by about 0.5 per cent of the Swiss population, but it is spoken nowhere outside of Switzerland. There is a decent case to be made that the very mountainous and predominantly rural Graubünden is the Connemara of Switzerland, and that its Romansh-speaking heartland is Switzerland's Gaeltacht. That would be the subject of another essay, though.
34  Desmond Fennell, 'Iosrael in Iarchonnachta', *Inniú* (28 February 1969): 9. This was reprinted in Fennell's pamphlet *Iarchonnacht Began* [1969] (Cill Chiaráin: Iarchonnachta 1985), 11–13, with some correspondence from *Inniú*'s letters pages reprinted directly thereafter. I'm going to refer here to the version in *Iarchonnacht Began*, which might be marginally easier to access than old copies of *Inniú*.
35  Fennell, *Iarchonnacht Began*, 11. See, for example, 'Cosaint Fhennell ar a Theoric "Iosrael in Iarchonnachta"', *Inniú* (5 June 1970): 1–2, which details arguments Fennell was having with Conradh na Gaeilge's Maolsheachlainn Ó Caollaí over whether rural areas could really be a vanguard for social change, and with Riobard Mac Góráin (a founder of the Dublin-based language group Gael Linn) over the matter of the model's suitability for other Gaeltacht areas (he was speaking at an event in Gaoth Dobhair, Donegal).
36  Robinson, *Labyrinth*, 171.
37  Robinson, *Labyrinth*, 172–3.
38  Desmond Fennell, 'Language Revival: Is it Already a Lost Cause?', *Irish Times* (29 January 1969): 8.
39  Fennell, *Beyond Nationalism*, 142.
40  See chapter 8 of *Connemara: A Little Gaelic Kingdom*, 'The Long March'.
41  Robinson, *Gaelic Kingdom*, 99.
42  Robinson, *Gaelic Kingdom*, 98.
43  Robinson, *Gaelic Kingdom*, 103.
44  Fennell, *Iarchonnacht Began*, 12.

# 11

# 'About nothing, about everything': listening in/to Tim Robinson

## Gerry Smyth

And now as if the cleaning and the scrubbing and the scything and the mowing had drowned it there rose that half-heard melody, that intermittent music which the ear half catches but lets fall; a bark, a bleat; irregular, intermittent, yet somehow related; the hum of an insect, the tremor of cut grass, dissevered yet somehow belonging; the jar of a dor beetle, the squeak of a wheel, loud, low, but mysteriously related; which the ear strains to bring together and is always on the verge of harmonizing, and at last, in the evening, one after another the sounds die out, and the harmony falters, and silence falls.[1]

– Virginia Woolf

There is no question that we belong to what is, and that we are present in this respect. But it remains questionable when we are in such a way that our being is song, and indeed a song whose singing does not resound just anywhere but is truly a singing, a song whose sound does not cling to something that is eventually attained, but which has already shattered itself even in the sounding, so that there may occur only that which was sung itself.[2]

– Martin Heidegger

In the preface to *Listening to the Wind*, the first volume of his *Connemara* trilogy, Tim Robinson describes a range of sounds he associates with the landscape in that part of the world: 'the sough (which we should not delude ourselves is a sighing) of the Ballynahinch woods, the clatter (not a chattering) of the mountain streamlets, the roar (not a raging) of the waves against the shore' (see Figure 23).[3] He goes on to discuss the 'toneless bulk noise' and the 'vast, complex sounds' characteristic of the Connemara soundscape, comparing these to 'the auditory chaos and incoherence that sound engineers call white noise'.[4] At this point, Robinson makes a discursive leap from the actual to the metaphorical, as he likens the chaotic noise of the contemporary world to the 'agonistic multiplicity' produced by 'the sound of the past'.[5] History may be musical, comprised of 'rhythms, tunes and

*Figure 23* Robinson 'listening to the wind' in the woods of Connemara (from Pat Collins's film *Tim Robinson: Connemara*, photo by Colm Hogan).

even harmonies', he writes; the actual 'voices' productive and constituent of those musical qualities, however, are random, chaotic and increasingly faint.

In the last three paragraphs of the Preface, Robinson goes on to compare what he calls 'true writing' and 'true speech' as discursive forms. The former is the 'outcome of the potentially endless reshapings of sense and intricate adjustments of word to context', whereas the latter is 'spontaneous, [uttered] out of the unfathomable depth of personhood'.[6] This creates a dilemma for someone in Robinson's position:

> How can writing, writing about a place, hope to recuperate its centuries of lost speech? A writing may aspire to be rich enough in reverberatory internal connections to house the sound of the past as well as echoes of immediate experience, but it is also intensely interested in its own structure, which it must preserve from the overwhelming multiplicity of reality.[7]

This dilemma haunts the Connemara project throughout its three volumes: how can one use writing to represent that (the past, place, the everyday lives of ordinary people) which resists representation?

If such a problem emerges in the first instance from Robinson's experience, at the same time it has implications for the ways in which readers may relate to his work and, wider still, for the ethics of interpretation itself – specifically, for those forms of interpretation that we call critical analysis. For if Robinson intuits a problem with his own project, there is at the same time a question mark over the existence at this time in the Western academy of a reading technique equal to the challenge of Tim Robinson's writing.

In the meantime, the author presses on and asserts an intention to organise his study of Connemara in terms of 'three factors whose influences permeate the structures of everyday life here: the sound of the past, the language we breathe, and our frontage onto the natural world'.[8] As its title suggests, *Listening to the Wind* privileges the first of these three (clearly overlapping and interconnecting) factors. It is in fact a book all about listening: about the various kinds of listening of which humans are capable, and about what it means to be a listener.[9] As such, it represents a key statement of the author's core themes – themes which themselves resonate in relation to a variety of philosophical and political systems.

'Listening' is a relatively small word for a dauntingly complex series of propositions and possibilities. There are numerous ways in which humans may listen, and numerous disciplines and theories for which listening constitutes a key category within a wider discursive system. Among other things, 'listening' connotes an ability and a practice that is at once physiological, psychological, philosophical, sociological, technological, musicological and cultural-historical. It is closely linked with an array of related human abilities such as speaking, thinking, understanding, acting, co-operating and creating. Despite an ocular bias embedded within human history,[10] ours is a species and a civilisation in which 'listening', in all its multifarious ramifications and connotations, remains of central significance.

One of the most influential engagements with the concept of listening was produced by the twentieth-century German philosopher Martin Heidegger. Among much that is arcane to the point of obscurity, in his famous philosophical treatise *Being and Time* Heidegger argues that 'Dasein, as a Being-with which understands, can *listen* to Others.'[11] Elsewhere in the volume he writes:

> Listening to … is Dasein's existential way of Being-open as Being-with for Others. Indeed, hearing constitutes the primary and authentic way in which Dasein is open for its ownmost potentiality-for-Being – as in hearing the voice of the friend whom every Dasein carries with it. Dasein hears, because it understands … Being-with develops in listening to one another.[12]

Listening is thus the mode whereby that which we may for the sake of convenience call 'the human subject' achieves presence, both in itself and in relation to the totality of existence (human and otherwise) within which it assumes a position. For Heidegger, in other words, it is by listening, not speaking, that Being is realised.

There is more. In a lecture of 1950, eventually published in 1959 as an essay entitled 'Language', Heidegger writes:

> Mortals speak insofar as they listen. They heed the bidding call of the stillness of the dif-ference even when they do not know that call. Their listening draws from the command of the dif-ference what it brings out as sounding word. This speaking that listens and accepts is responding … Man speaks in that he responds to language. This responding is a hearing. It hears because it listens to the command of stillness.[13]

The 'command of stillness' calls us to listen and respond, and in responding to confirm an affirmative relationship with that silence. Heidegger also writes of 'hearkening', 'heeding' and 'caring for' silence, while in another context (the essay 'What Calls for Thinking?') he traces the meaning of 'to call' through a range of equally benign and affirmative verbs: 'to commend, entrust, give into safe-keeping, to shelter'.[14] Not only must we learn 'to hear the peal of stillness afterward', moreover: true listening means 'to hear it even beforehand, and thus as it were to anticipate its command'.[15] 'To hear', in this regard, is to seek to master the voice of the Other in discursive terms; 'to listen', however, is to seek a position outwith discourse in respect of that which cannot be said, something that is literally unheard of.

Such gnomic statements form part of a complex system which Heidegger spent his entire career elucidating and refining. There is a readily available inference, however, which suggests that listening represents a crucial aspect of human experience, one that is ineluctably linked with a number of equally important abilities including understanding, speaking, dwelling and thinking. Such a profile renders Heidegger highly amenable to the kind of project undertaken by Tim Robinson in relation to the landscapes of the Aran Islands, the Burren and Connemara – what he himself has described as 'an existential project of knowing a place'.[16]

After Heidegger, the phenomenology of listening is taken up by a number of other philosophers, most notably Eugène Minkowski, Gaston Bachelard and Jean-Luc Nancy. In his book *Vers une cosmologie* (1936), Minkowsky discussed

> a new dynamic and vital category, a new property of the universe: reverberation ... It is as though the sound of a hunting horn, reverberating everywhere through its echo, made the tiniest leaf, the tiniest wisp of moss shudder in a common movement and transformed the whole forest, filling it to its limits, into a vibrating, sonorous whole.[17]

'Reverberation' refers not (or not just) to a sensory or physical experience, Minkowski suggests; it draws, rather, on 'the dynamism of the sonorous life itself which by engulfing and appropriating everything it finds in its path, fills ... the slice of the world that it assigns itself by its movement, making it reverberate, breathing into it its own life'. '[Although] not sonorous in the sensory meaning of the word', he explains, such reverberation is not 'less harmonious, resonant, melodic and capable of determining the whole tonality of life'.[18]

The notion of a form of non-sensory 'reverberation' is taken up by Bachelard and Nancy. In his phenomenological study of domestic space (itself a response to the possibility of Heideggerian 'dwelling'), Bachelard discusses 'a voice so remote within me, that it will be the voice we all hear when we listen as far back as memory reaches, on the very limits of memory, beyond memory perhaps, in the field of the immemorial'.[19] The human soul responds to such a voice by '[experiencing] the kind of reverberation that ... gives the energy of an origin to being'.[20] This is essentially how phenomenology differs from psychology or psychoanalysis: whereas the latter are always oriented toward the meaning (symbolic or real) of an actual sound, however faintly heard, the former reverberates in response to energies that that have no apparent 'sensory or physical' resonance. In such an analysis it is the listening, not the hearing, that is all important.

Jean-Luc Nancy develops this distinction between 'listening' and 'hearing' by self-consciously resisting the tendency of philosophy to assimilate the former merely as a mode of understanding. 'If "to hear" is to understand the sense,' he explains, 'to listen is to be straining toward a possible meaning, and consequently one that is not accessible.'[21] Nancy proposes that 'listening' connotes a different attitude toward the subject / object relationship, one oriented not toward 'sense', but toward 'resonance' – 'the tendency', that is, 'where sound and sense mix together and resonate in each other, or through each other'.[22] 'Sense' here does not equate to a 'meaning' apprehensible by the subject who hears: rather, '[when] one is listening, one is on the lookout for a subject, something (itself) that identifies *itself* by resonating from self to self, in itself and for itself, one in the echo of the other, and this echo is very like the very sound of its sense'.[23]

According to Nancy, then, 'hearing' is oriented toward that which is always already encoded: sounds and voices which may be processed as 'meaning' within a range of discourses. At the same time, we retain the capacity to 'resonate' – as he writes, 'to envisage "the subject" as that part, in the body, that is listening or vibrates with listening to – or with the echo of – the beyond-meaning'.[24] The burden of his thesis is to suggest that if we cannot do without the first (as patently we cannot), then we cannot afford to do without the second: we must resonate as well as hear / understand, and our existence must be oriented as much toward sound as sense, for it is only by opening ourselves to resonance that we achieve subjectivity.

There is one phenomenon, Nancy claims, one freely available everyday human activity in which the complementary orientations of meaning and resonance may be easily observed: music. For if music is always pulling us back to 'the' moment, or rather toward a series of moments (composition, performance, consumption, etc.), it also implicates us in *every* moment, when it converts the hearing subject into a subject who listens, a subject in and for whom sound resounds: 'Music is the art of making the outside of time return to every time, making return to every moment the beginning that listens to itself beginning and beginning again. In resonance the inexhaustible return of eternity is played – and listened to.'[25] In Nancy's terms, music demands a kind of sensory engagement (listening) that sensitises the subject to the Otherness that inheres within the Self, and to the ways in which the 'here and now' is symbiotically enmeshed with the 'always and everywhere'. Such an understanding recalls Robinson's methods which, as we shall see, are likewise concerned to experience the absolute interpenetration of past and present, object and subject. My suggestion at this stage is that Robinson's work is musical inasmuch as he is a listener who encourages listening among his readers – 'listening' here understood as that which resonates in relation to that which cannot achieve representation within discourse.

As might be expected, there is a political as well as a philosophical dimension to listening. In his analysis of the ecological implications of Heidegger's work, Charles Taylor differentiates between two principal forms of contemporary protest against 'the unreflecting growth of technological society'.[26] The first, which he describes as 'shallow', represents an attitude toward the natural world which is 'grounded ultimately on human purposes'.[27] The second is the 'deep ecology' movement which argues in essence 'that we have reason to set limits to our domination of Nature, which go beyond our own long-term flourishing, that Nature or the world can be seen as making demands on us'.[28]

Taylor is neither the first nor the last to identify a correspondence between certain aspects of Heidegger's thought and the concerns of the modern ecological movement. Although not entirely equivalent, the latter's work tends toward the 'deep' perspective, Taylor maintains, in so far as it develops a conviction that 'something beyond the human makes demands on us'.[29] This 'something' is identified by deep ecologists as a form of subjectivity to which mainstream Western discursive practices are simply not amenable:

> [We] require a viable environmental ethics to confront the silence of nature in our contemporary regime of thought, for it is within this vast, eerie silence that surrounds our garrulous human subjectivity that an ethics of exploitation regarding nature has taken shape and flourished ... Recognizing this need, some strains of deep ecology have stressed the link between listening to the nonhuman world (i.e., treating it as a silenced subject) and reversing the environmentally destructive practices modern society pursues.[30]

Heidegger's anti-intuitive understanding of language as a call to listen to silence rather than to utter or to articulate individual consciousness qualifies him to be described as a resource for 'deep ecology'. A willingness to listen to silence, moreover, is linked to the possibility of 'dwelling' – which, in one of his most suggestive essays, Heidegger tracks through old German etymology before eventually defining it as an ability 'to be at peace, to be brought to peace, to remain in peace':

> The word for peace, *Friede*, means the free, das *Frye*; and *fry* means: preserved from harm and danger, preserved from something, safeguarded. To free really means to spare. The sparing itself consists not only in the fact that we do not harm the one whom we spare. Real sparing is something *positive* and takes place when we leave something beforehand in its own nature, when we return it specifically to its being, when we 'free' it in the real sense of the word into a preserve of peace. To dwell, to be set at peace, means to remain at peace within the free, the preserve, the free sphere that safeguards each thing in its nature. *The fundamental character of dwelling is this sparing.*[31]

For Heidegger, 'History is the chronicle of man's concern for "place".'[32] The consequences of this are, firstly, that learning to dwell in (listen to) a place represents the perennial human plight, and, secondly, remembering that this is so constitutes the task of proper 'thinking'. This latter practice is linked with a whole series of concepts, attitudes and perspectives – preserving, commending, safeguarding, cherishing, freeing, sheltering, sparing – all of which orient us toward 'something deeper than an instrumental calculation of the conditions of our survival'.[33] Heidegger's 'thinking' is highly suggestive, in fact, of a form of 'deep ecology' that 'wants harmony with nature and recognition of its intrinsic value, limits on both consumption and the technologies that encourage consumption, and a rediscovery of local, community and human-scale concerns'.[34] At the same time, it also clearly suggests the profile of someone (Robinson) who has written that: 'Seeking out place-names, mapping, has become for me not a way of making a living or making a career, but of making a life; a mode of dwelling in a place.'[35]

So, developing from what has been broached thus far, I would like to suggest a philosophical distinction between 'hearing' and 'listening', and to suggest further that such a distinction may be linked to a range of specific political and critical considerations.

Through a process of immersion in the life, language and landscape of the west of Ireland, Tim Robinson has trained himself to 'hear' a range of Irish voices – what at one point he describes as 'the echo of that terrible shouting from the past'.[36] Whether literal or metaphorical, such voices continue to 'reverberate' in the present in a number of ways: in place names and place lore; in local song and story; in the memories of local people; in a great variety of textual traces, including old maps and books; and so on. These are 'real' voices which Robinson has dedicated the latter part of his life and career to retrieving and preserving – voices relating stories of real people negotiating the larger movements of history through the processes and the rituals (and the noises) of everyday living.

'Hearing' is oriented toward intended sound; listening, as we have seen, is oriented toward something else – a 'something' that may be silence or eternity or nature itself. There is that in Robinson's work, as well as in his attitude and outlook, which hearkens to this 'something else' – a resonance beyond the intended 'voices' of the past, compelling as they are. It is this hearkening which implicates Robinson in a discourse of deep ecology.[37] If he is concerned to 'hear' place names, for example, with all their specific temporal and spatial associations, he has also opened himself up to listening, to becoming a subject who reverberates and resounds in relation to *all* time and *all* space. This in itself connotes a different attitude toward landscape, toward the people who inhabit it and the forces claiming an interest in its use and disposal.

In the remainder of this essay I want to consider the economy of hearing and listening in Robinson's work, particularly in that volume which he has specifically identified as an exercise in listening. As I prepare to do so it strikes me that there are, resident within that body of work, theoretical and methodological implications for my own practice. What I mean by this is that if we consider Robinson's not inconsiderable output as representing a kind of discursive 'landscape', the question occurs: is there not a possibility, indeed an onus, to resist the drive to 'master' it all, and to focus instead on a limited aspect in order to enable the kind of engagement that only persistent, extended attention can produce? 'Reading', after all, is a kind of 'mapping' in which the subject attempts to actuate a complex, multifaceted text in terms of a range of pre-established signifying capacities. A text may be 'read', just as a landscape may be 'mapped', in relation to a range of specific interests and desires. At the same time, just as every map is a function of the cartographical technology that is brought to bear on the landscape, so every 'reading' is a function of the critical technology that is brought to bear on the text. Texts, like landscapes, are 'produced' according to what we want and how we represent them. Robinson teaches us that close, prolonged, patient attention to a restricted part of the text / landscape produces the possibility of an engagement which is deeper and richer than that afforded by 'comprehensive' responses.[38]

Robinson has written with insight and empathy about the indigenous music of Connemara, and about the people who use that music as part of their everyday engagement with the landscape. *Connemara: A Little Gaelic Kingdom* is perhaps the most obviously musical volume he has published, containing material on a range of songs (such as 'The Song of Wonders', 'The Keening of the Three Marys', 'The Song of Maínis') and figures (Joe Heaney, Sorcha Ní Ghuairim and Máire Ní Chlochartaigh, for example). Robinson tends to hear these songs and these singers with a folklorist's ear, listening for clues to local history and place names, noting archival and stylistic genealogies. Music in this sense is locked in to the landscape, its traditional function being to provide emotional expression and reportage to those lacking other means and other media.

Music also retains a rhetorical function throughout Robinson's work, much of it gleaned from an art tradition located at some remove from the traditional music which, as observed above, is embedded in a much more functional sense in the Connemara landscape. *Stones of Aran* he described as 'through-composed' in terms of the thousands of steps he took when walking the island.[39] In *Pilgrimage* he writes: 'The persistency, recurrency and interpenetration of images in a composed book ... are to be as modulable as those of themes in music.'[40] And in *Listening to the Wind* Robinson alludes to the pinnacle of the Western art tradition in order to try to encapsulate his impression of the view:

> Once when I was lying on the terrace of our house overlooking the bay, listening to music from the room behind me and watching a summer night subvert the scale of all things, I felt I could raise my hands and spread my fingers over the mountain range, solidly dark against the still wine-flushed sky, as if over the keyboard of a piano, and produce one tremendous, definitive Connemara chord. But Connemara tends to undefine itself from minute to minute, and this Beethoven moment quickly passed.[41]

Extended exposure to the landscape and culture of the west of Ireland has afforded Robinson an important insight: that *sean-nós* is capable of articulating the truth of Connemara in ways that art music cannot. Although there remains a world of difference between Joe Heaney and Beethoven, Robinson continues to employ musical metaphors from the art tradition associated with the latter; and if such images enable him to develop complex insights in relation to the landscape, at the same time they place him (along with Beethoven) at somewhat of a tangent to that landscape. There is a sense in which he remains an outsider with the knowledge of an insider, and that such a position has consequences for the way in which he relates to the landscape.

From my perspective here, the real interest of music in Robinson's work is that it provides a special instance of the phenomenology of listening. It is my sense that what at one level constitutes both a theme (music and the landscape) and a useful metaphor (listening to the landscape) comes in time to represent a much more active, enabling, and eventually literal, trope with which to conduct the search for 'true place'; and that music – and in particular the form of listening demanded and afforded by music – provided Robinson with both an example and a means to develop his particular understanding of what it means to dwell.

After a short introduction (from which I have already quoted above), Robinson begins *Listening to the Wind* with a chapter called 'Scailp'. This is an Irish word translating (among other things) as 'a cleft or small cliff, a clump of bushes or briars, a hut roofed with sods of turf'.[42] In this instance it refers to a particular place in Roundstone Bog, near the seaside village of Roundstone, located on the southern coast of that undefined territory in the west of Ireland known as Connemara – that place in which Robinson has made his home since 1984. Thus, like most Irish place names 'Scailp' is a multi-purpose signifier: a description that is also a location, a trace of previous landscape usage, a negotiation between nature and culture.

Describing his preferred mode of access to Scailp, Robinson introduces an auditory motif that is in keeping with the book's stated mode: 'Ideally, I feel, a walk should be undertaken with the respect for its own timescale and structures and ceremonies of mood one brings to the hearing of a piece of music.'[43] Perhaps the impersonal 'one' in this passage needs to be substituted by the personal 'I', for not everyone will listen to music with the same respect and attention that Robinson himself does. '[A] piece of music' is likewise hostage to the fortunes of taste and fashion: we are all familiar with the phenomenon of background music washing over us without anything resembling due care and / or attention, and we are probably all guilty at one time or another of disrespecting 'a piece of music' valued by someone else. Nevertheless, it is a striking and significant analogy, for in this opening paragraph Robinson uses music to describe a mode of engagement with the landscape which privileges balance (literally and figuratively) rather than mastery, one in which Scailp's essential emptiness comes, eventually, to resonate within his own being.

It would probably be a mistake to infer a particular Robinsonian 'methodology' as such, but he himself is quite open regarding a reliance on local knowledge, much of it stored in the language and the memory of those who dwell in the region. Such knowledge constitutes one aspect of the 'voice' of Connemara: a willed, intentional, human-oriented voice which relates the story of the species' historical engagement with the landscape. We encounter one instance of that voice on page six, when Robinson relates an anecdote concerning an elderly man complaining about the proliferation of weeds in a drainage channel by a field on the approach to Scailp: 'This is the last time I cut rushes and leave them to grow after me!' It is an innocuous, even banal, moment culled from millennia of textual traces. As an instance of the human attempt to modify the landscape in line with basic needs (survival) and desires (sustenance), however, it both symbolises and contains within itself all the other voices which have at one time or another imprinted themselves on the landscape.

This is an instance of what Heidegger refers to as 'Being-with'. It is a form of hearing oriented toward the Other, toward understanding and toward preservation. The voice – everything it symbolises and contains – is lost if Robinson is not there at that particular time, in that particular place, to hear it. In hearing it, he affirms a caring relationship with it, and he 'spares' the existential space-time within which it was uttered. Moreover, in hearing and accepting it, Robinson himself begins to 'reverberate' as a subject. He has, as it were, been called into being by the process of hearing.

Other 'human' voices proliferate throughout the remainder of 'Scailp': some actual and named – such as 'Tommy O'Donnell, a sheep farmer, from whose nearly forgotten Irish and obscurely mouthed English I tried to make out place-names and local histories'[44] – some lost and anonymous, such as the ancient shepherds who (in the author's imagination) shouted to each across the bog. In fact, Robinson's entire oeuvre maps a complex landscape of the identifiable and the anonymous, the named and the unnamed, the real and the imagined; and while he is not chary in naming the names, he is at the same time adamant that the unnamed should have their say, even those – like the obscene numbers who died during Ireland's various famines – whose only contribution to the conversation of history is the mound, of greater or lesser prominence, in the landscape where they are buried: 'these demanding dead whose voices I have promised to hear'.[45]

Famine graves constitute a ubiquitous, if often invisible, presence throughout Connemara. More obvious evidence of landscape modification is discernible in practices such as turf-cutting and field enclosure, cairns and shelters. Such traces tell their own story. The word 'booley' (pp. 12ff.), for example, an anglicisation of the Irish *buaile*, translating as 'milking pasture', embodies a range of economic, social, cultural, political and technological connotations, all lying silent within the signifier until they are invited into discursive presence by the hearer. If we listen to the word with Robinson's ears we hear a range of voices, ancient and modern: voices identifying the possibility and advisability of temporary pasturage; voices addressing legal questions of land ownership and usage; voices pursuing interpersonal dramas of courtship and community; and voices insisting on the expedience of transliteration from one language to another.[46]

Some might argue that all this implicates Robinson in a particular form of ecological consciousness – one set in a remorselessly declensionist mode, while at the same time primed to hear only the sounds of a human engagement with the landscape of Connemara. Given his commitment to and sensitivity toward that landscape, however, I think it would be disrespectful – and incorrect – to refer to such a response as 'shallow'. For if Robinson is a 'hearer', he is also a 'listener', and such 'listening' draws him (and his readers) toward other modes of being and other forms of politics.

Auditory images and references accumulate as the chapter proceeds: 'tall ranks of reeds … bow and scrape and whisper among themselves'; an 'invisible' stream is 'increasingly audible'; an exposed holly tree produces an 'amnesiac whispering'.[47] As before, such sounds are readily apprehensible in terms of various discourses – botany, geography, geology, etc. – as well as a number of sub-disciplines concerning, for example, vegetation proliferation and landscape morphology. All these disciplines are designed to 'discipline' these sounds in ways that answer a range of specifically human concerns. The effect is to interpellate the subject into a dialogue with an interlocutor who says: 'These sounds mean these things, don't they?' And of course, once positioned thus as an agent of disciplinary knowledge, it is very difficult to find a position outwith discourse from which to resist the power of that insistent, reasonable voice.

By virtue of his previous career as a mathematician and designer, allied to an insatiable thirst for knowledge about the landscape in which he lives, Robinson appears to be reasonably well versed in such disciplines: over the course of 'Scailp' he adopts by turn the language of the archaeologist, the botanist, the cartographer, the chemist, the climatologist, the conservationist, the county councillor, the economist, the entomologist, the folklorist, the geographer, the historian, the linguist, the literary critic, the musicologist, the ornithologist, the philosopher, the physicist, the politician and the zoologist – and probably others also which I have not noticed. It is with reference to these languages – their discrete vocabularies and grammars and syntaxes – that Robinson works 'to ensure that [he] has heard the tale of time and [has] taken it down correctly'.[48] And like the landscape for which it is named, the chapter entitled 'Scailp' is a complex 'network' of 'interested' discourses[49] – 'interested' in the intellectual sense, certainly, but also in the sense of the author's personal investment in the disposition of a precious resource which is symbiotically enmeshed with his own identity.[50]

All these discourses have an 'interest' in the whispering of the reeds and the trees, and in the sound of the invisible stream; they all 'hear' those particular sounds with reference to the regimes and the styles of knowledge which they represent. But what is the meaning of these 'unintended' sounds? How can they be listened to in ways that are not reclaimable by disciplinary knowledge, and what resonances do they set echoing in Robinson and, potentially, in the reader of his texts? What is it that the reeds and trees whisper among themselves?

One valid answer to this final question might be: 'nothing'. From the perspective of disciplinary knowledge, 'whisper' is a figurative signifier chosen by the author to describe a peculiar sonic effect produced by the action of certain properties (wind force) upon certain materials (reeds and trees). From such a perspective, the inference of willed communication or exchange encoded in the word 'whisper' is entirely unwarranted. The subject *hears* nothing because there is nothing to be heard; and to 'hear' nothing is to confront the limits of disciplinary knowledge, and to confront also the existential absence at the heart of the subject who wields that knowledge.

Robinson cultivates a different relationship with 'nothing', one oriented toward resonance rather than (or as well as) sense. This relationship is broached in the opening paragraphs of 'Scailp':

> Sometimes ... after one of these almost ceremonial or ritual walks I am disappointed to find very little in my mental knapsack; I have taken the distance only in my stride and not in my mind. But perhaps that is for the best in the case of a walk with a goal like Scailp, where there is nothing, or almost nothing; I go out there to wrestle with emptiness, and success would be to bring exactly nothing home with me, not even a catalogue of finds and observations or that rather exciting ego-whiff of sweat and wilderness.[51]

The theme is picked up again a few pages into the next chapter: 'Sometimes I come back from such a walk with my head so empty it seems not a single thought or observation has passed through it all day, and I feel I have truly seen things as they are when I'm not there to see them.'[52] Or, we may add, in a variation on the old chestnut, 'truly heard the sound of a falling tree when I'm not there to hear it'. Described thus, the author's engagement with the landscape represents a state or a disposition rather than an insight or a fact – a state in which *every* moment inheres within *the* moment. Robinson goes on to describe how he will on occasion use a mobile telephone to call his partner at their Roundstone home and talk to her 'about nothing, about everything – out of this vast space brimming with changeable light and pristine breezes and murmurous silences'.[53]

Critics will point to the fact that, as in relation to the word 'whisper' above, the language of such passages is highly rhetorical, using devices such as metaphor ('wrestle with emptiness'), repetition ('*about* nothing, *about* everything'), assonance ('pristine breezes'), onomatopoeia ('murmurous silences'), polysyndeton ('with ... and ... and ...') and so on. The reason for this is that it is only through the use of figurative language that one may broach a kind of thinking capable of responding to the 'everything' that inheres within the 'nothing'. I am referring here to a kind of Heideggerian 'thinking' which, as quoted above, would 'heed the bidding call of the stillness', eliciting a kind of response that 'hears because it listens to the command of stillness'. To engage in such thinking is (as described in *The Last Pool of Darkness*) to set 'an ear to the earth' – to suspend the insistence of discursive sense, and to open oneself instead to the resonance of 'nothing' in order to become the subject who resonates, who listens.[54] It is to adopt an essentially musical attitude toward living, one in which *all* time inheres within *every* time. It is to implicate oneself in the immemorial, in 'everything', and to learn to reverberate in response to the still and the silent.[55] It is, finally, to spare, to preserve, to set at peace, to learn 'to dwell' rather than merely to occupy or to inhabit.

Robinson is a listener as well as a hearer – someone who, although implicated in the rational world produced by disciplinary knowledge, has learned to open himself and his texts up to the resonance of silence. Since coming to Ireland he has oriented himself toward 'deep places, places that demand fidelity to their truth … Aran, the Burren, Connemara'.[56] Such an orientation draws him toward a version of deep ecology characterised by an accommodation between the dualities which bear upon the modern subject: local / global, lore / knowledge, intuition / sense, feeling / thinking, etc. The possibility of such an accommodation emerges in relation to a philosophical tradition (articulated variously in a range of phenomenological, existential and poststructuralist discourses) which has been concerned to orient the subject toward the absolute reality of 'the Other', whether that otherness be manifested as a form of absence or stillness or silence. 'Orientation' in this sense connotes a particular attitude in which the listening subject 'resonates' in sympathy with that 'otherness'. It is precisely by listening, moreover, by opening himself up to the possibility of reverberating energy, that the subject becomes a subject.[57]

If 'listening' in this sense is implicated in the question of political engagement with the landscape, it also raises a range of issues relating to the fate of those imaginative landscapes that we call 'books'.[58] To compare textuality with landscape is to observe phenomena undergoing widely divergent experiences in the modern era. In the age of digital proliferation there are vastly more books – and more of them available, and in new formats – than ever before in human history. The fact is that modern textuality is virtually limitless, in which condition it contrasts profoundly with a physical landscape that is under increasing pressure from rationalising and exploitative technologies. More books, less land: that is one of key equations ordering human experience in the early twenty-first century.

There are aspects of certain existing critical systems that recall aspects of Robinson's work in relation to the landscape of the west of Ireland. Formalism (of different varieties) tends to be concerned with what one might describe as the morphology of the text – which is to say, how a text coheres as a functioning system of meaning. Close reading fosters an advanced appreciation of how a text orchestrates the array of binaries of which it is comprised. Both deconstruction and psychoanalysis, on the other hand, typically operate by identifying and scrutinising elements – gaps, silences, paradoxes, marginalia – which threaten to expose the text's essential incoherence: the fact that it is comprised of all manner of conflicting, incompatible elements, and that its coherence is an illusion maintained by the will (always politicised to a greater or lesser extent) of the reading subject. Marxist criticism, meanwhile, has always incorporated an element of the process of 'cognitive mapping' identified by Fredric Jameson (1988) as a necessary element of any critical system with radical socio-political pretensions – the need for the subject to know where and how they stand in relation to the energies that order the world in which they live.

From my perspective, all these systems suggest interesting parallels with Robinson's work. And so far as I can see, such parallels as do exist emerge in the first instance from the influence of phenomenology, in particular the early Heideggerian focus on what he called 'the thingness of the thing'.[59] But the questions remain: how would a reading practice based on Robinson's methods function? Does the ability 'to listen' as well as 'to hear' function as part of any existing critical system? And is there any such system that would encourage adherents to be content to practise their art on a relatively tiny portion of the available corpus of texts?

Another time, another essay. In the meantime, I believe that it is important to begin to imagine a critical system equal to the challenge of Tim Robinson's work – one inspired by the same levels of commitment and focus as may be discovered in his books and maps. An 'ecocriticism' worthy of the name would have to find ways to balance its disciplinary inheritance – criticism as a 'masterful' discourse which looks to 'map' the text in terms of one or other 'interested' technology – with a willingness to attune itself to the inaudible music of that same text. Yes, in the face of an increasingly instrumental rationality it remains vitally important to know when and where, how and why. At the same time, the ecocritic has to learn to reverberate in response to the resonance of the text, and to create opportunities for the reader to do likewise. Such represents both the example and the challenge set us by Tim Robinson's life and work.[60]

### Notes

1. Virginia Woolf, *To the Lighthouse* [1927] (Oxford: Oxford World's Classics, 1992), 192.
2. Martin Heidegger, 'What Are Poets For?' [1950a], in *Poetry, Language, Thought*, trans. by Albert Hofstader (New York: Perennial Library, 1971), 89–142 (139).
3. Tim Robinson, *Connemara: Listening to the Wind* (Dublin: Penguin, 2006), 1.
4. Robinson, *Listening*, 2
5. Robinson, *Listening*, 2.
6. Robinson, *Listening*, 3.
7. Robinson, *Listening*, 3.
8. Robinson, *Listening*, 3.
9. In so far as this is the case, one of Robinson's intertexts – a 'precentor' (one who sings before), so to speak – is the chapter entitled 'Sounds' in Henry David Thoreau's *Walden*. See Henry David Thoreau, *Walden and Civil Disobedience* [1854, 1849] (New York: Norton, 1966), 74–86.
10. See Don Ihde, *Listening and Voice: A Phenomenology of Sound* (Albany: State University of New York Press, 2007), 3–17, and Martin Jay, 'In the Empire of the Gaze: Foucault

and the Denigration of Vision in Twentieth-Century French Thought', in David Couzens Hoy *(ed.)*, *Foucault: A Critical Reader* (Oxford: Blackwell, 1986), 175–204.
11 Martin Heidegger, *Being and Time* [1927], trans. by John Macquarrie and Edward Robinson (Oxford: Blackwell, 2008), 315. Original emphasis.
12 Heidegger, *Being and Time*, 206.
13 Martin Heidegger, 'Language' [1959], in *Poetry, Language, Thought*, trans. by Albert Hofstader (New York: Perennial Library, 1971), 187–210 (209–10). Original hyphenation. The reference to 'difference' here, and indeed the overall thrust of the argument, anticipates Derrida in a number of respects, at which point we recall the latter's apprenticeship as a student of phenomenology. Derrida's work on the relations between verbal language (as articulated by the voice) and writing in *Of Grammatology* (1967) represents his elaboration of a poststructuralist rejoinder to phenomenology, based in part on Heidegger's late work. Despite its suggestiveness, a consideration of that project is beyond my remit here.
14 Martin Heidegger, 'What Calls for Thinking?' [1971], in David Farrell Krell *(ed.)*, *Martin Heidegger: Basic Writings* (London: Routledge & Kegan Paul, 1978), 341–68 (364).
15 Heidegger, 'Language', 209.
16 Tim Robinson, *Setting Foot on the Shores of Connemara and Other Writings* (Dublin: Lilliput Press, 1996), 76.
17 Eugène Minkowski, *Vers une cosmologie* (1936), quoted in Gaston Bachelard, *The Poetics of Space* [1958], trans. Maria Jolas (Boston: Beacon Press, 1994), xvi–xvii.
18 Minkowski, *Vers une cosmologie*, xvii
19 Bachelard, *The Poetics of Space*, 13.
20 Bachelard, *The Poetics of Space*, 14.
21 Jean-Luc Nancy, *Listening* [2002] (New York: Fordham University Press, 2007), 6.
22 Nancy, *Listening*, 7.
23 Nancy, *Listening*, 9. Original emphasis.
24 Nancy, *Listening*, 31.
25 Nancy, *Listening*, 67.
26 Charles Taylor, 'Heidegger, Language, Ecology', in Herbert L. Dreyfus and Harrison Hall (eds), *Heidegger: A Critical Reader* (Oxford: Basil Blackwell, 1992), 247–69 (247).
27 Taylor, 'Heidegger', 267.
28 Taylor, 'Heidegger', 267. For further discussions of deep ecology, see Warwick Fox, 'Deep Ecology: A New Philosophy of Our Time' [1984], in Andrew Light and Holmes Rolston III (eds), *Environmental Ethics* (Oxford: Blackwell, 2003), 252–61 and Arne Næss, 'The Deep Ecology Movement: Some Philosophical Aspects' [1998], in Light and Rolston III (eds), *Environmental Ethics*, 262–74.
29 Taylor, 'Heidegger', 247.
30 Christopher Manes, 'Nature and Silence', in Cheryll Glotfelty and Harold Fromm (eds), *The Ecocriticism Reader: Landmarks in Literary Ecology* (Athens, GA: University of Georgia Press, 1996), 15–29 (16).
31 Martin Heidegger, 'Building Dwelling Thinking' [1954], in *Poetry, Language, Thought*, trans. by Albert Hofstader (New York: Perennial Library, 1971), 145–61 (149). Original emphasis.
32 Martin Heidegger, 'An Ontological Consideration of Place', in *The Question of Being*, trans. by William Kluback and Jean T. Wilde (London: Vision, 1956), 18–26 (24).
33 Taylor, 'Heidegger, Language, Ecology', 266.
34 Christopher Belshaw, *Environmental Philosophy: Reason, Nature and Human Concern* (Chesham, Bucks: Acumen, 2001), 181.
35 Robinson, *Setting Foot*, 164. In a gesture that is now commonplace, Taylor points out that while Heidegger was alive to the 'tremendously positive uses' of words

retrieved from silence, he was deaf to the 'terrifyingly dangerous' uses – in the cult of Nazism, for example – such words could also come to represent. Terry Eagleton refers to the 'sinister' implications of Heidegger's thought in *Literary Theory* [1983] (Oxford: Blackwell, 1988), 64.
36  Robinson, *Listening*, 123.
37  It is this 'something' that also leaves Heidegger's work – and deep ecology itself – open to accusations of a potentially dangerous irrationalism. Rather than 'a global attack on reason', however, Heideggerian listening suggests a 'need to dismantle a particular historical use of reason, a use that has produced a certain kind of human subject that only speaks soliloquies in a world of irrational silences'. See Manes, 'Nature and Silence', 25.
38  This does not mean that other times and other places have no bearing on the landscape/text before us; and I shall of course be drawing on Robinson's wider oeuvre when/where I deem it useful or necessary.
39  Robinson, *Setting Foot*, 213. 'Through-composition' refers to a mode of song composition in which the music 'reflects' or adheres to the lyric in terms of its emotional burden, in which respect it differs markedly from 'strophic' forms such as may be found in ballad and popular song.
40  Tim Robinson, *Stones of Aran: Pilgrimage* (London: Penguin, 1986), 160.
41  Robinson, *Listening*, 362.
42  Robinson, *Listening*, 12.
43  Robinson, *Listening*, 5.
44  Robinson, *Listening*, 7.
45  Robinson, *Listening*, 202.
46  Robinson's ambition to hear the lost voices of the west resonates with the similar one pursued (on the advice of W. B. Yeats) by J. M. Synge to express a form of life that had never found expression. For Robinson's response to Synge and his work see the long chapter entitled 'Place/Person/Book', *Setting Foot*, 108–50.
47  Robinson, *Listening*, 6, 10, 16.
48  Tim Robinson, *Connemara: A Little Gaelic Kingdom* (Dublin: Penguin, 2011), 152.
49  Robinson, *Listening*, 20.
50  In an essay entitled 'Listening to the Landscape' (first delivered as a lecture in 1992), Robinson discusses the connection between the Irish language and the physical environments (sky, sea and land) in and against which that language evolved. In order that they may gain access to what he describes as 'true place, with all its dimensions of subjectivity, of memory and the forgotten' (163), Robinson invites his audience 'to hear the language as if it were spoken by the landscape' (153). See the reprint of the essay in *Setting Foot on the Shores of Connemara*.
51  Robinson, *Listening*, 6.
52  Robinson, *Listening*, 26.
53  Robinson, *Listening*, 26.
54  Tim Robinson, *Connemara: The Last Pool of Darkness* (Dublin: Penguin, 2009), 136.
55  The combination of eclectic knowledge, conservationist energy and identification with the local establishes Robinson's profile as a 'deep mapper'.
56  Robinson, *Last Pool*, 342.
57  Volume 2 of the Connemara trilogy – *The Last Pool of Darkness* – takes its title from a description of the region by Ludwig Wittgenstein. It is interesting that the great philosopher's major work should end (famously) with the invocation of the very phenomenon – silence – with which I am concerned here: 'Whereof we cannot speak, thereof we must remain silent' (quoted in Robinson, *Last Pool*, 31). This is no

coincidence, of course, as 'what mattered [to Wittgenstein]', according to Robinson, 'was exactly that which cannot be said' (32). Wittgenstein's relevance to deep ecology awaits exploration.

58 In the essay entitled 'Four Threads', Robinson writes that for the people who lived there the Connemara countryside 'was like a book in fineness of detail, closeness of print; every corner of it conveyed a message, held a memory'. See *Setting Foot*, 166.

59 Martin Heidegger, 'The Thing' [1951], in *Poetry, Language, Thought*, trans. by Albert Hofstader (New York: Perennial Library, 1971), 163–82 (167).

60 Such a programme suggests neither the 'passiveness' nor the 'slavish abnegation' invoked by Terry Eagleton in his description of Heideggerian phenomenology in *Literary Theory*, 63.

## 12

# 'Another half-humanized boulder lying on unprofitable ground'?: the visual art of Tim Robinson/Timothy Drever

## *Catherine Marshall*

Some years ago, introducing Views From an Island – an exhibition of contemporary Irish Art from the collection of the Irish Museum of Modern Art – to an audience in Beijing, and wondering, as one does, where to begin, I found myself explaining, with some difficulty for the listeners, that Ireland is as small as it is, that it floats on the westernmost edge of a large land mass, and that art from this geographically peripheral place presents, inevitably, a view from an edge. Tim Robinson, too, is interested in marginal places. In his introduction to J. M. Synge's *The Aran Islands*, he observed, 'If Ireland is intriguing as being an island off the west of Europe, then Aran, as an island off the west of Ireland, is still more so; it is Ireland raised to the power of two.'[1] I was aware, however, in Beijing in 2004, that the little spot on the global map that represented Ireland could also be read as the centre of a wider global network, depending on your value systems and your perspective. I had not then seen the fascinating unattributed 1929 map, *The World in the Time of the Surrealists*, in which the United States of America has been wiped out and Britain reduced to a tiny spot, while Ireland – as a country perceived to be supportive of Surrealism – has ballooned to a significant land mass.[2] The Surrealists were, of course, interested in more imaginative geographies and the writings of Jonathan Swift and Synge had encouraged André Breton to believe that Irish influence in this area was bigger than that of some other cultural centres. In discussion of the Surrealist map, Luke Gibbons points out that Breton and his friends, unlike the Futurists and other modernists who sought to eliminate the past, looked to history for the voices that were silenced in it, or vanquished in the name of progress.[3]

Robinson's maps of the Aran Islands, Connemara and the Burren could not be described as Surrealist but they share that movement's understanding of the importance of historical voices, especially the voices of those who have been overlooked or diminished in the thrust toward economic progress. Robinson's maps and writings have been widely discussed as important tools in the preservation of important residues from an environment and a history threatened by modernity and in the rediscovery of much of what has already been eroded by history or the banalities of modern convenience. Much less has been said about them as artworks and it is this narrower view of Robinson's work that I wish to consider in this essay.[4] In doing so, I am mindful that the artist is the sum of all his parts, and that it makes no sense to think of Robinson only as an artist, or only as a writer or map-maker. My excuse for this exercise is that his work has not received as much attention from the art world as it has from literary critics, geographers, historians and other writers. Just how does this body of work, as art, contribute to our world? How does it accommodate itself within the wider context of visual art, in particular, the art of the landscape? How does it compare with other projects to represent this part of Ireland? What impact has his mapping project had on contemporary visual arts?

There is a well-established artistic context for maps and cartography. Bartolomeo Facio, the fifteenth-century Italian writer and humanist, saw a map made by Jan Van Eyck for Philip the Good, now lost and known only through his description from around 1456:

> There is a circular representation of the world painted by him for Philip, prince of the Belgians, and it is thought that in our own time no one has made a more perfect work. It is not only possible to see the location of various places and continents on it, but to measure the distance between them.[5]

Elizabeth Dhanens argued that Van Eyck's map was neglected by historians of cartography, not only because it disappeared but because it was not made by a professional.[6] A hundred and fifty years later, artists in Antwerp introduced a new strand in painting, as a strategy to raise the profile of painting by creating images of the well-stocked cabinets of the cultured collector. In addition to wide selections of paintings from Venice, Germany and the Netherlands, antique sculpture and instruments for science and measurement, maps and globes are an essential feature of these collections, frequently along with a beautiful woman and some exotic animals. The best-known illustration of the elitist connections between art, travel and status is Hans Holbein's painting *The Ambassadors* from 1533.[7] The link was, therefore, well established centuries before Timothy Drever/Robinson used his practice of map-making and art to undermine notions of exclusiveness and authority.

In Ireland, mapping projects from those of the medieval Welshman Giraldus Cambrensis to the nineteenth-century Ordnance Survey maps have always been associated with colonisation and power. Letters to the *Freeman's Journal* and *The Times* show that the suppression of the Ordnance Survey Memoir in 1840 in Ireland was clearly understood as a way of denying a subjected people access to their history and identity. Postmodern artists, such as Jimmie Durham in the United States and Tim Robinson/Drever, have drawn on the concept of the map

*Figure 24*   *To the Sun* (oil on canvas, 1969, Timothy Drever).

as a highly subversive force, calling attention to other cultural views ignored if not totally destroyed under the colonial regimes of the past.

One of the consistent drives in Robinson's work has been the political one of empowering those outside his particular disciplines to feel that they have a right to interact with them. Through his art practice he sought to overcome the segregation of art from the wider public through the elitist art market and a critical discourse directed in a language that excludes those who do not have the kind of education enjoyed by its writers. Although he studied and taught mathematics, Robinson turned to painting under the name Timothy Drever in Vienna in the early 1960s. His work from this period is highly abstract, and mathematical, but unlike the optical experiments of Bridget Riley and others to whom his work bore a superficial resemblance, Drever's work was all about how the viewer can position himself/herself in relation to the image. It led him from painting to installations of rods of different lengths and thickness, scattered on the floor or even suspended from the ceiling of the gallery, each one suggesting the potential of scientific measurement, but seeming to point in all directions, temporally as well as spatially, at the same time (see Figure 24 for an example of Drever's relational and spatial artwork). The questions they raise go beyond the image of the artwork in the gallery, and invite bigger examinations about how art is itself positioned in contemporary culture.

On his return to London, where similar questions to his own were being raised by Stuart Brisley, Hamish Fulton and Richard Long, Drever/Robinson took part in the campaign to prevent the Minister for Art, Lord Eccles, from introducing admission charges to the National Gallery. But neither the performative environmentalism of Long, Fulton nor even Joseph Beuys satisfied his need for social change. He showed his anti-elitist credentials again when, as Timothy Drever, he and Peter Joseph challenged the art gallery as the locus of authority in the visual arts in two exhibitions at the grounds of Kenwood House. On those occasions Drever invited the public to complete his work by selecting and placing his large abstract forms according to their own preference. With Peter Joseph, he authored a text in *Studio International* in which they outlined their dissatisfaction with current practice.[8] It was following this, and feeling disillusioned with the London art world, that he came, with his partner Máiréad, to visit the Aran Islands in 1972 and began his first map, *Oileáin Árann*.

This project came about as a response to a request from Máire Bn. Uí Chonghaile (the postmistress of Cill Mhuirbhigh) to draw up some kind of a map that would be useful to visitors to the island.[9] It was the perfect task for an artist who wanted to work away from the gallery system, and who was interested in environmental issues and community participation. In Ireland, in the late 1970s, Kathy Prendergast, then a sculpture student, was beginning to experiment with maps as a means of undermining the tyranny of sculpture in representing the three-dimensional. Robinson's maps of the Aran Islands, the Burren and Connemara, the first of which appeared to acclaim at that time, offer a strategy for representing yet another dimension – Robinson's 'mysterious and neglected fourth dimension of cartography which extends deep into the self of the cartographer'.[10]

For Drever/Robinson cartography was the perfect vehicle to add a new element to a widely understood visual tradition. He combined that tradition with a conceptual element that required the fullest co-operation with those who lived in the area to be mapped or who had written about it, with its folk history and its dream world, as well as with documented history, science and culture in the widest sense. Most of all, it paved the way for an exploration of the language, especially the Irish language so eroded by Ireland's colonial history, and sounds of the area that are as important to the visual understanding of a place as anything offered in more conventional strategies within fine art.

Robinson called his publishing company Folding Landscapes; the name is important here. While maps have existed since classical antiquity, the idea of the landscape as a fit subject for artists only really began to emerge in European art in the sixteenth century as artists produced paintings and drawings of ideal or realistic scenes from nature in significant numbers. Ernst Gombrich took the new genre very seriously, claiming that 'of all the "genres" which the 16th century specialists began to cultivate in the North, Landscape painting is the most revolutionary'.[11] Every age has its particular concerns. Gombrich identified landscape as the only genre in late Renaissance Northern Europe that was not subjugated to religious or moral content, no matter how subtly disguised, as he felt history, portraiture and still life continued to be. In a modernist context, such considerations have little relevance. Instead, it is the elitism surrounding easel paintings and the power of the gallery, irrespective of subject, that Robinson's maps initially subvert. The concept of the folded map as a landscape that can be purchased at petrol stations and tourist shops is as subversive of the modernist order as the non-religious subject painting was in the sixteenth century. And following Paul Klee's tenet – that art reveals the visible rather than imitates it – Robinson's maps go on to unveil secrets hidden in the landscape and voices and sounds long forgotten.

The landscape of the West of Ireland first emerged as a popular and increasingly saleable strand within Irish art with the publication of Edmund Burke's *A Philosophical Enquiry into the Origin of Our Ideas of the Sublime and Beautiful* (1757) and Jonathan Fisher's views of the Lakes of Killarney, as the West of Ireland proved itself to be a less climatically difficult terrain from which to observe the wilderness than the Highlands of Scotland or the Alps in the early nineteenth century. Wild nature in the eighteenth and nineteenth centuries was a taste confined to those who could contemplate it from the safety and comfort of the drawing room rather than the struggling farm labourer. Gradually, however, a new attitude along with a new audience began to appear, especially as artists like Aloysius O'Kelly honed in on the dignity and the hardships endured by Irish peasants in such paintings as *Mass in a Connemara Cabin*.[12]

By the early twentieth century, and with growing momentum following the Irish War of Independence and the establishment of the Free State, the West of Ireland and its occupants became the practical equivalent in visual art terms of the figure of Éire with her harp and wolfhound as a symbol of Irishness.[13] Yvonne Scott has pointed out that in the counties of the western seaboard the fusion of Irish language, traditional rustic lifestyle and predominantly Catholic religion appealed to nationalists and also to the families of those who had emigrated in large numbers from the west and to the continuing flow of economic migrants, who wanted to remember the country as they or their forebears had known it. Seán Keating, J. Crampton Walker, Laeititia Marion Hamilton, Charles Lamb and Paul and Grace Henry all produced canvas after canvas in which, to reference Scott, the thatched cottage had become one of the country's 'most defining motifs'.[14] The darling of the thatched cottage school was undoubtedly Paul Henry, whose painting *Connemara* was chosen as the frontispiece of the *Saorstat Eireann Handbook*, published in 1932 to mark ten years of independent Irish government. In over two thousand catalogued paintings and drawings, apart from a brief dalliance with commissions for portraits and book illustrations, Henry devoted himself to paintings of the landscape of Achill Island and Connemara, with occasional forays as far as Kerry and Wicklow.[15]

Despite their beauty, the subject matter in these paintings was almost formulaic, and had to include some, or all, of the following: thatched houses or villages of thatched houses, high mountains, turf stacks and a body of lake or seawater, with very occasionally in his earlier work, men or women fishing or digging, usually seen from the rear, or in the mist. Henry's paintings were popular beyond all expectations. In a country where there was little opportunity to show or sell art, and only a small purchasing clientele, he was outstandingly successful. His paintings were also economically useful as the tools to attract tourists to an environment that was clearly 'other' to the rest of Europe and America. The cosy-looking cottages, sheltered by their high mountains and reassuring turf stacks at the gable end, offered the new city dwellers the promise of a return to a tried and tested lifestyle that was native and not contaminated by modernity. At the same time, Henry's paintings gave them a sense of achievement at now being able to live in more modern accommodation elsewhere. The element of nostalgia in this was as old as the genre of landscape painting itself, having been a powerful impetus in the development of Dutch landscape painting in the sixteenth and seventeenth centuries. Just as the Dutch countryside shrank or was brutalised to make way for new towns, canals and industrial developments, and the new burghers sought to fill their houses with images of what they had left behind, so the Irish, in their flight eastward to Dublin, or Britain, sought to keep the memory of the West, although not its lifestyles, alive. Irish art became so synonymous with the landscape of the West that as recently as 1979 Robert Ballagh, who introduced Pop art to Ireland, was forced to insist:

> I never had access to the culture that many people think is the Irish culture, the rural Gaelic tradition. I can't paint Connemara fishermen ... My experience of Ireland is an urban, Dublin one and I paint that. It would be dishonest of me to paint anything else. But being Dublin is also being Irish.[16]

For Robinson, the pull of the West initially was simply that it was not London. Nonetheless, his response to it was to open up wide-ranging alternatives to older representations of the area.

Paul Henry lived for nearly ten years on Achill Island and writes with obvious love for the place, and some of the people he met there, in his autobiography, *An Irish Portrait* (1951). Yet his paintings and his writings employ a level of selectiveness that is hard to equate with the apparent openness and simplicity of both. His autobiography makes no reference whatsoever to his wife, Grace, who not only accompanied him to Achill but who also painted the life and landscape and shared every exhibition he had during their Achill years, while the people he does write about are frequently treated as picturesque characters, rather than as equals. However, it is in the painting that the selectiveness is particularly interesting. As a body of work representing the West of Ireland between the years 1910 and 1958 (when the artist died), it seems strange that there are almost no images of the derelict houses that have dotted the western countryside ever since the mass evictions and widespread depopulation of the area following the famine in the nineteenth century. Henry was a personal witness to that dereliction, noting that 'it is extraordinary how quickly an untenanted thatched cottage falls to pieces ... The thatch falls quickly down and the rafters follow, and in a few months there is nothing left except the empty shell.'[17] The emphasis in his paintings, however, is always on snugness, on a primitive and simple comfort that eschews signs of desperation just as it avoids images of churches and shebeens (places where alcohol is sold without a licence). An advertisement for the London Midland and Scottish Railway Company's steamers offers one explanation for this. Below a reproduction of Henry's painting *Connemara*, it states as boldly as advertisements are wont to do that the painting 'illustrates in a striking manner a typical phase of Irish scenery. Ireland – the new post-war Ireland – is now ready to welcome the English tourist, and wishes you to explore the beauties of her coast and countryside.'[18]

The war referred to here could refer to the so called 'Great' War of 1914–18, but is far more likely to refer to the war of independence in Ireland between 1919 and 1921, and about which English tourists might need the reassurance that the painter's formula was designed to offer. Henry's paintings were similarly used by the Irish Tourist Board, Coras Iompar Éireann (Irish national transport system) and Aer Lingus, as well as by various travel companies around the world, offering the same promise of peace and tranquillity, with little trace of the unacceptable poverty that strewed the countryside with abandoned cottages. The local occupants, if they appear at all, are shown digging their potato ridges or pushing their fishing boats out, never catching our gaze, never imposing themselves into the privileged viewer's ownership of the dream contained in the image. Anything suggestive of social or religious division or the often contested histories of his picturesque hamlets has been carefully avoided. The local communities carried that history in their imagination (as Robinson was to prove so effectively), and the tourists preferred to be shielded from it. As Scott has also pointed out, there is little to indicate that the governments of the Free State cared much for economic improvement in those very areas that they encouraged Henry to paint, and nothing in Henry's writing or painting demanding government action. Henry's influence was so pervasive that in 1996, almost a century after he first visited Achill, Caroline McCarthy (as a recent graduate from the National College of Art and Design) questioned her ability to locate herself within a tradition so largely defined by his work. *Greetings* is a short video work in which a jumping figure (the artist herself) tries vainly to position herself in a typical Irish 'picture', only to repeatedly fall out of view.[19] The title of the work suggests that this landscape is only for temporary visits, not for lifetime engagement.

All artists have to be selective and Robinson is no exception. But unlike Henry, Robinson states his preferences from the outset: 'I am aware of the selectivity of my written response to living in Connemara. I concentrate on just three factors whose influences permeate the structures of everyday life here: the sound of the past, the language we breathe, and our frontage onto the natural world.'[20] In committing himself to living in the West of Ireland for over four decades now, Robinson works from the position of a permanent inhabitant, not a visitor, to make the special features of the place, whether they are conceptual and emotional or purely physical, visible. Whereas Henry's approach was to generalise, Robinson's personal challenge was to find a way of linking the particular to the global and the mythic. It is on its physical and human histories that the individuality of a place largely depends, and that is what Robinson's maps and writing continually reveal. His written imagery is rooted in his earlier, abstract artworks, the clusters and arrangements of wooden rods that made up his art installations *Winged Victories*, *To the Centre*, *Autobiography*, *Inchworm* and *Ghost Staircase* – in Vienna (1963), London (1972 and 1973), the Irish Museum of Modern Art (1996) and Dublin City Gallery The Hugh Lane (2011) – with their tantalising promise of measurement and scientific stability amid apparent randomness and potential change. By mapping the Aran Islands he could take the ideas explored in those early works and apply them to a very specific place with its physical make-up, its history, its inhabitants and its visitors who became both end users and empowered participants in the process itself. Because, like Van Eyck, Robinson was not a professional cartographer, he was not constrained by existing mapping conventions or training. His response to the task was immediately the artistic one, of how best to represent all that he found interesting and useful for himself and the public – his 'audience' in conventional art terms, determined not to allow anything to come between himself and the land to be mapped. He writes, 'I came to the practice of cartography largely ignorant of its specific techniques, theories and received ideas, not to say deeply suspicious of its technological and organizational structures that distance the drawer of the map ever farther from the place to be drawn, alienating the hand from the foot.'[21] Like a performance artist, his body becomes the medium while the process of the journeys he makes as he maps, and the people that he meets, all colour the final object.[22] The fusion of artist, medium and subject could not be more complete so that as he says, 'while walking this ground I am the pen on the paper; while drawing this map, my pen is myself walking the land'.[23]

Drawing on his extensive knowledge of mathematics and his artistic fluency with a pencil, Robinson produced maps using signs and symbols suited to the unique locations of the islands, with the shapes of cliff faces clearly and beautifully articulated because of their importance to the fisherfolk, and other features of interest to the islanders themselves that had not been shown on older Ordnance Survey maps. For a conceptual artist to have to add verbal language in the form of place names was a natural next step, but here the pressure to understand the stories that the place names revealed, especially in their older forms as learned from the islanders, became both a challenge (he had to learn the Irish language) and a real commitment. 'The landscape here speaks Irish,' he declared, and goes on to write, 'while it is true everywhere that linguistic change forces on us a constant effort of retrieval of lost meaning, it is clear that the death of spoken Irish in its last native environments would be a tragedy'.[24] While Henry had made frequent reference in his autobiography to his inability to understand or speak Irish, Robinson's approach to landscape is driven by the underlying conviction that you cannot claim to represent a place without knowing its people or its written and oral histories, and a thorough understanding of their language and folklore is central to that. For this reason, he points out, 'In recording, largely from the lips of Aran itself, as best I can these place-words, and expressing them in today's media of map and book, I want, not to insure against that tragedy [the death of the language] but to inspire the rebellion against its seeming inevitability.'[25]

The case for the language, however, should not be seen as a support for nationalism. Having opened himself fully to the lessons of geological history and the 'underearthly' as well as the 'unearthly' as revealed in the folklore, Robinson is very explicit that his representations of Aran, Connemara and the Burren are an argument against petty concepts of ownership. In the chapter 'Who Owns the Land?', from *Connemara: The Last Pool of Darkness*, he argues that 'we need ... a shared sense of the Earth's surface as a palimpsest, the compiled and over-written testimonies of all previous generations, which it is our right and duty to read'.[26] And concluding that same chapter – drawing from his multidisciplinary studies of language, folklore, history, prehistory and geological remains of volcanic and glacial periods, all of which are mirrored in Doon Hill/Cnoc an Dúin, Errismoe – he insists, 'If we know in our hearts that mountains and oceans have their day then our proprietorial attitude to patches of land appears in perspective, as a littleness.'[27]

When the mapping and the place names inevitably led to writing, Robinson's exploration of the 'frontage onto the natural world' leads to wonderful imaginative passages, informed by an artist's eye at all times: 'Stroll down the boreens, and you go arm in arm with the Atlantic, for their pattern is that of the fissures caused by the forces that separated Europe from America sixty or seventy million years ago.'[28] Robinson's embrace of the history of place – combining the mega-view derived from his deep respect for the geology and climatic conditions of the areas his maps represent, with the smaller, but also land-changing impact of cultures and more recent historic struggles – makes the landscape visible in a way that more traditional landscape painters could not consider, even if Robinson ultimately communicates this verbally. This becomes very clear in essays such as 'The Bay of Doubt', where his sensitivity to the mood of the bay at Cill Rónáin leads to discussions about very limited but recorded sightings of rare botanical specimens, which to his educated eye can only be explained by seawater penetrating a hundred yards beyond the high tide mark through fissures in the limestone and speculation about the correct name of Carraig an Bhanbháin, Carraig an Mharbháin and Carraig na Mara Báine, evoking the story of a tragedy that gave rise to another place name, Aill na mBád, in 1899. The imagery conjured up by this discussion (much of it carried by the place names) is powerful, but difficult to capture in a single artwork in language other than something of such a symbolic order that it would lose much of its accessibility. So, Robinson resorts to writing. This, however, is all about the visual. 'Aran', he says, 'is one of civilisation's loftiest windows onto its own origins in the past and the natural world.'[29] It invites us to look through that window, not at a pictorial formula, but at a singular, living thing with a past, present and future in which we are participants with responsibilities, not privileged viewers. His window does not offer the idealised view of the Italian Renaissance but rather the multiple perspectives and nominalist embraces of the individual in the cosmos of Van Eyck and his contemporaries. Robinson's alternative approach calls on the skills of the artist, the writer, the scientist and the mathematician to do so.

Lest anyone should suspect that Drever the artist has been lost in Robinson the writer, it is useful to consider the uses to which the maps have been put. *The Distressed Map of the Aran Islands*, so-called after it, as a much enlarged version of the Aran map, was shown in Vinyl – an exhibition curated by Simon Cutts for Cork's European City of Culture Year in 2005. The work was defiantly anti-elitist and subversive of the preciousness of the art market. In this exhibition, in a manner reminiscent of Timothy Drever's work in Kenwood House in the 1960s, it was sited in an old school playground, with a notice saying:

> The original Aran map is published in a paper edition, and if a copy wears out one can get another. The present enlarged version on vinyl is singular, like the islands themselves, and what becomes of it remains to be seen. You are invited to walk on it, write on it, dance on it, treat it as you see fit.

In its new 'distressed' state, bearing new marks in the form of graffiti and footprints, it was included in Hans Ulrich Obrist's Map Marathon at London's Serpentine Gallery in 2010, where it rubbed noses with work by artists from all over the world like Louise Bourgeois, Alighiero Boetti, Jimmie Durham and Ed Ruscha, among others. Aran, albeit distressed, was no longer peripheral.

Robinson's work is a passionate plea for the individual that is everywhere challenged by global economics and the blandness of mass communication, a plea for each one of us. He writes, 'Individually, none of the names I have mentioned is of much intrinsic interest. But if we think of all the placenames that humanity has applied to the surface of this planet as constituting a single vast fingerprint, can we neglect even its most minute particularities in trying to identify ourselves?'[30] Extending this argument outwards, he says, 'If islands loose their singularity, the world becomes smaller.'[31] This is the message that now drives a younger generation of artists, like Alan Counihan and Gypsy Rae, to devote themselves to their Townlands project in Co. Kilkenny and Deirdre O'Mahoney to reopen a former post office in Kilnaboy (North Co. Clare) as an arts centre, where new models of community building, rooted to the local environment, their place names and their histories, are now emerging. Their work, like Robinson's, poses growing difficulties for those who want art to serve elites and support exclusion.

Internationally the empowerment potential of the map can be seen at its most radical in the District Six Museum in Cape Town, South Africa. Instead of showing artefacts, as conventional museums do, the floor of the main exhibition space is filled with a map of the area from which 60,000 former inhabitants of the original mixed community of District Six were forcibly ejected and their homes demolished to make way for occupation by white people. The original occupants are invited to add their written memories of their place. As Tim Robinson has shown, the map without the personal stories is meaningless. His description of his practice as a map-maker is entirely consistent with his paintings and installations from four decades ago: 'My task is to establish a network of lines involving this dimension [the mysterious fourth dimension of cartography], along which the landscape can enter my mind, the unfragmented and undistorted, to be projected into a map that will be faithful to more than the measurable.'[32]

## Notes

1. Tim Robinson, 'Place/Person/Book: Synge's *The Aran Islands*, in J. M. Synge, *The Aran Islands* [1907] (London: Penguin, 1992), xvii.
2. *Variétés: Le Surréalisme en 1929* (Brussels: Editions Variétés, 1929), 26–7.
3. Luke Gibbons, 'Peripheral Visions: Revisiting Irish Modernism', in Enrique Juncosa and Christine Kennedy (eds), *The Moderns* (Dublin: Irish Museum of Modern Art, 2011), 88–101 (91).
4. For another perspective on the Robinson/Drever relationship, see Nessa Cronin's essay in this collection, '"The Fineness of Things": The Deep Mapping Projects of Tim Robinson's Art and Writings, 1969–72' (Chapter 3).
5. Elizabeth Dhanens, *Hubert and Jan Van Eyck* (Antwerp: Art Books Int. Ltd, 1980), 170.
6. Dhanens, *Hubert and Jan Van Eyck*, 172.
7. See Hans Holbein, *The Ambassadors* [1533] (London: National Gallery).
8. Timothy Drever and Peter Joseph, 'Outside the Gallery System: Two Projects for Kenwood', *Studio International* (June 1969).
9. Tim Robinson, *Setting Foot on the Shores of Connemara and Other Writings* (Dublin: Lilliput, 1996), 76.
10. Robinson, *Setting Foot*, 19.
11. Ernst Hans Gombrich, *Norm and Form: Studies in the Art of the Renaissance* (London: Phaidon Press, 1966).
12. Aloysius O'Kelly, *Mass in a Connemara Cabin* [1883] (Dublin: National Gallery of Ireland).
13. For a discussion of this, see Yvonne Scott, *The West as Metaphor* (Dublin: Royal Hibernian Academy, 2005).
14. Scott, *The West*, 28.
15. See S. B. Kennedy, *Paul Henry: Paintings, Drawings, and Illustrations* (New Haven, CT: Yale University Press, 2007).

16  Robert Ballagh, 'A Sense of Ireland', unpublished Arts Festival Catalogue, ed. Simon Oliver (London, 1980), 50.
17  Paul Henry, *An Irish Portrait* (London: B.T. Batsford, 1951), 98.
18  See J. Crampton Walker, *Irish Life and Landscape* (Dublin, Talbot Press, 1926), n.p.
19  Caroline McCarthy, *Greetings* (Dublin: Irish Museum of Modern Art, 1996).
20  Tim Robinson, *Connemara: Listening to the Wind* (Dublin: Penguin, 2006), 3.
21  Robinson, *Setting Foot*, 77.
22  For more on map-making as a form of performance art, see Derek Gladwin's essay in this collection, 'Documentary Map-making and Film-making in Pat Collins's *Tim Robinson: Connemara*' (Chapter 4).
23  Robinson, *Setting Foot*, 77.
24  Tim Robinson, *Oileáin Árann: A Companion to the Map of the Aran Islands* (Roundstone: Folding Landscapes, 1996), 19.
25  Robinson, *Oileáin Árann*, 19.
26  Tim Robinson, *Connemara: The Last Pool of Darkness* (Dublin: Penguin Ireland, 2009), 290.
27  Robinson, *Last Pool*, 313.
28  Robinson, *Oileáin Árann*, 84.
29  Robinson, *Oileáin Árann*, 84.
30  Tim Robinson, *Stones of Aran: Pilgrimage* [1996] (London: Penguin, 2008), 246.
31  Robinson, *Oilean Árann*, 84.
32  Robinson, *Setting Foot*, 19.

# 13

# 'An ear to the earth': matrixial gazing in Tim Robinson's walk-art-text practice

*Moynagh Sullivan*

Tim Robinson's writing cannot be considered as separate from his map-making, nor indeed from the powerful physical practice of his walk. Each of these acts is distinct, yet intimately interlinked to form his own unique aesthetic, and in this essay I consider each of these elements as co-emerging, co-poetic and co-extensive aspects of his oeuvre. Such a practice calls on a deep feminine structure, but not in ways we have come to expect from other well-rehearsed instances of feminising the land, especially the western seaboard, in Irish culture. For instance, Jim Sheridan's film *Into the West* (1992) opens and closes with a visually striking scene of a white horse running across a shoreline near Roundstone with the partial image of a foetus discernible in the sonography of the moon's reflection in the wake of the sea, and it constructs the meeting of wave and earth as a maternal space, which forms the deep background against which the father–son relationship that drives the plot of the film can be dramatised. In the film's plot it appears as if the maternal is both foreclosed (the mother is already dead) and sublimated into the landscape which comes to stand in for the mother. The visual moment of the maternal is easily missed because of the powerful theatre of the film's story as an allegory of Ireland's place in modernity. And yet, the critical staining of the visual text, the opening up of this sonar system on the coast, seems to evoke what artist, philosopher and psychoanalyst Bracha L. Ettinger calls the feminine or matrixial sphere. This essay considers how Robinson activates the matrixial in his walk-art-text practice.

Ettinger conceives of a supplementary feminine matrixial substratum that is the other of what the psychoanalyst Jacques Lacan refers to as the phallic stratum. Taking the late prenatal stage as its psychic origination, it creates another model of relationship. Traces of this experience persist throughout our lives in shared partial encounters, compassionate wit(h)nessing and through encounters with art. In the phallic stratum we learn to experience ourselves as subjects, as 'one', as a unified self, yet there is much to being human that cannot be accounted for by this understanding of the monad. Ettinger's theorisation of the matrixial borderspace posits another substratum, which both precedes and sits alongside the phallic. In other words, while we experience ourselves as 'one' we also experience ourselves as partially related to many, as part of a weave of connections. Our experience of co-becoming with(in) our mothers contributes to this understanding of existence-in-relation (not existence-in-opposition) and distance-in-proximity (sharing a borderspace, yet distant). This trans-subjective understanding of the process of co-becoming involves what Ettinger calls metramorphosis – a process in which 'borderlines and thresholds' are 'transgressed or dissolved, thus allowing the creation of new ones'.[1] Metramorphosis becomes available to us through artwork:

> Art bears the traces of the phallic and matrixial objet(s) of its creation. The work of art we create, and the work of art in which we take part as viewers is not only the gaze approaching us but is what metramorphoses us into part-objects and partial subjects in a matrix larger than our separate one-selves.[2]

The matrixial gaze is a core part of metramorphosis, this experience of border and threshold transgression or refiguration, and it allows the matrixial to be represented, to come into the symbolic, to become legible. The matrixial gaze almost sees what isn't there to see, or perhaps better, what the viewers have not themselves witnessed:

> The matrixial gaze conducts imprints from 'events without witnesses' ... the artist in the matrixial dimension is a wit(h)ness without event in com-passionate wit(h)nessing. The viewer, and this partially includes the artist in its unconscious viewer position, is the wit(h)ness without event *par excellence*. The viewer will embrace while transforming traces of the event and will continue to weave metramorphic borderlinks to others, present and archaic, cognized and uncognized, future and past ... is carried by an event s/he did not necessarily experience, and through the matrixial web an unexpected transformation and reaction to that event arises.[3]

The wit(h)nessing matrixial gaze calls up imprints from traumatic events and encodes an understanding of a matrix larger than our one-selves.

Robinson's understanding of being partial-subjects, part-objects in a matrix larger than our 'separate one-selves' is clear in the text he wrote for the artist Siobhan McDonald's solo exhibition Eye of the Storm (2012), and noted that 'the artist observes, records, relates. Since the Cosmos and all that's in it were born of a singularity, all things are related. The task of the artist is to trace the lines of this universal cousinage.'[4] Robinson's writing, artwork and walking practice can be read as a metramorphosis that brings into the symbolic the lines of consciousness that striate the matrixial stratum as elaborated by Ettinger. For Robinson, the mathematician Benoît Mandelbrot's 'collation of two ideas, self-similarity and fractional dimensionality, opened up new vistas of thought'.[5] According to Robinson, this notion

> gives a precise mathematical definition of a fractal, but also allows for a looser use of the term to cover the sorts of things considered above, such as coastlines, that exhibit a degree of self-similarity over a range of scales and are therefore too complicated to be described in terms of classical geometry which would indeed regard them as broken, confused, tangled, unworthy of measure.[6]

These ideas underwrite much of Robinson's work, which considers the intimate spaces as 'worthy of measure', or at least of representation. His work, however, also reveals, like the matrixial gaze, a substratum of fractal dimensionality that can be made legible in another system of understanding of what can and cannot be seen or gazed upon and he is the wit(h)ness without event *par excellence* to traumas in the landscapes and the lives of those he meets. Robinson writes that Mandelbrot's work 'surprises us yet again with the unfathomable depth and richness of the natural world; specifically it shows us that there is more space, there are more places, within a forest, among the galaxies or on a Connemara seashore, than the geometry of common sense allows'.[7] Ettinger notes how the limits from the symbolic order – from what Robinson calls the 'geometry of common sense' – construct 'evocations of and irruptions from the feminine/prenatal encounters and emergence of matrixial cross-scription of imprints' as 'psychotic-like when they have no symbolic access whatsoever in a culture that takes them for non-sense'.[8] Robinson's own work reforms the geometry of common sense just as the matrixial opens the matrixial to legibility within the symbolic order. Such cross-scriptions, according to Ettinger, 'are a ground for thinking the enigma of the imprints of the world on the artist and of the inscriptions of the artist on the world's hieroglyphs'.[9] In other words, Robinson's 'inscriptions on the west's hieroglyphs' involve seeing the shore not as a line but as markings indicating awareness of spherical space.

When walking shorelines, Robinson treads in waves, in the space that holds ocean and land apart yet brings them together, and this represents an important aspect of the matrixial. Further, the encounter between sea and land echoes the encounter between the mother whose land surrounds the child's Atlantean home, and the push and pull of the amniotic aerials of the tide that call the mother into being as the child washes in and through her. *Stones of Aran: Pilgrimage* is described as 'portray[ing] the inner and outer life of a landscape and its inhabitants', and the work abounds with deep listening, with imaginative and visionary gazing and the synesthetic overlap of the two: 'this is the scheme emerging under my hands, at the moment of writing'.[10] In walk-art-text practice, Robinson accepts his results from a 'geometry of excess' that do not 'converge on a true answer, the real length of the shoreline in question; they just get bigger and bigger, to infinity … It is now known that many natural phenomena have this sort of geometry of excess.'[11]

In understanding the shorelines of the western coast of Ireland as having a 'geometry of excess', Robinson's walk-art-text practice represents a movement 'from a phallic structure to a matrixial sphere'.[12] Robinson's writing and map-work involves in-between internal and external states, and translates the reverberations of the land and sea on the tympanic membrane of his footfall, or under his hands in his transformative traces on and tracing of the landscape. Robinson 'listens' to the landscape, to the wind and to the ocean; he is heard by it, sees and is seen by it, and makes a visual and textual record of the event of bringing it to his life by walking it, an event that creates an encounter horizon:

> The shoreline, as I could see even from a small-scale map, was exceedingly complicated, with many headlands and islands, deep bays and long ramifying inlets. I soon discovered that the islands had other smaller islands off them, the inlets had lesser creeks opening into them, the headlands bifurcated, trifurcated, polyfurcated; the only regularity was irregularity. This coastline presented itself to me as a challenge; it boasted of being unmappable. And that suggested to me that I should walk its entire length, which I did, over some years, piece by piece, in a curious mixture of ritual and research.[13]

No map can ever represent the landscape as experienced, as seen from high or low, or in weathers that descend upon or lift your view. Connemara's shoreline, as must all other space, remains always unmappable, as 'even the best map would not show every little point and crevice of a coastline'.[14] But the maps that Robinson produced, as well as his writing, proceeded from what Ettinger has referred to as 'scopic erotic antennae in-tuning and borderlinking with the trauma of the Other', which occurs through his attentive looking with the geology and listening with the history of human imprints, including his own, on a place.[15] Thus, the shoreline is brought into a different historical, aesthetic and ethical life, similar to how Ettinger's theory of a matrixial feminine evokes psychic dimensions at the threshold of consciousness.

This idea of the feminine is not rehabilitation of Mother Ireland, nor of a Mother Earth, but as Ettinger explains below, is constitutive of the link and border between subjects themselves:

> This feminine different-difference is not a configuration of dependency derived from disguising oneself in a phallic masque (Joan Rivière's femininity as masquerade), or parody and irony (in Judith Butler). Nor is it a revolt or a struggle with the phallic texture (the feminine as the moment of rupture and negativity in Julia Kristeva). We can advance in this way of thinking only if we free ourselves from the compulsion not only to disqualify as mystical or psychotic whatever lies beyond the phallic border, but also to grasp that the borderline itself can become transgressive and should not be perceived only as a castration, a split and a bounding limit – and if we distinguish between subjectivity and the individual.[16]

The *Connemara* trilogy and *Pilgrimage* both demonstrate an affective record of a deepening understanding of the relationship between Robinson and the coastlines he journeys, or what Ettinger calls 'a co-poietic metramorphosis'.[17] Robinson writes that the

> shoreline cannot be conceptualized as a one dimensional line, however contorted. Any such representation of it omits detail. On the other hand, a shoreline is not quite an area, with two dimensions. It is something in between the one-dimensional and the two-dimensional. This seems a counterintuitive idea, but Mandelbrot makes sense of it, and proves that the average coastline has a dimensionality of about 1.2; I've done my own crude calculations and estimate that the complex shoreline of south Connemara has about one and a third dimensions, and some stretches of it must be of even higher dimensionality.[18]

The matrixial similarly transforms borders, or 'one dimensional lines' – the shoreline between part-subjects and part-objects – into borderspaces that do not translate into two full subjects, or one subject one object, but create horizons of encounter. Ettinger theorises in her elaboration of matrixial theory how the interweaving of sound in visual art and poetics, resonance and sight in the matrixial borderspace, expresses a dimension of experience and being that is subjacent to the instrumentalist and logical operations of language, and which is derived from experience of the first landscape – the sensorium of the mother's body. The traces of this time 'are remembered without being recollected and are revealed in a phantasm saturated with imprints of the trauma of a partial and shared subjectivity'.[19] Traces of these imprints circulate 'by affects and by waves', which Ettinger calls 'a parallel psychic activity, not of drives as internal and autonomous, but of the erotic antennae of the psyche. This process 'engenders a trans-subjective psychic sphere she calls "matrixial"'.[20] Robinson's keen listening to waves on many frequencies, resulting in mapping and texts, works to produce an erotically charged matrixial dimensionality of the spaces he walks, different from the Ordnance Survey maps, from the narrative histories, from the curves of post-Oedipal logic. He writes,

> These three localities, of which I have made maps, are the Aran Islands, which stand in the mouth of Galway Bay, the Burren in County Clare on the south side of the Bay, and Connemara on the north side of the bay. I'll enhance the echoic properties of these terrains by giving each an extra dimension as well as the normal three of length, breadth and height.[21]

The feminine in matrixial theory is conceived of as a relation, a generative borderlinking based on unprocessed memories of the mother-to-be/subject-to-be encounter felt by those who will later become gendered. In the matrix, the feminine is not in opposition to the masculine. Post-Oedipal representations and practices of masculinity and femininity are unimaginable in the matrix because it is predicated upon borderlinkings, trace-connections, not the divisions that for psychoanalysts

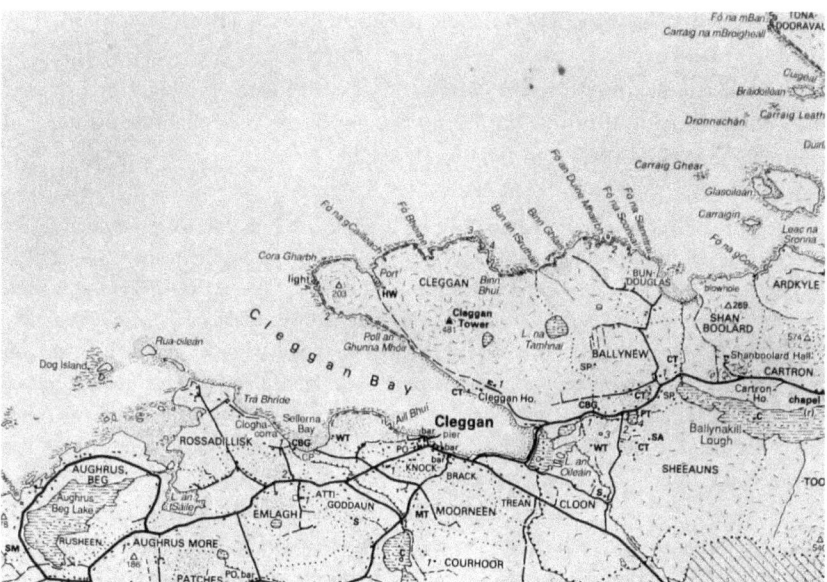

*Figure 25* Map of Cleggan Head, Connemara, by Robinson (cropped from original sized map).

Sigmund Freud and Jacques Lacan create sexually individuated subjects. As a theory of sexual difference, or subjectivity, it does not seek to displace what is understood to be the touchstone of psychoanalytical theory, the Oedipal scene, but rather seeks to expand the psychic dimensions of our phenomenological experience. Robinson's work recalls both the separated and sexed individuals created from 'castrating cuts' that see sexual difference, and yet traces of the matrixial can be powerfully discerned in its patterns and evocations. In the following description of an encounter with Constantin Brancusi's heads (close in style and execution to his 1915 piece 'The Newborn'), in which they appear to draw a thousand-year-old breath, Robinson is prompted to imagine Cleggan Head listening to the breathing of a mother earth (see Figure 25). He thus sees the earth as gendered – as maternal:

> once in an art gallery I found myself alone with five or six of Brancusi's sculpted marble heads, each lying on its side in its own glass shrine. There was no ugly truncation of their necks; they were rounded off as naturally as eggs, complete and content in themselves –and profoundly they slept! One could imagine their taking perhaps one breath in a year, and turning over, gently, so as not to wake themselves, once in a thousand years. Similarly the huge bulk of Cleggan hill is weighed down under sleep and the force of gravity, not a dead but a living weight, its ear pressed to the earth, lulled by her breathing, her heartbeat, her snoring.[22]

But far from recalling an idealised state of 'oceanic feeling' Robinson's work is deeply ambivalent and the horizon created is an often violent and cruel intrusion into any possible easing into a nostalgic relationship with a romanticised or feminised Irish landscape.

In the chapter titled 'Into the Mist', from the final part of the trilogy *Last Pool of Darkness*, Robinson begins with lines that recall the matrixiality of water and white horse in Sheridan's film *Into the West*:

> Aughrus – Eachros, horse peninsula, in proper Irish – is the broad, low-lying lobe of land at the head of the peninsula south-west of Cleggan bay. Its main topographical feature is a large lake set centrally within it, so that its flowery hay-fields and marshy pastures form a rim only 200 or 300 yards wide between lake and sea; quiet country lanes serving a scattering of small farmhouses and holiday homes make a loop around this rim, which I think a horse could trot in less than half an hour. I connect Aughrus with childhood and have wanted to write about it with that theme in mind ever since, a decade or so ago, an elderly lady from there called on me with the typescript of her autobiography, looking for my advice on publication.[23]

The suggestive note of childhood in these lines cues us to the pre-Oedipal, pre-natal processes of being a part-subject and part-object that saturate this chapter. The 'typescript' of 'autobiography' is linked to this, and yet we find as we read on that some years later when Robinson tried to get in touch again with the elderly lady whom he had met, one Mary Walsh, *née* de Lappe, Robinson discovered that she had passed away and that 'the fate of the typescript seemed to be unknown'.[24] Thus a process of co-creating begins, in which Robinson works from 'a few pages' of her reminiscences of 'this amphibious life in the raw edges of Connemara that he had photocopied', which included 'the piglet who used to go swimming with her sister, the geese with quills from their own wings pushed through their beaks to stop them rooting up the potatoes, the white-faced baby calves "with an almost angelic quality", her mother taking off her coat to cover a badly mutilated corpse retrieved from the sea'.[25] Robinson evokes an account of Mary's younger life in a Connemara cottage in the early 1920s, and in doing so, looking through the lens of Ettinger's matrixial, we see that he opens up the possibility of transforming the means of connection to the traumas of others:

> In an era of events-without-witness, whether you take/find/your materials with-in or with-out should not make a difference at not-all; the problems fade out and move elsewhere. We participate in the traumatic events of the other. What makes the difference is the awareness, and through it, the possibility to transform the ways we join in the traumatic events of others. Under the matrixial gaze then, different aesthetic and ethical problems arise.[26]

While Mary's later life events may have been without a witness in Robinson's account, they were not without his wit(h)nessing of the traces and gaps. Robinson works as Mary's mother had done to weave the fabric of their family life with the material he had with-in and with-out: 'Mary's mother was always at work "making something out of nothing".[27] Mary's lost typescript links to a question Robinson continually asks in his writing: 'how can writing, writing about a place, hope to recuperate its centuries of lost speech?'[28]

In this chapter, the house as a place to hold writing serves as one of the many connections in the multicursal patterns between Robinson, Mary Walsh, the actress Olwen Fouéré and the dramatist and writer Antonin Artaud in a horizon of trauma. Robinson raises his self-consciousness of this act of writing and its comparison to a house in *Listening to the Wind*: 'A writing may aspire to be rich enough in reverbatory internal connections to house the sound of the past as well as echoes of immediate experience, but it is also intensely interested in its own structure, which it must preserve from the overwhelming multiplicity of reality.'[29] In 'Into the Mist', the first house is Mary's cottage:

> with its three-foot-thick walls of stone … its roof of thick, rough trusses and rafters and wooden slats, covered by a heavy layer of sods, grassy side down, and then a layer of thatch, fastened down with *sugan* ropes of plaited straw tied to heavy stones lying on the ground, and old fishing nets thrown over it to keep the wind from plucking at it.[30]

This cottage, the crucible from which this woman – who left behind only fragments of her past – emerged 'reads, in her account of it, as if it were designed by the elements themselves', comes to stand for the matrixial substratum shaped by the feminine/prebirth encounter.[31] In a series of ever-intensifying centres, the kitchen, 'centrally placed within the house', was 'never without a cradle, for "new arrivals came in rapid succession"'.[32] The cradle is centred as the magical space in the core of the cottage in which precious objects were placed and from which lost objects could be drawn. If anything was missing – odd shoes, cutlery, school satchels, the grandmother's bone corset – the cry was 'did you search the cradle?'[33]

The storyline takes us from Mary to Olwen through the narrative device of the land sold by Mary's mother to Olwen's father to house the lobster pots for the shellfish company with which Mary's father also worked. After a summary of Fouéré's impressive career, Robinson introduces the fact that her 'performance-theoretical pole star is Antonin Artaud, originator of the Theatre of Cruelty'.[34] Robinson at this point becomes linked to the relationships and sub-connections between Artaud and Fouéré, when he explains that he 'had written a few pages on the visit Artaud had made to the Aran Islands in 1937, when he was already on the cusp of madness … in search of the last descendants of the druids', and wishing 'to return to Ireland a staff that he believed to have belonged to St Patrick'.[35] Cinnzia Hardy, a director of the Dialogues project, which is a UK-based organisation that put artists together 'who might otherwise never make contact',[36] had contacted Robinson suggesting a collaboration with Fouéré. He writes that at this time he was

> nursing an obsession with the labyrinth of strange, high-walled fields in Inis Meain, one of which, called *Balla an Tiarbh*, the bull's wall, had the layout of a stage, on which no one had ever appeared but which seemed to hold some theatrical experience in suspension in its emptiness, such as perhaps, an Artaudian treatment of the legend of the Minotaur's labyrinth in Crete. I therefore began to write an outline of a film with nobody in it, and for which Olwen Fouéré seemed the right protagonist.[37]

Robinson's wonderfully mischievous yet powerful proposal for a 'film with nobody in it' (which did not proceed 'because an actual film would be no advance' on the proposal) depended 'solely on our awareness of her past successes in embodying a bizarre range of passions ... her non-appearance in *The Bull's Wall* draws its strength from all her past appearances, her commitment to the art of immediacy, the ambiguities of the physical'.[38] Robinson's account here echoes Ettinger's theory of the 'matrixial *objet a*', in which she explains:

> Being shared, the flickering of the matrixial gaze for one partial subject is not its exhaustion for the other, and therefore is never a complete atrophy, not even when one of the partial subjects totally disappears. Thus, the matrixial subjectivity may carry from one to an unknown other, and also from one generation to another, the 'trace of the trace' (Derrida) and the 'mother of the mother' (Fedida).[39]

Robinson's evocation of the sound of shorelines suggests a matrixial surfing of unknown others. He writes that

> these indefinite but enormous noises are part of Connemara. Sometimes from my doorstep on a still night I become aware that the silence is set in a velvet background like a jewel in a display case, a hushing that, when attended to, becomes ineluctable. It is compounded of the crash of breakers along distant strands, variously delayed, attenuated, echoed and re-echoed.[40]

His proposal for Fouéré was underwritten by such an understanding of 'indefinite but enormous noises' comprised of a 'hushing' and 'compounded' of the crash of her other waves breaking, other traces of her past performances. These traces seem to depend on us willing her to life through awareness of the traces of her past performances, an act in which, according to Ettinger, 'artwork incarnates traces of traces, mothers of mothers, and "events without witness" ... the emotional and mental conduct of an artwork may reflect on far away matrixial unconscious events'.[41]

Their meeting appears to have taken on some of the qualities of Robinson's idea, for although he had been greatly looking forward to meeting Fouéré, and was interested especially in her childhood in Aughrus, Robinson did not leave her company with such knowledge about her childhood. He observes that he must have forgotten to ask about her memories of the lake, nor did he ascertain what the anchor on the beach 'meant to Olwen as a child'.[42] He writes that 'trying to understand afterwards how it was that I came home with little or nothing of interest about her own childhood in Aughrus, I realise that my concept of a walk, which almost precludes talk except of the immediate data of landscape consciousness, had imposed itself'.[43]

It appears to be more than a missed encounter, but a matrixial one, with all the tonal quality of being caught up in the gaze of the artwork and the other where they 'experience a similar metramorphosis and will momentarily co-emerge with the gaze that is caught in the artwork'.[44] Robinson writes, 'Perhaps she had brilliantly performed the role of nobody, as in my first conception of her. Looking back on our circuit of Aughrus and its central lake, I only know that I had felt the presence of a lake of stillness within her.'[45] The language of conception, and an inner lake of stillness, as well as Robinson's 'nursing' an obsession, suggests a matrixial connection to Fouéré, which becomes apparent on the 'peculiar day' they spent together in which 'a sea mist had come into existence and annulled all distant prospects'.[46] In this enveloping mist, both Robinson and Fouéré appear to have entered a 'matrixial borderspace' in which 'a specific aesthetic field comes into light, with metramorphosis as a poietic-artistic process with ethical implications'.[47] The sea mist is described as if the land has migrated to the sea, as it 'lay in vague tufts and wads just offshore'.[48] The gaze into this merging of land and seawater is as if it is surrounded by the spectral traces of a structure with no sides: 'from the vanished gazebo we looked out to sea, or rather into the mist, which still surrounded the land with a soft obscurity, so that it felt as if we had been walking around the interior of a diorama, or were ourselves mere projections onto an illusory and depthless background'.[49] Here the scopic is impressed upon by the archaic traces of the matrixial gaze, which Ettinger describes as

> not purely visual, and it enters visuality, disturbs it and change the tableau, because it penetrates and alters the scopic field, which by definition is im-pure – inseparable from other unconscious dimensions of the psyche (oral, anal), informed by different sources of sensibilities (like changes in pressure, movement, touch etc.) and also connected to the unconscious of others in inter-subjective and trans-subjective spheres.[50]

In this metramorphic state, the sense of the sacred matrixial occupancy of the shore through their shared matrixial gaze appears when Fouéré, describing how Tony White was buried in Omey graveyard 'in the same corner as the cross commemorating the Cleggan disaster victims', mentions, seemingly as an aside, how her own two children are also buried there. However, Robinson directs us to other sources for the details of this trauma in her life when he writes that this was 'a tragedy [he] knew of, but not about, from a TV film of her life'.[51]

This suffering becomes story in an interview with Ciara Dwyer, when Fouéré talks movingly about the heartbreak of the loss of her two children within the space of six years, her daughter Morgane at a day and a half after birth, her son Jo Jo at five months in the womb. Cherishing the prenatal relationship with them, they remain present to her in her understanding that 'maybe they were given to me for a reason, like I was given them to be touched on a physical level. So they're a kind of gift.'[52] Here she honours this time so often symbolised and foreclosed in our understanding of being a human subject – from both the perceptions of the child and the mother, and from the wider social and political fabrics in which we live. Talking about Omey she says, 'we often go there. It's a lovely peaceful place, a real place of healing. They're buried in the sand, which is beautiful and if I get to choose I'd like to be there.'[53] When Robinson asks, 'what is the connection between the great calm and her theatre life?', a clue might be found in how she continued to rehearse for a Marina Carr play for two weeks when she knew that Jo Jo had already passed away.[54]

This stillness, this matrixial trace of the archaic tranquillity that she carries, surfs under the symbolic understanding of Fouéré as an Artaudian, a disciple of Artaud, the originator of the Theatre of Cruelty, the work of whom prompted the connection between Fouéré and Robinson in the first place. Robinson asks, 'what would be the Paradox of the Artaudian?' and answers by speculating 'that only the deeply sane can present extremes of the instinctual and irrational convincingly and without being harmed by them'.[55] Perhaps Fouéré's work as a channel for the partial imprints of others is more matrixial than paradoxical:

> In the matrixial psychic sphere, my imprints will be trans-scribed in the other, and to begin with in the m/Other, thus my others will process traumatic events for me, like my m/Other processed archaic events for my premature and fragile subjectivity. Thus, female bodily specificity is the site, physically, imaginatively and symbolically, where a feminine divergence emerges, through which a 'woman' is interlaced as a figure that is not confined to the one-body, but is a hybrid 'webbing' of links between several subjectivities, who by virtue of such a webbing becomes partial.[56]

The notable stillness of Fouéré's body as a cradle not only for her children, one of whom she rocked inside her after death even as she made herself available to the trauma of others, imprints itself on Robinson powerfully, in a way he finds hard to cognitively process or name. In a day full of impression and half-sights, of glimpses and mists, his own connection to Artaud is partially and incompletely called up. Fouéré acts as the 'hybrid webbing'. The lake of stillness inside her allows her to open a passage from Artaud to Robinson, and contrasts with the internal agitation associated with Artaudian dramatic principles.

Robinson's own relationship with the landscape and weather systems of Connemara have often placed him in the theatre of the wild, assaulted by sounds of nature like the audience member in the Theatre of Cruelty:

> The ear constructs another wholeness out of the reiterated fragmentation of pitches, and it can be terrible, this wide range of frequencies coalescing into something approaching the auditory chaos and incoherence that sound engineers call white noise: zero of information-content, random interference obliterating all messages,

utterly dire, a metaphysical horror made audible, sometimes dinned into prisoners' heads to drive them mad in the cells of their brains.[57]

Artaud's soundwork – where the language is pre-symbolic, and comprises guttural howls, cries, noises, shrieks, rack screams – is designed to produce the affect of trauma. Robinson's work also listens for the sound of the past and to the traumas inflicted on the land and the people who cared for it. For Artaud, this was an unbearable place, and his avoidance of it culminates in his theorisation of the 'organless body', which was so influential for the French philosophers Gilles Deleuze and Félix Guattari. This appears to travel some of the same pathways as Robinson, Mandelbrot and Ettinger, sharing some metaphors of geology, flora and geometry, but it is not the same creature, and arrives at a different destination. In 'Antonin Artaud and Freud's "Family Romance": The Transgressive Sublime', Lynn Hughey Engelbert writes,

> In his description of 'how I lost my mind,' Artaud correlates a variety of natural phenomena with his rational degeneration in order to support this venture, so that 'joints in stones,' as well as 'arboreal bunches of mind's eyes ... set into glossaries and stone aeroliths will fall' (Artaud 25). This 'fall' would then 'effect a general understanding of "non-spatial geometry".'[58]

This finally arrives at radical body that also for Artaud, as Hughey Engelbert notes, was the maternal body, which is described in terms of misogyny and fear:

> This radical body returns the individual to their birth, and the mother, and for Artaud, this maternal source was a 'shapeless female principle' which 'suffocates and engulfs the rational masculine mind' (Greene 84). She is also the immanent yet omnipotent reproducer for Artaud: 'The cunt without the mother' (Watchfiends 67). She is the 'Center-Mother and Boss-Pussy' (111) and Artaud's 'tongue' becomes her counterpart as a masculine metaphor that 'tries to pierce the earthly crust of soft matter which threatens to subdue and annihilate it' (Greene 84). This is the organ that literally defines him as well, as 'this tongue between four gums, this meat between two knees, this piece of hole for madmen'.[59]

The matrixial is reduced to a hole: 'Out of the cunt without mother I will make an obscure, total, obtuse and absolute soul.'[60]

In Artaud's symbolic world of corporeal brutality, feminine and masculine are opposed, keeping such thinking on the phallic plane and not accessing the matrixial substratum where sexual difference is not yet established, as would be developmentally understood later on. Artaud's audience is helpless, whereas Robinson's walk-art-text practice is a process of 'co-poetic metramorphosis', in which nature responds 'in more intimate and less patriarchal terms'. Robinson goes on to say that 'these landscapes are the epitome of what I have called the Echosphere, the zone in which a balance is maintained between culture and the wild, so that, through daily frequentation and the communal memory of placelore, nature answers to the human voice'.[61] For Robinson, the sound of the past is less overwhelming and more a co-emerging encounter. He writes,

> Similar too is the sound of the past, the wreck of time's grand flow in torturous passages. It includes and sometimes drowns the sound of history. History has rhythms, tunes and even harmonies; but the sound of the past is an agonistic multiplicity. Sometimes, rarely a scrap of a voice can be caught from the universal damage, but it may only be an artefact of the imagination, a confection of rumours.[62]

In 'Into the Mist', the sound of the past and the sound of history are interwoven via the matrixial, evoking the trauma of Europe's interwar and post years, which Artaud was also so clearly imprinted by (as well as, of course, by a brutal childhood). Ettinger posits that 'in the matrix up to a certain extent, there is an impossibility of not sharing trauma'. This is most evident in art practices which 'in our post-Duchampian era' dissolve 'the oppositions between ready-made as a textual appropriation on the one hand and materials emerging from the self as originally "mine"'. In this way, 'art may lead us to discover our part of shared responsibility in the events whose source is not "inside" the One-self'.[63] Robinson takes other 'ready-made' materials as his source events whose source is not inside his one-self – the 'erotic antennae of the psyche engender a trans-subjective psychic sphere', who herself calls on sources outside her one-self in her 'gruelling' installation as 'Artaud incarcerated in a glass box' – where his art sought to evacuate the one-self, a few pages of a typescript, the shore, ruins and sea, and creates a matrix for the sharing of trauma, however elusively it may appear to be articulated.[64]

The question of loss circulates here from the loss of Mary's cottage cradle to Fouéré's lost children, from the lost typescript – Mary's brother asked 'was that book ever published?' – to the unwritten texts of Artaud's plays.[65] Judith Butler, writing about Ettinger's work, notes that 'the work of borderlinking' is to 'ask about historical losses, the ones that are transmitted to us without our knowing, at a level where we cannot hope to piece it together, where we are, at a psychic level, left in pieces, pieces that might be linked together in some way, but will not fully "bind" the affect'.[66] Feeling and opening to the affect of these 'unbound' stories is, according to Butler,

> part of the work of borderlinking that Bracha writes about, and it is, in her view, prior to identity, prior to any question of construction, a psychic landscape that gives itself as partial object, as grains and crumbs, as she puts it, as remnants that are, on the one hand, the result, the scattered effects of an unknowable history of trauma, the trauma that others who precede us have lived through and, on the other hand, the very sites in which a new possibility for visual experience emerges, one which establishes a temporality in which the past is not past, but is not present, in which the present emerges, but from the scattered and animated remains of a continuing, though not continuous, trauma.[67]

In contrast to the 'unbound' affect of these matrixial remnants, *Towards a Federal Europe* (1968), written by Olwen's father Yann Fouéré, functions as a text of identity politics, and in doing so focuses on the socio-political causes in which much of the trauma binding occurs: a concentration that can lead to over-determination. Politically inflected, this book argued for a federal Europe of 'region-states', in which, Robinson notes, are not mentioned 'the vast immigrant populations in the big states; nor are the Jews, a nation central to a cultural definition of Europe but denied even a region in it; not those of us who do not want to be defined by the culture we were born into, or by any other'.[68] This book, as well as other recordings, gave Robinson some sense of 'the cauldron of ideologies out of which Olwen's family had blown like a bubble to this Atlantic Shore' – Fouéré's father had in fact come seeking refugee, having been wrongly accused of collaboration with the Vichy regime, a charge he was later cleared of.[69]

We finally return to the idea of a house in the title of Yann's latest book, *La Maison du Connemara*, which with its European cross-scriptions tells the story of his journey from Brittany to the west coast of Ireland and of how his family is washed up on the shores on Connemara.[70] In Yann, Robinson finds a man with a 'rather Cartesian regard for exactitude in space and time', which contrasts perhaps to Robinson's own exactitude, which can be characterised more as a fidelity to the excesses of space and time he finds in Connemara. Robinson explains,

> the invisible vortex spinning across the hillside according to equations of fluid dynamics I've never been able to master, spilling its energy into smaller vortices, and those into still smaller ones, dissipating itself fractally into a pocket of turbulence and the mad gyrations of individual molecules of air, feels in memory like a friendly nod and pat on the back from the incomprehensible system of all things.[71]

In interweaving the 'ready-made' from Mary's lost cottage and Yann's books about his *maison*, by suggesting other threads to unravel, through artistic and classical allusion, through the theatre of the coast, by receiving traces of trauma in cast-away words and body language, Robinson's walk-art-text practice opens him – and us – to the sphere prior to the identity marked as singular. His ear, pressed to the earth like Cleggan Head, and felt in his foot in synesthetic prenatal transmission of affect, invites another order into our lines of sight. His particular mode of 'transportation' approaches the extended aesthetic of *fascinance*, a mode of gazing that activates the seam of trauma running beneath his feet and the lives of those he has encountered. Of the relationship of art and trauma Ettinger has written:

> The place of art is for me the transport-station of trauma; a transport-station that, more than a place, is rather a space that allows for certain occasions of occurrence and encounter, which will becomes the realization of what I call *border-linking* and *borderspacing* in a matrixial *trans-subjective* space by way of experiencing with an object or process of creation.[72]

Robinson's own practice, like Fouéré's, calls forth the traces of other's'trauma, shapes it and experiences it – but not as one might expect through mapping, coding or codifying, but through *wit(h)-nessing* with the borderspace of the coast and the lives he encounters.

## Notes

1 Bracha Lichenberg Ettinger, 'Matrix and Metramorphosis', *Differences* 4 (1992): 201.
2 Bracha Lichtenberg Ettinger, *The Matrixial Gaze* (Leeds: Feminist Arts and Histories Network, 1995), 48.
3 Bracha Lichtenberg Ettinger, 'Wit(h)nessing Trauma and the Matrixial Gaze: From Phantasm to Trauma, from Phallic Structure to Matrixial Sphere', *Parallax* 7:4 (2001): 109
4 Tim Robinson, 'Seism: Essay Written for Siobhán McDonald on the Occasion of Eye of the Storm', *Eye of the Storm* (Galway: Galway Arts Centre, 25 May 2012), 8–9.
5 Tim Robinson, *Connemara: A Little Gaelic Kingdom* (Dublin: Penguin Ireland, 2011), 252.
6 Robinson, *Gaelic Kingdom*, 249.
7 Robinson, *Gaelic Kingdom*, 252.
8 Ettinger, 'Wit(h)nessing', 104.
9 Ettinger, 'Wit(h)nessing', 104.
10 Tim Robinson, *Connemara: Listening to the Wind* (Dublin: Penguin Ireland, 2006), 4.
11 Tim Robinson, 'The Irish Echosphere', *New Hibernia Review / Iris Éireannach Nua* 7:3 (Autumn 2003): 16.
12 Ettinger, 'Wit(h)nessing', 90.
13 Robinson, 'Echosphere', 15.
14 Robinson, 'Echosphere', 15.
15 Ettinger, 'Wit(h)nessing', 101.
16 Bracha Lichtenberg Ettinger, 'Weaving a Woman Artist with-in the Matrixial Encounter-Event', *Theory Culture Society* 21:1 (2004): 73.
17 Ettinger, 'Wit(h)nessing', 102.
18 Robinson, 'Echosphere', 16.
19 Ettinger, 'Wit(h)nessing', 109.
20 Ettinger, 'Wit(h)nessing', 90.
21 Robinson, 'Echosphere', 11.
22 Tim Robinson, *Connemara: The Last Pool of Darkness* (Dublin: Penguin Ireland, 2008), 137.
23 Robinson, *Last Pool*, 181.
24 Robinson, *Last Pool*, 181.
25 Robinson, *Last Pool*, 184.
26 Ettinger, *The Matrixial Gaze*, 51.
27 Robinson, *Last Pool*, 183.
28 Robinson, *Listening*, 3.
29 Robinson, *Listening*, 6.
30 Robinson, *Last Pool*, 182.
31 Robinson, *Last Pool*, 182.
32 Robinson, *Last Pool*, 182.
33 Robinson, *Last Pool*, 182.
34 Robinson, *Last Pool*, 184.
35 Robinson, *Last Pool*, 185.

36 Robinson, *Last Pool*, 184.
37 Robinson, *Last Pool*, 185.
38 Robinson, *Last Pool*, 185–6.
39 Ettinger, *The Matrixial Gaze*, 49.
40 Robinson, *Listening*, 1.
41 Ettinger, *The Matrixial Gaze*, 49.
42 Robinson, *Last Pool*, 189.
43 Robinson, *Last Pool*, 193.
44 Ettinger, *The Matrixial Gaze*, 48.
45 Robinson, *Last Pool*, 193.
46 Robinson, *Last Pool*, 187–8.
47 Ettinger, 'Wit(h)nessing', 104.
48 Robinson, *Last Pool*, 188.
49 Robinson, *Last Pool*, 190.
50 Ettinger, 'Wit(h)nessing', 90.
51 Robinson, *Last Pool*, 189.
52 Ciara Dwyer, 'Olwen Fouere's Fire and Ice', *Independent.ie* (25 January 2009): n.p. Accessed 16 February 2014, www.independent.ie/woman/celeb-news/olwen-foueres-fire-and-ice-26508461.html.
53 Dwyer, 'Olwen Fouere's Fire and Ice', n.p.
54 Robinson, *Last Pool*, 193.
55 Robinson, *Last Pool*, 194.
56 Ettinger, 'Wit(h)nessing', 103–4.
57 Robinson, *Listening*, 2.
58 Lynn Hughey Engelbert, 'Antonin Artaud and Freud's "Family Romance": The Transgressive Sublime', *Mosaic: A Journal for the Interdisciplinary Study of Literature* 47:2 (June 2014): 28.
59 Engelbert, 'Antonin Artaud', 28.
60 Engelbert, 'Antonin Artaud', 23.
61 Robinson, *Last Pool*, 146.
62 Robinson, *Listening*, 2. For more on sound in Robinson's work, see Smyth's essay in this collection, '"About Nothing, About Everything": Listening in/to Tim Robinson' (Chapter 11).
63 Ettinger, *The Matrixial Gaze*, 95.
64 Robinson, *Last Pool*, 193.
65 Robinson, *Last Pool*, 189.
66 Judith Butler, 'Bracha's Eurydice', *Theory Culture Society* 21:1 (2004): 97.
67 Butler, 'Bracha's Eurydice', 97.
68 Robinson, *Last Pool*, 191.
69 Robinson, *Last Pool*, 193.
70 Robinson, *Last Pool*, 192.
71 Robinson, *Gaelic Kingdom*, 253.
72 Bracha Lichenberg Ettinger, *The Matrixial Borderspace* (Minneapolis: University of Minnesota Press, 2006), 91. Original emphasis.

14

# Essayist of place: postcolonialism and ecology in the work of Tim Robinson

## Eóin Flannery

In his 1993 study of cartography and folklore, *Mapping the Invisible Landscape*, Kent C. Ryden underscores the necessary interdisciplinarity of what he terms 'the essayist of place'.[1] Impelled by a desire to do justice to the complexity, or 'thickness', of place histories, of place – visual and textual – for Ryden, 'the essayist of place is at once a cartographer, a landscape painter, a photographer, an archivist, and a folklorist, as well as a storyteller ... [and] a conscientious and dedicated chronicler of landscapes and of the ways of life that those landscapes witness and support'.[2] Ryden's emphasis on interdisciplinarity, firstly, intersects with calls within current ecocritical studies for just such a modus operandi within that field, as well as gesturing to the inevitable difficulty of documenting the sedimented layers of place identities.[3] Equally, Ryden's catalogue of disciplines betrays an acquisitive dynamic in the relationship it constructs between place and those that strive to essay its features. His argument seems, therefore, to privilege a medley of representationalist engagements with the physical exteriorities of place. Yet, as vexed debates within ecocriticism testify, recourse to signification is not always seen as inevitable and often adjudged to be entirely anthropocentric.

In this light, Ryden subsequently tempers the apparent representationalism of his earlier point by disaggregating factual records of place from the living processes of these locations. Again he refers to the essayist of place, whose task it is to go 'beyond the *facts* of place to emphasize the *life* of place, exploring in depth its symbiotic relationship with his or her own life or with the lives of the residents of the place to which the essayist of place turns a keen eye and an empathetic imagination'.[4] Ryden's methodology tends toward a dynamic relation with place rather than any tendency to abstract physical realities; and there is a palpable recognition of the need for authorial self-consciousness on the part of the essayist of place in Ryden's scheme. In addition, his final point alights upon the notion of 'empathetic imagination', an idea that seems to preserve the living dignity and agency of the physical topographies and cultural geographies under scrutiny. Not

*Figure 26* Essayist of place – Robinson at his desk juxtaposed against the Connemara landscape (from Pat Collins's film *Tim Robinson: Connemara*, photo by Colm Hogan).

only should the essayist of place display requisite self-reflexiveness in tracking their own rapport with place, but Ryden's communal and empathetic vision also highlights that interdependence and mutuality are components of the essayist's ecological vision.

The field of Irish cultural studies has yet to exploit fully the critical and analytical resources of ecological criticism. Indeed very little sustained and enabling historical or critical writing has emanated from the field that might productively contribute to international conversations on the political and cultural implications of global environmental change. There have always been creative and critical engagements with the Irish landscape – a trend partly occasioned by the country's protracted history of colonialism (a prime concern of ecological criticism). While there have been isolated interventions in both literary studies and economic history, these creative and critical legacies have yet to yield a body of ecological critical writing worthy of the name.[5] Perhaps, in some measure, on foot of the cynical, and ultimately self-defeating, transvaluation of the Irish landscape under the yoke of the Celtic Tiger, and its accompanying property 'boom', there have been belated and sobering critical responses that have taken impetus from some international ecological criticism. Combining a plurality of disciplinary fields – authorial self-consciousness and a lateral, transhistorical empathetic vision – Tim Robinson's cartography and prose writing are, this essay will contend, explicit incarnations of Ryden's profile of the essayist of place. Robinson's pioneering visual and textual narrations of place from the West of Ireland are exemplary maps of the visible and invisible landscapes of these regions (see Figure 26).

The accumulated knowledge and deeply felt sensitivity to place – in terms of its ecologies, its social histories and its contemporary patterns of living – evident in Robinson's works make them salutary reminders and embodiments of the significance and import of ecological consciousness in the contemporary moment. Reading across his works as, inter alia, postcolonial, ecological and interdisciplinary, this discussion will position Robinson's career and publications as potentially informative of contemporary debates within ecocritical and postcolonial ecocritical studies, as well as potent reminders of the currency of place and landscape in post-Celtic Tiger Ireland. In respect to this latter point, Robinson's works restore a sustainable ethical relationship with place in the Irish context: place as a historically rooted and valued context, marked by an informed interaction with the cultural histories of that locale. In other words, Robinson reminds us of, and exemplifies, the mutually sustaining bonds between the materialities of place and landscape, and representations of such locations. Too often in the Irish context both place and landscape have been abstracted into nostalgia and/or relative financial value. In these respects the argument will be centred on foregrounding Robinson's work as ecocritical *avant la lettre* in an Irish context, as well as suggesting that his work harbours continuing dynamic potentials for future ecocritical thought and praxis in Ireland and elsewhere. As Robinson concludes in his 1996 essay 'Listening to the Landscape', 'Enquiring out placenames, mapping, has become for me not a way of making a living or making a career, but of making a life; a mode of dwelling in a place.'[6] Such remarks might well echo those of an ecological politics that is rooted in the local, a brand of politics that abjures the national or the global, and that can mutate into insular conservatism. Yet, as we shall see, both the forms and the contents of Robinson's works are, in fact, directed toward more global ethical endpoints.

Reviewing Robinson's *Stones of Aran: Pilgrimage*, Seamus Deane pinpoints the sensibility of the author, with its intimate attachment to the endured vagaries of island life encountered on his pedestrian navigation of the coastal extremities of Árainn Mhór (Inishmore). Deane characterises Robinson's work as

> a marvellous act of retrieval which is not marred by a sentimentality that fails to take into account the plight of the people who live on the island. His Árainn Mhór remains a place of habitations, not a museum-piece in which interesting relics can be observed under the white light of a severe scholarship.[7]

Deane's comments are loaded with postcolonial irony – he is well aware of and intellectually invested in the critique of British imperial rationalisations of Ireland's 'otherness'.[8] The Aran Islands were an isolated fastness of the British Empire and, subsequently, of the Irish State, and its inhabitants have populated waves of external representations. As a resolutely peripheral Celtic outpost of Europe, the Aran Islands have attracted both scientific and anthropological investigators, and legions of visitors attracted to, and presuming, its remote spiritual difference from metropolitan life. Thus, as Deane intimates, Árainn Mhór has assumed limited reiterated identities, more often imposed by exogenous visitors. And through this profusion of reinventions and reimaginings of its geographies and its histories, much of the reality of Árainn Mhór's prehuman and human histories have been elided or disfigured. Granted Robinson's work could be enumerated among the dense skeins of visual and verbal representations of Árainn Mhór, but what is also implicit in Deane's assessment are the virtues of Robinson's methodologies in pursuing his versions of Árainn Mhór. Robinson would never claim 'truth' and 'objectivity', but there is the sense in which the almost encyclopaedic generosity of his narrative style, its undiscriminating embrace of minor tales and major political events, are indicative of what Ryden called earlier 'an empathetic imagination'. Or, in his own words, Robinson lays bare the practices that informed the composition of *Pilgrimage*:

> I have walked the islands in companionship with such visiting experts as well as the custodians of local lore whom I sought out in every village, and have tried to see Aran through variously informed eyes – then, alone again, I have gone hunting for those rare places and times, the nodes at which the layers of experience touch and may be fused together.[9]

Robinson's anatomisation of his practice reveals the interfaces with and debts to the fields of academic, scholarly research, and the accreted strata of local folklore expounded by his native interlocuters on Árainn Mhór. The description of his routine of research, significantly, alerts us to the basic physicality of the mutually informative enterprises of mapping and of writing the island. Robinson's perambulations – lengthy and interactive – are committed and arduous immersions in the physical pasts and presents of the island. Such a form of investigation, in fact, partakes of an ethics of 'slowness', a practice, a disposition, a temperament, that favours a decelerated and rooted attachment to the density of place histories. But what is equally striking, at a more general level, is Robinson's willingness to explicate the specifics of his authorial creative processes. Right across his writings, Robinson reveals his anxieties and his pleasures in acts of writing about the natural and forgotten human histories of the Atlantic seaboard of the West of Ireland. Robinson is fully cognisant of his own implication in these histories as cartographer and as prose chronicler, and he is sensitive to the limits of both subjective experience and the mechanics of 'realistic' representationalism.

In a recent interview with Brian Dillon, Robinson further exposed some of the authorial self-reflexiveness that is characteristic of his work, as well as further gesturing to the tensile relations that can exist within ecological writing between the author and the environment, between human history/time and prehuman and geological time frames. Referring to the time spent resident on Árainn Mhór, Robinson reflects,

> Living on an island, a little habitable space in the midst of a rampant wilderness, forces a metaphorical or allegorical dimension into everyday space ... At least, so it did for me, though many an islander might have been amused by the idea. But I also relished the island's deep anteriority to all human interpretation of it.[10]

Again, the extremity of location occasions a humbled response from the author, but, equally, compels him to respond to such extremity through language and through signification. And there is a sense in which Robinson is eager to stress the limits of such representation, but also the integrity of language and landscape. In other words, while his, or any, linguistic registration of Árainn Mhór's sublimities are always already failures, the temporary, lapsed representational effort stems from a material relationship between place and language. What is significant is not just Robinson's recognition of the boundaries of writing, but also the novelty of human engagement with the gnarled geological forms of the island. Time is a concept we cannot elude when confronted with the prehuman histories of geological formations and evolution of these Atlantic islands.[11]

The anteriority of the Aran Islands generates awe and insignificance in Robinson; the physical and cultural imprints of humanity on the islands have been brief yet often violent and destructive. But Robinson dismisses any presumption that language and writing can somehow contain the innumerable phases of ecological development on the islands. His authorial self-consciousness, then, is conditioned by the limits of human history and by the related fact that, as a consequence, language/representation are ill-equipped to house the millennial materialities of the landscape's evolution. Yet, this recognition of the overwhelming excessiveness of prehuman history is also true when we consider the lives of Árainn Mhór's human histories. For Robinson, when he turns to the stories of individuals and communities in the West of Ireland, 'the truest of writing about the past can hardly offer more than an appalled recognition of the injustice of time, its brutally hasty recourse to mass graves for irreplaceable individualities and to landfill for delicate discriminations'.[12] Writing and textual record can, then, do some material justice to a thin selection of subalternated historical voices, but not to any more than that of history's plurality of discarded stories. Even within the confines of a small geographical region, Robinson is confounded by narrative evasion, contradiction and forgetting. Despite these obvious limitations, which are features of all ecological-historical narration, Robinson evinces a material attachment to place, one steeped in a desire to do justice to and to dignify the lives and the lost histories of remote and plundered landscapes of the Aran Islands and Connemara, and their generations of populations.

Quite apart from the finished textual products of Robinson's years of living in the West of Ireland, it is just as important to highlight the labourious physicality he invested in these creative processes. Walking and interpersonal, face-to-face conversation are two of the primary modes through which Robinson gleans his folkloric, botanical, geological, etymological and historical information. Thus, there is a deep material symbiosis between the author, the cartographer, and the people and the places that come to animate his works. Again, in interview with Dillon, Robinson confesses to the physical rigours of his life's work: 'It was an exhausting struggle against rain and wind, but by the end of it I felt that the islands had been so deeply etched into my very being that I could have printed off an image of them by rolling on the paper.'[13] The bodily exertions endured by Robinson during his researches and observations are, once more, indexical of a decelerated ecological vision – a mode of discovery and living that privileges immersion and depth in its relationship with the living landscape's storehouse of natural and cultural imprints. In addition, Robinson's methodologies are exhaustive and exhausting in their pursuit of a tentative accuracy in historical, botanical, linguistic and folkloric information. But what such a regime of research suggests is the embeddedness of human living with the concrete materialities of the Aran Islands and, in his later work, of Connemara. Robinson's metaphorisation of the scripting of the island's histories and topographical features onto his body, thus, is indicative of the ethical animus of his project – an ethics that impresses the elusive dynamism of Irish cultural landscapes, and that cherishes the enduring cultural energies of the folkloric and submerged historical artefacts and voices of Ireland's experiences of British colonialism. When we come to study Robinson's narration of these contested cultural terrains, a further geological figuration by the historian Kevin Whelan seems to resonate:

> The Irish cultural landscape displayed igneous or metamorphic, not sedimentary historical layering ... In such circumstances, there could be no easy partitioning of the past from the present in Ireland. The landscape itself was a palimpsest, containing contested narrative of history and culture. Its monuments and traces reached from the present down into earlier layers from which the derived their power and presence, their aura.[14]

Building on his critiques of the deformative cultural politics of colonialism and latter-day neocolonialism in earlier works such as *Orientalism*, *Culture and Imperialism* and *Covering Islam*, Edward Said's more recent essay 'Invention, Memory and Place' centres on the misappropriation and manipulation of the past as both history and as memory in contemporary crucibles of political domination – most notably as part of the Israeli–Palestinian conflict. For Said, 'the art of memory for the modern world is both for historians as well as ordinary citizens and institutions very much something to be used, misused, and exploited, rather than something that sits inertly there for each person to possess and contain'.[15] In his estimation, the modern world is to be distinguished from, for example, classical antiquity, where memory was an ordered, and ordering, art form. Perhaps what aggravates Said more than anything is the wholesale cynicism characteristic of institutional and popular (mis-)appropriations of memory toward oppressive and exclusionary ends. Said's temporal specificity on what constitutes the 'modern world' is ambiguous. The 'nation' and 'nationalism' – particularly in virulent ethnic forms – are the most culpable in these regards. In constructing his argument – one that is, at the same time, both uncontroversial and highly contestable – Said dwells on the adjacent disciplines of history and geography; and, in the current context, it is his attention to the geographical exertions of imperial conquest that seems most germane. Again, in dealing with the implication of geography within the exercise of imperial authority, Said foregrounds both its provisionality and resulting susceptibility to misemployment.

'Geography', for Said, is 'a socially constructed and maintained sense of place', and as such, 'geography can be manipulated, invented, characterized quite apart from a site's merely physical reality'.[16] Such an assessment of the potential distortionary effects of the mishandling of 'geography', of place and landscape, is, of course, most readily apparent in the protracted and variegated histories of global conquest and imperialism. Comprehensively documented by scholars such as Paul Carter, we see how colonised space is perceived and understood as *terra nullius*, blankly awaiting inscription and assimilation through cartography, travel writing and visual and literary arts.[17] While Said's intervention is profoundly political in its temper and motivation, as it registers the bluntness of the modes through which geography and history are managed as discourses in the fashioning of systems of national and international remembrance, his attention to the 'socially constructed and maintained sense of place' actually speaks to contemporary ecocritical and postcolonial-ecocritical projects. As mentioned, Said's argument in this essay is continuous in many respects with his previous work, but particularly with *Orientalism*. And this is just one of the ways in which Said's writing here, and in *Orientalism*, falls short of recent postcolonial engagements with the ecological imprints and legacies of global imperialism. Both of Said's texts labour the institutional co-option of the discourses of history and geography to their self-interested ends. And, in doing so, the arguments neglect the possibility of resistance or complexity arising out of these contested and conquered geographies and histories. If he remains convinced of the 'constructed' cast of geography, then, of necessity it can be read as culturally contestable, as transient. Though he conjoins history and geography to the exploitative politics of memory under imperial and neo-imperial regimes, his argument is limited by a blindness to the cultural complexity that is layered upon and that emanates from a diversity of local geographies.

Reflecting on the 'postcolonial' motivations underwriting his various labours, Robinson dwells upon the linguistic denuding of the Irish landscape and of Irish place name lore: 'So it became my intention to try and undo this historical insult to the [Irish] language and its speakers; there was an element of post-colonial reparation in undertaking this task.'[18] And such an attentiveness to the vernacular landscape chimes with recent work within postcolonial ecocriticism. Invoking the writings of the Guyanese novelist and poet Wilson Harris, Elizabeth DeLoughrey and George B. Handley introduce their volume *Postcolonial Ecologies* seeking to demystify the historical legacies of colonialism.[19] They argue,

> In order to engage a historical model of ecology and an epistemology of space and time, Harris suggests that we must enter a 'profound dialogue with the landscape' … This historical dialogue is necessary because the decoupling of nature and history has helped to mystify colonialism's histories of forced migration, suffering, and human violence.[20]

While Harris's immediate context is the protracted, and variegated, experiences of slavery, plantations and indentured migration in the Caribbean, the authors seek to progress from Harris's local vision to a working methodology for the discourse of postcolonial ecocriticism. Given that our discussion strives to situate Robinson's oeuvre within this latter discourse, DeLoughrey and Handley's critical manifesto offers relevant, and enabling, arguments on just how Robinson's work coalesces with extant critical and creative projects within postcolonial ecocriticism. Indeed, DeLoughrey and Handley provide a statement of what this critical enterprise or series of critical resources should entail: 'A postcolonial ecocriticism, then, must be more than a simple extension of postcolonial methodologies into the realm of the human material world.' More ambitiously, they suggest, 'it must reckon with the ways in which ecology does not always work within the frames of human time and political interest. As such, our definition of postcolonial ecology reflects a complex epistemology that recuperates the alterity of both history and nature, without reducing either to the other.'[21] The implication in the first part of this assertion is that by merely importing the reading strategies of postcolonial studies into postcolonial ecocriticism, the critic is restrained by a residual anthropocentrism, as well as by temporal 'presentism' in terms of their relative perspective on human–nature relations. This is not to impute that a progressive postcolonial ecocriticism, or any form of ecocriticism, needs to be, or ought to be, anti-humanist in approach, but that the influences of 'deep-time' history – evident in Robinson's work – together with a world-ecological perspective must be incorporated in the central workings of postcolonial ecocriticism. Introducing such frameworks into postcolonial criticism, as DeLoughrey and Handley suggest, as viable critical foci does not preclude the parallel employment of eco-materialist criticism, or eco-Marxian thought, within a postcolonial-inflected ecocriticism.[22] Nevertheless, the final contention above by DeLoughrey and Handley is open to critique, as they appear to disaggregate history and nature, a point that has been effectively contested in much recent eco-Marxist writing, by scholars such as Jason Moore and John Bellamy Foster.[23] But what is provocative about their conclusions on history and nature is the idea that postcolonial ecocriticism, at its best, should alienate us from preconceived and complacent assumptions about our relationships with history and non-human nature. In a sense, what unites their conjunction of postcolonial studies and ecocriticism, and our reading of Robinson in terms of both these critical discourses, is a commitment to 'the inextricability of environmental history and empire building'.[24]

Prior to their outline of what an integrated postcolonial ecocriticism should involve, DeLoughrey and Handley alight upon a key area that once more explicitly aligns their critical project with those undertaken by Robinson. Without losing sight of the global, tentacular reach of historical and contemporary imperialisms, the authors refocus their attention on the multiplicities of the local.[25] If the temporal logic of capitalist imperialism has proceeded in a destructive linear vector – too often across 'unmarked' spatial planes, or landscapes that require suitable 'civilisation' or 'homogenisation', so as to be co-opted into the cultural and social economies of capitalist imperialism – then an interrogative postcolonial ecocriticism must orient itself so as to contradict and confound such violent spatial and temporal simplifications. And one of the ways of addressing these dominant, and dominating, world views is through a renewed attention to, and understanding of, 'place'. Specifically,

> Place has infinite meanings and morphologies: it might be defined geographically, in terms of the expansion of empire; environmentally, in terms of wilderness or urban settings; genealogically, in linking communal ancestry to land; as well as phenomenologically, connecting body to place … postcolonial studies has utilized the concept of place to question temporal narratives of progress imposed by the colonial powers.[26]

Notwithstanding the constructive 'infinity' of signification proposed in the opening sentence, this reassertion of the fecundity of place is, perhaps, one of the keynotes of Robinson's entire visual and verbal archive. The geographical, environmental, genealogical and phenomenological definitions cited here are all abundantly present in his work, and indeed, the glaring omission of language – indigenous versus imported and acquired – by DeLoughrey and Handley can be added to this catalogue encountered across Robinson's works.

But Robinson also unearths many of the material realities of Ireland's history of feudal colonialism. History and oral cultural records memorialise the punitive excess and subversive actions characteristic of the colonial feudal system on Árainn Mhór, but there are other, more 'ecological' exhibits that testify to the influence of landlordism when Robinson traverses Connemara, specifically the village of Roundstone in Co. Galway, where he resides. As Robinson's pedestrian excursions reveal, landscape speaks of the past and to the present, it is revelatory of some but not all of the human drama and interactions played out within its contours. Recalling the architectural and demesned decorum of the 'Big House' novel genre of Irish literary history, Robinson provides a telling portrait of one such domicile and property on the outskirts of Roundstone. Despite the background of the Land War across the 1880s, the property encountered by Robinson here seems to indicate a family enjoying prosperity during this time of struggle and impoverishment:

> The mansion, unostentatious and comfortably-off in appearance, that Henry Robinson built in 1885 on a rise above the road just beyond the northern limits of Roundstone village must have been a more commanding presence in that era of cottages and hovels, until it was hidden from intrusive eyes by his plantation of pines, sycamores and beeches, still known as Robinson's Wood.[27]

Robinson's opening signals a typically dichotomous situation of tenanted poverty shadowing muscular, if somewhat paranoid, expressions of feudal authority. But there is a complex interplay of landscape, class, history and place name enacted in Robinson's cursory description. The plantation of trees, the imposition of an imported 'natural' screen, on this land is figurative of the orderliness witnessed in the Ordnance Survey (OS) reinscription of the place names of Ireland as part of the concerted discursive operations of British colonialism in Ireland. In this physical act of husbandry, 'nature' becomes a 'guard' against indigenous human eyes; non-human ecology is deployed against a constituency of the human population.

By extension, this arboreal implantation not only has physical effects on the topographic, but assumes toponymic significance as 'Robinson's Wood'. Thus Robinson's microcosmic portrait of landlordism in Ireland, with its 'big house', its demesned lands, its screen of planted trees and, finally, its toponymic legacy, throws into relief the ecological watermarks of this local colonial experience. Class exploitation and class-based management of the Irish landscape, then, have left physical and toponymic legacies across the local geographies of Robinson's Connemara: 'Only gentlemen could afford to plant trees and to protect them from cattle, goats and sheep by well-maintained fences, and in Connemara such a wood is a certain sign of a history of landlordism'.[28] What is obvious is that Robinson's works can be read as postcolonial. His methodologies, his interpersonal encounters, his environmental advocacy, his linguistic interrogations and retrievals, and his immersion in, and sensitivity to, the contradictions and excesses of British colonialism along Ireland's western seaboard are in alignment with the critical ambitions and spirit of DeLoughrey and Handley's projected critical enterprise. In Robinson's words, and referencing one of the principal routes through which he navigates the local histories and genealogies of Irish colonial and folkloric inheritances, place names are crucial to the efficacy of ecological consciousness:

> Placenames, whether they exist in the mind of the Irish *seanchai*, the custodian of traditional lore, or in the memory banks of a database, are only the anchor points of a discourse of place. To create a language for the secular celebration of the Earth, with the height and power of the religious tradition but purged of supernaturalism, can be seen as the task of ecoliterature, tracked and made conscious of itself by ecocriticism.[29]

The colonial context is one that informs our readings of Robinson's visual and verbal texts, and is a context that is implicit and, quite often, explicit in Robinson's engagements with the histories and geographies of property, economics, religious faith (pagan and Christian), myth and folklore, and culture in Ireland. 'Locality' and 'dwelling' are cornerstones of Robinson's authorial economy, and one of his primary vehicles for attempting to grasp the changing relationships between history and geography, place and population, and land and language in Ireland is the discourse of place names. The proximity of place names and colonisation in Ireland clearly centres on the tensions and accommodations between the Irish and English languages.[30] But, as Catherine Nash outlines in her progressive postcolonial critique of the potential inflammatory effects of narrow-gauge nationalistic deployments of place names, place names reveal cross-cultural negotiations more often than they assert the existence of unsullied cultural authenticity. The importance of place names for Nash – and in this she draws on, and extols, Robinson's work in the field – is that they link 'language and geography'.[31] Furthermore, they are 'at once both material and metaphorical, substantive and symbolic – read, spoken, mapped, catalogued and written in the everyday intimate and official bureaucratic geographies of road signs, streetnames and addresses – are all about questions of power, culture, location and identity'.[32] Given their potential for powerful symbolic use and their centrality to questions of communal, ethnic and national identities, it is understandable that 'modern Western states throughout the nineteenth century consolidated their authority and eased their governance through archives and registers of people, places and things, places were mapped, censuses taken and lists compiled'.[33] And, crucially, 'In Ireland, this process of modernization was profoundly complicated by a colonial relation, the rapid decline of the Irish language, partition of the island and Southern independence.'[34]

As we have mentioned, Nash's situation of place name discourse within the epistemic and political lineages of imperialism chimes with Said's coupling of empire and geography, and also aids us in setting Robinson's oeuvre within a broadly postcolonial ecocritical context. Nash's argument is predominantly centred on how place name projects in contemporary Ireland can productively inform non-sectarian, postcolonial understandings of Irish history. Her survey is well attuned to the divisive, and mendacious, idealisations and mythologies that adhere to 'place' and its naming when the said 'place' – as an historical and as a geographical space – is under contestation. And, of course, such potentially corrosive mythologisations are routinely exposed by the complexity unearthed by, and that frequently confound, Robinson in his pursuit of place name origins. With an anti-colonial, or narrowly conceived romantic nationalist agenda, according to Nash, '[a] critique of naming would seem to invite a celebration of cultural re-appropriation, resistance, recovery and reclamation'.[35] And Nash is surely correct in suggesting that while 'this perspective is critical of colonial forms of representation and relations of power, there are problems with this interpretation if it depends upon the notion of a pure, homogenous, pre-colonial culture suppressed by colonialism but ultimately recoverable'.[36]

The pitfalls of such an ethically solipsistic perspective are clear, and from an ecocritical standpoint an equal degree of insular self-identification, or cultural validation, is characteristic of many outdated and superseded forms of localism, which abjure the undeniable global imprints on their localities. But what is just as pertinent in Nash's argument, and just so in Robinson's work, is the disqualification of 'recoverability' as a telos. This is a necessary cultural and political corrective by Nash, as faith in the dreary recoverability of a pristine or authentic pre-colonial identity and heritage is nothing less than fuel for a politics of exclusion and division. And 'recoverability' as a pursuit of previously jettisoned or traduced cultural origins is anathema to Robinson's labours on the place names and broader cultural historical artefacts of Ireland's western climes. From a general critical viewpoint, then, where does the value of place name discourse reside? How can place name research prove useful to postcolonial and ecocritical readings of Irish and other global histories and geographies? If an ecological politics rooted in, but not tethered to, 'locality' and 'proximity' is a viable trajectory for current and future environmentalist thought and praxis, this can be realised through our appreciation of local difference, an understanding of the depths and complexities of all local micro-modernities.[37] In other words, through an informed engagement with and attachment to one local place, we can become more attuned to, and empathetic with, the idea of 'locality' on a global scale. Understanding the local becomes a means of appreciating the fragile fragments that make up the global; place name archaeology is one of the vital strands in generating a sense of truly informed residency in a particular location. But, as we have argued above, this is not a licence to revert to an undiluted myth of origin; as Nash notes, 'Like language, placenames have been interpreted as keys to neat and natural cultural categories, but also like languages, placename projects and policies present opportunities to rethink culture as a dynamic and social form of communication and meaning-making. Placenames speak of a sense of location and culture.'[38]

Key here are the notions of dynamism and of an ongoing process of identity-formation, each of which opens up the possibility, indeed the likelihood, of generative conflict in the construction of place identities. Difference and plurality across time in any one place are core features of Nash's recalibrated postcolonial place name discourse, as she concludes:

> At the same time the different versions of a single placename and the different cultural traditions they reflect can be read as pointers to cultural diversity and dynamism. It is this sense of multicultural traditionalism and pluralistic belonging that link attempts to reimagine culture and location. Identities and cultures can be imagined as simultaneously mobile and located, constantly being reshaped and at the same time specific and situated.[39]

Thus it is in Robinson's mining and recovery of the micro-histories of places and place names that we witness both the enactment and embodiment of a vital ecological consciousness. There is a recalcitrance to such local histories that verges on the subalternated, bearing the hallmarks of subaltern culture, as described by the cultural historian, Joep Leerssen, in his writing on history, remembrance and trauma.[40] For Leerssen, a subaltern culture is characterised by the fact that 'its historical memory remains inchoate, uncanonized, informal, a matter of folklore and local communities rather than of state-sanctioned political life or academic scholarship'.[41] Naturally this cannot mutate into a disabling critical-historical romanticisation of marginal cultural mores, a hypostatisation of so-called 'tradition'. In Leerssen's view, 'community remembrancing' is principally borne by orally transmitted folklore; thus, it is mutable and intimate, and 'persists by traditional renewal, self-repetition and re-enactment'.[42]

Robinson's physical and intellectual excursions into the histories of the Aran Islands and Connemara seek out and disinter precisely these forms of cultural artefacts. His materials and his methods depend upon – indeed exist as – living examples of Leerssen's 'community remembrancing'. The qualification being that Robinson is highly self-conscious about the ways in which he corrals such forms and contents into his own verbal and visual texts. This tension between the orally based elements of Robinson's methodologies and his commitment of these to written text is a fraction of a recurring authorial self-reflexiveness tangible across each of his written volumes. An openness to contingency and contradiction informs Robinson's place name work, and this underlies the re-enchantment and re-presentation of these landscapes in his work. Robinson disinters and rekindles meanings within these localities that are only legible within a particular geographical, linguistic or historical context, but whose exemplarity as ecological case studies is not bounded by their local contexts. There are forceful and enabling political, ethical and cultural lessons to be gleaned from Robinson's remarks on the agency of place name lore:

> Placenames then are the last faded ghosts and echoes of powers and words of power we have let lapse into oblivion ... But in the Irish names, and particularly in the West where the language still lives or at least has not long withdrawn, and the names are still pronounced properly even if not comprehended, we are surrounded by poetic acts as by the flowers of the fields. I always record local opinions of the meanings and origins of placenames, even though some professionals would regard this as naïve credulity.[43]

Place names, then, are rich and legible access points into the intimacies of local histories for Robinson, and they are vital bulwarks against cultural homogeneity and neglect, as well as against political complacency. Such resources are foundational to Robinson's ecological sensibility, a point emphasised again in a recent essay, 'A Land without Shortcuts', where he insists on the necessity and the durability of place names toward the retention of local identities and cultural difference. In his view, 'a placename summarizes the place's attributes and origins, asserts its excellencies and rights to respect. There the handing down and rehearsal of its placename is a place's first defence against neglect or exploitation, against its being regarded as a mere shortcut to some other more profitable place.'[44]

Utility and speed are the targets of Robinson's ecological critique here, as he rails against the framing of landscape as a mere blanked/denuded obstacle to be overcome and/or traded. Language and landscape are not idealised into abstraction by Robinson, but endure in tandem as resistant agents against the amnesia of a one-tracked futurity. This latest ecocritical broadside chimes with much of Robinson's earlier practice and writing on place and place naming, as he urges in 'Listening to the Landscape', 'placenames then are the last faded ghosts and echoes of powers and words of power we have let lapse almost into oblivion'.[45] Landscapes and the histories of their namings are communal creations and inheritances, and as these two related quotations from Robinson attest, they are communal responsibilities. From an ecological perspective, Robinson abjures any sense of culture and nature existing in a dyadic relation. Instead our living histories of engagement with, and dependence upon, landscapes mean that such potentially destructive symmetries are unsustainable for Robinson. Echoing the poet John Montague's figuration of the Irish landscape as a text we have forgotten how to decipher and to read, later in 'Listening to the Landscape' Robinson underscores the link between human and non-human ecology.[46] He highlights his concern with meaning and textuality as both abstract verbal, and deeply somatic, experiences: 'Thus we, personally, cumulatively, communally, create and recreate landscapes – a landscape being not just the terrain but also the human perspectives on it, the land plus its overburden of meanings.'[47] Curiosity and responsibility fuel Robinson's physical and creative projects, and it is precisely these motive forces that underpin the ecological politics of his work. His collection and explication of place name tales and genealogies is not designed as a process of museumisation; it is a series of acts of retrieval and restoration, but also, crucially, re-presentation. In many respects, Robinson's re-presentation of the myriad, fantastic, often contradictory narratives attached to the namings of local places on Árainn Mhór and in Connemara are acts of faith. His project re-presents the depths of local orally transmitted and textually preserved narratives in a textual form, and part of the legacy of these publications is surely the prospect, the hope, that they can inform, but also that they can provoke ecological and historical awareness. In other words, these voluminous histories of local Irish places may well compel readers to rethink, and to revalue, the complexities and contradictions of their own local places, and of the stories and personalities that help to shape, and are shaped, by these landscapes. For Robinson, it is about identifying a place, and identifying with a place, through which an invested mode of meaningful dwelling can accrue, and, once again, it is apparent in his own description of his working methods in *Connemara: Listening to the Wind*, where he emphasises his attentiveness to the micro-details of place:

> A tidemark of tales lies ready for my beachcombing along this way. They are extremely heterogeneous, but every one of them involves contention: between religions, between classes, between animals, between this world and the other … I hear them from people born in the area, and write them down to preserve the little asperities and particularities of the way, without which it is just a road one drives along too fast, commending the scenery in watercolour generalities.[48]

Though much of Robinson's writing is preoccupied with, and sustained by, the 'deep' historical and human historical traces and narratives of Connemara and Árainn Mhór, across his work he evinces a long-term environmental investment in the contemporary well-being of these landscapes. While his career is noteworthy for its recuperative and preservative labours, more recently he refers to Ireland as a land of shortcuts, and it laments a dearth of genuine lived affection for local place in

Ireland – an absence which eased the transactions and lifestyles of the Celtic Tiger property 'boom'.[49] And at junctures in his broader historical surveys, Robinson draws upon his own experiences of such avaricious and destructive transvaluations of land in Connemara. In *Listening to the Wind*, while dwelling upon the agencies of place names and the lingering marks of the mid-nineteenth-century Great Famine, among other topics, Robinson registers a specific concern about the marketisation of rural Irish landscapes, and at the same time, abstracts to state the case for the mutual implication of human and non-human nature. In other words, we witness the co-location of Irish ecological degradation and a broad ecocritical assertion of human responsibility for the environment. Pointing to the perilous alchemy of the property 'boom', Robinson notes that at the beginning of the twenty-first century 'the selling off of Connemara is immensely profitable to landowners and developers, and as a result planning regulations are flagrantly subverted, with the connivance of clientele-dependent politicians. It corrupts our eyes, we see every field as a potential house site, flaunting a price tag instead of its ragged hawthorn tree.'[50]

Robinson's allusion to the legal-political nexus that facilitated the manic consumerism of the Celtic Tiger period needs no development at this point, and, indeed, tensions between local communities along Ireland's western seaboard and exogenous property developers predates the exertions of the Celtic Tiger.[51] But what is provocative in the contemporary context in terms of ecological thought and environmental awareness is Robinson's reference to the fact that our very modes of perception, our temperaments, are remoulded and distorted by the white heat of economic modernisation and financial profiteering. There was, in this view, a radical and retrograde recalibration of how people 'lived' in nature; as we have noted the alchemical faculties of property speculation sundered tracts of the Irish landscape and they became concretised into commodities. But, as Robinson makes plain, the actual physical assault on the landscape is merely a symptom and is not the underlying disease. The pure promise of profit consumes, and urges consumption, and it was such a promise that re-enchanted the Irish landscape as a canvass for excess. One might argue that ecocriticism seeks to re-enchant landscapes that have been evacuated of affective currency, but without retreating to ahistorical nostalgia. As Robinson argues above, a form of corrupted enchantment prevailed, superseding history and myth with short-term presentism, which, of course, came buttressed with its own self-sustaining myths of prosperity. But Robinson's ecological position, which challenges this latter view, is not underwritten by an idealist's view of non-human nature. He does not subscribe to a version of non-human nature as wildly pristine and independent of human influence. Crucially, Robinson is dissatisfied with the neglectful complacencies afforded by a dyadic nature/culture framework.

To conclude, I wish to return to the present moment, and remain with Robinson. His work is not only exemplary in its historical ecological consciousness, but he is also acutely aware of the ongoing temperamental and attitudinal dangers to his local landscape and to the planet. Returning to the essay 'A Land without Shortcuts', Robinson bemoans the increasing lack of affective investment in the Irish landscape. Again, intersecting with extant critiques of the contemporary cult of speed, which diminishes place to traversable space, Robinson argues that this historical period trades 'space for time; technology is shortcuts, ever bridging the gap between intent and fulfilment. And, technology having hurried us into the present crux of global warming, it now offers to deliver us from it, in return for the ground beneath our feet.'[52] The immediate context here of Robinson's protests is his scepticism about many new technological panaceas for global climate change. Yet, what is more relevant, for the moment, is the devaluation and ignorance of place that these equally destructive technologies create and thrive upon. Speed conditions much of what passes for daily life in the developed world, the global North. Speed and desire are complicit in the relegation of place to commoditised space.[53] Rather than reflect upon, inquire about, or, ultimately, begin to understand and value, place, our attention spans have been disabled so that velocity stands as an indissoluble virtue of what it is to be 'modern' in the developed world and beyond. And even within Irish society, while we persistently protest and retain our historical consciousness of our language, landscape and culture, in Robinson's view, we have traded in these values. Contrarily, we retain a hypocritical and superficial interest in the toponymic (the study of place names) fabric of our country, one that has mutated in what he terms 'The Old Woman of the Four Green-Field Sites'.[54]

The degradation of the Irish landscape and its ecologies is implicit in Robinson's argument here – a matter of direct political activism we will summarily address below. But the more pressing issue is that if Irish people simply feign commitment to place and its naming, then the value of land and place is calculated in other terms and scales. The obvious consequence for Robinson, then, is that if place is valued according to the currency of the market, of the property exchange and access to raw materials, then all localities are evacuated of historical depth and difference – in this regard speed and homogeneity become complicit agents. Furthermore, the kinds of superficial attachments attributed to Irish people by Robinson are indicative of a society that has abdicated any sense of duty or of responsibility to their landscape of places, and the histories embedded within, and which emanate from, those topographies. In the end, this recent intervention by Robinson performs a couple of differential but aligned functions. In the first instance, as we have outlined, Robinson indicts a culture that chooses to by-pass the complexities of its proximate geographies, abandoning them to the logic of market capitalism and short-term economic opportunism. And secondly, this latest invocation of the worth attached to the study and comprehension of place, and most especially place names, acts as a justification for his life's work, as well as a reminder that this life's work is available to us yet as a potent resource for ecocritical ripostes to 'the age of the shortcut'.

Within Irish society, while we persistently protest our historical consciousness of the currency of our language and landscapes, in Robinson's experienced view, these have been pawned in the past decade and a half. Contrarily, we retain a disingenuous interest in the toponymic fabric of our country. But his more serious assessment is that 'this Irish fascination with placenames is too often purely nominal and does not extend to caring for the places themselves; nostalgia stands in for conservation and convenience trumps all'.[55] In this one national context, in these local coordinates, we see visual and verbal mapping, and, I would suggest, a resonant ecocritical voice is registered. To reprise Ryden, with whom we began our consideration of Robinson's work, 'Through its documentary and preservationist impulse, the essay of place stands as a brave and necessary attempt to stem this erosion, an effort to preserve much of what gives life value.'[56] Ryden's précis speaks to the combined postcolonial and ecocritical politics of Robinson's writing, which, as we have seen, weighs 'value' according to less materialistic calculus. Finally, across Robinson's work and methods we are witness to the variety of spatial scales – fractal, local, national, global – and we see the historical scales through which we think as human historical actors, in addition to the deep historical scales of our evolutionary development, and of the planet's geological formation. In these cursory ways, his career and oeuvre are exemplary of the kind of scalar consciousness that has begun to inform historical and cultural debate in the last decade, and that must inform the ways in which we map ourselves in the shadow of the global climate crisis.

## Notes

1 Kent C. Ryden, *Mapping the Invisible Landscape: Folklore, Writing and the Sense of Place* (Iowa City: University of Iowa, 1993), 248.
2 Ryden, *Mapping*, 251.
3 Glen A. Love writes, 'Ecocriticism urges its practitioners into interdisciplinarity, into science'. See 'Ecocriticism and Science: Toward Consilience?', *New Literary History* 30:3 (1999): 561.
4 Ryden, *Mapping*, 254–5. On debates within contemporary ecocriticism concerning literary realism and 'authenticity', see Lawrence Buell, *The Environmental Imagination: Thoreau, Nature Writing and the Formation of American Culture* (Cambridge, MA: Belknap Press, 1995); Timothy Morton, *Ecology Without Nature: Rethinking Environmental Aesthetics* (Cambridge, MA: Harvard University Press, 2007); and Dana Phillips, *The Truth of Ecology: Nature, Culture and Literature in America* (Oxford: Oxford University Press, 2003).
5 On Ireland, colonialism and ecocriticism, see my essay 'Ireland of the Welcomes: Colonialism, Tourism and the Irish Landscape', in Christine Cusick (ed.), *Out of the Earth: Ecocritical Readings of Irish Texts* (Cork: Cork University Press, 2010), 85–107. Three recent publications in this emerging, but still limited, field of Irish ecocritical studies include: Christine Cusick (ed.), *Out of the Earth: Ecocritical Readings of Irish Texts*; Eamonn Wall, *Writing the Irish West: Ecologies and Traditions* (Notre Dame, IN: University of Notre Dame Press, 2011); and Robert Brazeau and Derek Gladwin (eds), *Eco-Joyce: The Environmental Imagination of James Joyce* (Cork: Cork University Press, 2014). In the field of Irish economic history, see two publications by Eamonn Slater and Terence McDonough: 'Marx on Nineteenth-Century Ireland: Analyzing

Colonialism as a Dynamic Social Process', *Irish Historical Studies* 36:142 (November 2008): 153–72; and 'Bulwark of Landlordism and Capitalism: The Dynamics of Feudalism in Nineteenth Century Ireland', *Research in Political Economy* 14 (1994): 63–118.
6   Tim Robinson, *Setting Foot on the Shores of Connemara and Other Writings* (Dublin: Lilliput Press, 1996), 164.
7   Seamus Deane, 'Ultimate Place: Review of Tim Robinson's *Stones of Aran: Pilgrimage*', *The London Review of Books* 11:6 (March 1989): 10.
8   As is well established, Deane's critical work was pivotal to the emergence of an Irish postcolonial criticism. See especially *Strange Country: Modernity and Nationhood in Irish Writing since 1790* (Oxford: Clarendon Press, 1997).
9   Tim Robinson, *Stones of Aran: Pilgrimage* (Dublin: Lilliput Press, 1986), 11.
10  Brian Dillon, 'An Interview with Tim Robinson', *The Field Day Review* 3 (2007): 41.
11  On 'deep' time history and ecological criticism, see Eóin Flannery, '"The Buried Life": Ecological Criticism and the "Deep" Past', *Critical Quarterly* 54:4 (2012): 93–109.
12  Tim Robinson, *Connemara: Listening to the Wind* (Dublin: Penguin, 2006), 4.
13  Dillon, 'Tim Robinson', 34.
14  Kevin Whelan, 'Reading the Ruins: The Presence of Absence in the Irish Landscape', in Howard B. Clarke, Jacinta Prunty and Mark Hennessy (eds), *Surveying Ireland's Past: Multidisciplinary Essays in Honour of Anngret Simms* (Dublin: Geography Publications 2004), 300.
15  Edward W. Said, 'Invention, Memory and Place', *Critical Inquiry* 26 (Winter 2000): 179.
16  Said, 'Invention, Memory and Place', 180.
17  See Paul Carter, *The Road to Botany Bay: An Essay in Spatial History* (London: Faber and Faber, 1987) and *The Lie of the Land* (London: Faber and Faber, 1996).
18  Dillon, 'Tim Robinson', 34.
19  A similarly titled, and recent, volume is Alex Hunt and Bonnie Roos (eds), *Postcolonial Green: Environmental Politics and World Narratives* (Charlottesville: University of Virginia Press, 2010).
20  Elizabeth DeLoughrey and George B. Handley, 'Introduction: Towards an Aesthetics of the Earth', in Elizabeth DeLoughrey and George B. Handley (eds), *Postcolonial Ecologies: Literatures of the Environment* (Oxford: Oxford University Press, 2011), 4.
21  DeLoughrey and Handley, 'Introduction', 4.
22  See, for example, John Bellamy Foster, *Marx's Ecology: Materialism and Nature* (New York: Monthly Review Press, 2000).
23  See Jason Moore, 'Ecology, Capital, and the Nature of Our Times: Accumulation and Crisis in the Capitalist World-Ecology', *Journal of World-Systems Research* 17:1 (2011): 108–47, and 'The Modern World-System as Environmental History? Ecology and the Rise of Capitalism', *Theory & Society* 32:3 (2003): 307–77. See also John Bellamy Foster, *Ecology Against Capitalism* (New York: Monthly Review Press, 2002) and *Marx's Ecology*.
24  DeLoughrey and Handley, 'Introduction', 10.
25  For a recent intervention in Irish ecocritical studies that emphasises the agency of the local, see Michael Cronin, *The Expanding World: Towards a Politics of Microspection* (London: Zero Books, 2012).
26  DeLoughrey and Handley, 'Introduction', 4.
27  Robinson, *Listening*, 229.
28  Robinson, *Listening*, 229.
29  Tim Robinson, 'A Land without Shortcuts', *The Dublin Review* 46 (Spring 2012): 44.
30  Writing on the centrality of the Irish language to his work and to the landscapes he has surveyed, Robinson writes, 'The landscape here [Arann] speaks Irish. The cliffs,

rocks, fields and paths are named in words nearly all still alive, descriptive and rich in memories for the islanders themselves, Aran being one of the strongholds of the spoken language.' *Arann: A Companion to the Map of the Aran Islands* (Roundstone, Co. Galway: Folding Landscapes: 1996), 19.

31 Catherine Nash, 'Irish Placenames: Post-colonial Locations', *Transactions of the Institute of British Geographers* 24:4 (1999): 457.
32 Nash, 'Irish Placenames', 457.
33 Nash, 'Irish Placenames', 457.
34 Nash, 'Irish Placenames', 457.
35 Nash, 'Irish Placenames', 463.
36 Nash, 'Irish Placenames', 457.
37 Cronin's *The Expanding World* is also instructive on these ideas.
38 Nash, 'Irish Placenames', 476.
39 Nash, 'Irish Placenames', 476.
40 On subalterneity and Irish postcolonial criticism, see David Lloyd, *Ireland After History* (Cork: Cork University Press, 1999).
41 Joep Leerssen, 'Monument and Trauma: Varieties of Remembrance', in Ian McBride (ed.), *History and Memory in Modern Ireland* (Cambridge: Cambridge University Press, 2001), 214.
42 Leerssen, 'Monument and Trauma', 215.
43 Robinson, 'Listening to the Landscape', 161.
44 Robinson, 'A Land without Shortcuts', 42.
45 Robinson, 'Listening to the Landscape', 161.
46 Of interest here is John Montague's essay, 'Notes and Introductions', in *The Figure in the Cave and Other Essays* (Dublin: Lilliput Press, 1989), 42–57.
47 Robinson, 'Listening to the Landscape', 162.
48 Robinson, *Listening*, 277.
49 The criticism of (post-) Celtic Tiger Ireland is an industry in itself; some enabling texts include: Debbie Ging, Michael Cronin and Peadar Kirby (eds), *Transforming Ireland: Challenges, Critiques and Resources* (Manchester: Manchester University Press, 2009); Fintan O'Toole, *Ship of Fools: How Stupidity and Corruption Sank the Celtic Tiger* (London: Faber and Faber, 2009); and Peadar Kirby, *The Celtic Tiger in Collapse: Explaining the Weaknesses of the Irish Model* (Basingstoke: Palgrave/Macmillan, 2010).
50 Robinson, *Listening*, 110.
51 Robinson writes about his role in a campaign to prevent the construction of an airstrip across the archaeologically significant site at Roundstone Bog in *My Time in Space* (Dublin: Lilliput Press, 2001).
52 Robinson, 'A Land without Shortcuts', 41.
53 On speed and Irish culture, see Michael Cronin, 'Inside Out: Time and Place in Global Ireland', in Eamon Maher (ed.), *Cultural Perspectives on Globalisation and Ireland* (Bern: Peter Lang, 2009), 11–30, and Cronin, 'Speed Limits: Ireland, Globalisation and the War against Time', in Peadar Kirby, Luke Gibbons and Michael Cronin (eds), *Reinventing Ireland: Culture, Society and the Global Economy* (London: Pluto Press, 2002), 54–66. More recently, see Eóin Flannery, 'Ecology, Memory and Speed in John McGahern's Memoir', *Irish University Review* 42:2 (2012): 273–97.
54 Robinson, 'A Land without Shortcuts', 41.
55 Robinson, 'A Land without Shortcuts', 40.
56 Ryden, *Mapping*, 253.

# Epilogue: On the rocks road

*Andrew McNeillie*

*Figure 27* Bothar na gCreag (photo by Andrew McNeillie).

*There . . .*

Preserve us I say
From narrow-gauge minds
But not narrow roads
With green spines
Where the heart's affections

Put best foot forward:
Between two stone walls
Built to the rhythm
Of rock-form and contour
Of labour and time
Straight as a die
Dipping in and out of sight
Opening and narrowing
Ahead behind – behind ahead
In threadbare karst country
Grown where nothing grows
Better than light and lichen
Rare alpine, common thorn,
Atlantic gale and storm
Limestone, stone by stone
Advancing to delay
To the last angle and oval
With makeshift-erratic
Punctuation of granite
Relief work in stark relief –
As now at home recalling
I step up from Cill Rónáin
Over the top and down
To Gort na gCapall (a.k.a. West Cork)
The field of the horse
On my solitary walk
Unpicking as I go
The old formula:
Distance over speed and time –
Beyond recognition
In my mind-body economy
Of presences and memory
In and out of step
Balancing line on line
Not carelessly picked
Or casually piled
But as those men worked
With steady eye and hand.
But hold your step
As the Rocks Road
Has ever done
Since it began
Never to travel its own length
Nor time its progress
Never to see but to be
From end to end

Its centre of gravity where
But here and there?
In infinite recession
Beyond the sum of knowledge
Beyond botany book
And guide to birds
Studies of fossils
Or place-names
In the folds of a map
And where the wild rose
Blows in nothing's name.
What do I bring?
Nothing it knows.
What do I take?
Nothing it will miss.
Where am I bound?
To the field of the horse.

*And back . . .*

As now I am again
Stepping out of time
Working my way in
From the port
Of the fort's mouth
Through a giant jetsam
Of rock and boulder
Beside which sea-wall
The village of the horse takes shelter
Itself an island in a sea of stone
Safe as houses but no safer
As a FOR SALE sign tells
Wired to a garden gate
The rogee Time at work
Waiting on the highest bidder
No safer than mortality:
The O'Flaherty home
Of radical fame.
A story to be told there
But to the stranger more
One of stage directions
With pauses and no text between
Where no one seems to live
But when the little bus comes round
A woman disembarks

With shopping bags from 'SUPERMAC'S'
To disappear indoors
And at evening an old man
Emerges to Flymo the lawn
And two brothers further on
Look up startled as
I take them by surprise
Calling as I go: 'Fine evening!'
And at once bend back
To hack at briars
As if lifetimes ago
Drifting into silence behind me
In deepening shadow
As the day fades
And evening sharpens
To monochrome
Then dims to a glimmer of lights
Beside the sea's fire
And the nightwatch begins
Behind drawn curtains
Via dish and aerial
As I start the steep climb
Back the way I came
Out of sight out of mind
Out of mind into vision
And I pass myself
Coming back in silence
The way not the same
My step different
Though not my passion
For the Rocks Road.
Do not ask life's purpose
But live every step of the road
(The time it takes knows
Nothing of distance or speed)
In the world of matter
And mind brought together
As in making's invention
I write this for you:
Call it ordinariness
Call it best of all love
As the walls ahead touch
At their vanishing point
And keep opening.

# Bibliography

Aalen, F. H. A., Kevin Whelan and Matthew Stout (eds). *Atlas of the Irish Rural Landscape*, 1st edn. Toronto: University of Toronto Press, 1997.
Abbey, Edward. *Desert Solitaire* [1968]. New York: Touchstone, 1990.
Adorno, Theodor. *Minima Moralia: Reflections from a Damaged Life*. Translated by E. F. N. Jephcott. London: Verso, 1974.
Agamben, Giorgio. *The Idea of Prose*. Translated by M. Sullivan and S. Whitsitt. Albany: State University of New York Press, 1995.
Andrews, J. H. *A Paper Landscape: The Ordnance Survey in Nineteenth-Century Ireland* [1975]. Dublin: Four Courts, 2001.
Andrews, J. H. 'Paper Landscapes: Mapping Ireland's Physical Geography'. In *Nature in Ireland: A Scientific and Cultural History*. Edited by John Wilson Foster. Dublin: Lilliput Press, 1997, 199–218.
Andrews, J. H. *Plantation Acres: An Historical Study of the Irish Land Surveyor and His Maps*. Belfast: Ulster Historical Foundation, 1985.
Antonioni, Michaelangelo. *That Bowling Alley on the Tiber: Tales of a Director*. Translated by William Arrowsmith. New York: Oxford University Press, 1987.
Arthur, Chris. '(En)Trance'. In *Best American Essays 2008*. Edited by Mary Oliver. New York: Houghton Mifflin, 2009, 3–16.
Arthur, Chris. *Irish Elegies*. New York: Palgrave Macmillan, 2009.
Arthur, Chris. *Irish Willow*. Aurora, CO: Davies Group, 2002.
Arthur, Chris. *On the Shoreline of Knowledge: Irish Wanderings*. Iowa City, IA: University of Iowa Press, 2012.
Atkins, G. Douglas. *Reading Essays*. Athens, GA: University of Georgia Press, 2008.
Atwan, Robert. 'Notes Towards the Definition of an Essay'. *River Teeth* 14:1 (2012): 109–17.
Baber, Benjamin. *The Death of Communal Liberty: A History of Freedom in a Swiss Mountain Canton*. Princeton: Princeton University Press, 1974.
Babine, Karen. '"All the Sky Were Paper and All the Sea Were Ink": Tim Robinson's Linguistic Ecology'. *New Hibernia Review* 15:4 (Geimhreadh/Winter 2011): 95–110.
Bachelard, Gaston. *The Poetics of Space* [1958]. Translated by Maria Jolas. Boston: Beacon Press, 1994.
Bakewell, Sarah. *How to Live, or A Life of Montaigne in One Question and Twenty Attempts at an Answer*. New York: Other Press, 2011.
Bellamy Foster, John. *Ecology Against Capitalism*. New York: Monthly Review Press, 2002.

Bellamy Foster, John. *Marx's Ecology: Materialism and Nature*. New York: Monthly Review Press, 2000.
Belshaw, Christopher. *Environmental Philosophy: Reason, Nature and Human Concern*. Chesham, Bucks: Acumen, 2001.
Berger, John. *Pig Earth*. New York: Pantheon, 1980.
Berger, John. *Ways of Seeing*. London: Penguin Books, 1990.
Biggs, Iain. '"Deep Mapping": A Brief Introduction'. In *Mapping Spectral Traces 2010, Exhibition Catalogue*. Edited by Karen E. Till. University of Virginia Tech College of Architecture and Urban Studies, 5–8.
Biggs, Iain. 'The Spaces of "Deep Mapping": A Partial Account'. *Journal of Arts and Communities* 2:1 (2011): 5–25.
Blau DuPlessis, Rache. *The Pink Guitar: Writing as Feminist Practice*. Tuscaloosa: University of Alabama Press, 2006.
Brazeau, Robert and Derek Gladwin (eds). *Eco-Joyce: The Environmental Imagination of James Joyce*. Cork: Cork University Press, 2014.
Buell, Lawrence. *The Environmental Imagination: Thoreau, Nature Writing and the Formation of American Culture*. Cambridge, MA: Belknap Press, 1995.
Buell, Lawrence. *The Future of Environmental Criticism: Environmental Crisis and Literary Imagination*. Oxford: Wiley Blackwell, 2005.
Butler, Judith. 'Bracha's Eurydice'. *Theory Culture Society* 21:1 (2004): 95–100.
Carter, Paul. *The Lie of the Land*. London: Faber and Faber, 1996.
Carter, Paul. *The Road to Botany Bay: An Essay in Spatial History*. London: Faber and Faber, 1987.
Casey, Edward S. *The Fate of Place: A Philosophical History*. Los Angeles and London: University of California Press, 1997.
Casey, Edward S. *Getting Back into Place: Toward a Renewed Understanding of the Place-World* [1993]. Bloomington, IN: Indiana University Press, 2009.
Chatwin, Bruce. *In Patagonia*. New York: Penguin, 1978.
Conley, Tom. *Cartographic Cinema*. Minneapolis: University of Minnesota Press, 2007.
Connolly, S. J. *The Oxford Companion to Irish History*. Oxford: Oxford University Press, 1998.
Corlett, Christiaan and John Medlycott (eds). *The Ordnance Survey Letters: Wicklow*. Roundwood, IR: Roundwood and District Historical and Folklore Society, 2000.
Corlett, Christiaan and Mairéad Weaver (eds). *The Liam Price Notebooks: The Placenames, Antiquities, and Topography of County Wicklow*, Vol. 1. Dublin: Dúchas, 2002.
Cresswell, Tim. *In Place/Out of Place: Geography, Ideology and Transgression*. Minneapolis: University of Minnesota Press, 1996.
Cresswell, Tim. 'Landscape and the Obliteration of Practice'. In *Handbook of Cultural Geography*. Edited by Kay Anderson, Mona Domosh, Steve Pile and Nigel Thrift. London: Sage, 2003, 269–81.
Cronin, Michael. *The Expanding World: Towards a Politics of Microspection*. London: Zero Books, 2012.
Cronin, Michael. 'Inside Out: Time and Place in Global Ireland'. In *Cultural Perspectives on Globalisation and Ireland*. Edited by Eamon Maher. Bern: Peter Lang, 2009, 11–30.
Cronin, Michael. 'Inside Out: Time and Place in Global Ireland'. *New Hibernia Review* 13: 3 (2009): 74–88.
Cronin, Michael. 'Speed Limits: Ireland, Globalisation and the War against Time'. In *Reinventing Ireland: Culture, Society and the Global Economy*. Edited by Peadar Kirby, Luke Gibbons and Michael Cronin. London: Pluto Press, 2002, 54–66.
Cronin, Michael. *Translating Ireland: Translation, Languages, Cultures*. Cork: Cork University Press, 1996.

Cronin, Nessa. '"Disciplinary Ghettoes": Irish Studies and Interdisciplinary Negotiations'. *Journal of Nordic Irish Studies* 6 (2007): 1–16.

Cronin, Nessa. 'Lived and Learned Landscapes: Literary Geographies and the Irish Topographical Tradition'. In *Irish Contemporary Landscapes in Literature and the Arts*. Edited by Marie Mianowski. London: Palgrave Macmillan, 2011, 106–19.

Crowley, Caroline and Denis Linehan (eds). *Spacing Ireland: Place, Society and Culture in a Post-Boom Era* (Manchester: Manchester University Press, 2013).

Cusick, Christine, 'Mapping Placelore: Tim Robinson's Ambulation and Articulation of Connemara as Bioregion'. In *The Bioregional Imagination: Literature, Ecology, and Place*. Edited by Tom Lynch and Cheryll Glotfelty. Athens, GA: University of Georgia Press, 2012, 135–49.

Cusick, Christine, 'Mindful Paths: An Interview with Tim Robinson'. In *Out of the Earth: Ecocritical Readings of Irish Texts*. Edited by Christine Cusick. Cork: Cork University Press, 2010, 205–11.

Cusick, Christine, (ed.). *Out of the Earth: Ecocritical Readings of Irish Texts*. Cork: Cork University Press, 2010.

Deane, Seamus. *Strange Country: Modernity and Nationhood in Irish Writing since 1790*. Oxford: Clarendon Press, 1997.

Deane, Seamus. 'Ultimate Place: Review of Tim Robinson's *Stones of Aran: Pilgrimage*'. *The London Review of Books* 11:6 (16 March 1989): 9–10.

Dee, Tim. 'Nature Writing'. *Archipelago* 5 (Spring 2011): 20–30.

DeLoughrey, Elizabeth and George B. Handley (eds). *Postcolonial Ecologies: Literatures of the Environment*. Oxford: Oxford University Press, 2011.

Derrida, Jacques. *Of Grammatology* [1967]. Translated by Gayatri Chakravorty Spivak. Baltimore: Johns Hopkins Press, 1976.

DeSilvey Caitlin. 'Observed Decay: Telling Stories with Mutable Things'. *Journal of Material Culture* 11:3 (November 2006): 318–38.

Dhanens, Elizabeth. *Hubert and Jan Van Eyck*. Antwerp: Art Books Int. Ltd., 1980.

Dillon, Brian. 'An Interview with Tim Robinson'. *The Field Day Review* 3 (2007): 33–41.

Doherty, Gillian M. *The Irish Ordnance Survey: History, Culture and Memory*. Dublin: Four Courts Press, 2004.

Drever, Timothy. 'Field Work 3: A Structured Arena'. *Bulletin of the Computer Arts Society* (February 1971): n.p.

Drever, Timothy and Peter Joseph. 'Outside the Gallery System: Two Projects for Kenwood'. *Studio International* (June 1969): 255.

Eagleton, Terry. *Literary Theory* [1983]. Oxford: Blackwell, 1988.

Ellis, Jack C. and Betsy A. McLane. *A New History of Documentary Film*. New York: Continuum, 2005.

Emerson, Ralph Waldo. *Nature* [1836]. Project Gutenberg, 2009. Accessed 4 July 2014, www.gutenberg.org/ebooks/29433.

Engelbert, Lynn Hughey. 'Antonin Artaud and Freud's "Family Romance": The Transgressive Sublime'. *Mosaic: A Journal for the Interdisciplinary Study of Literature* 47:2 (June 2014): 17–32.

Ettinger, Bracha Lichenberg. 'Matrix and Metramorphosis'. *Differences* 4 (1992): 176–208.

Ettinger, Bracha Lichenberg. *The Matrixial Borderspace*. Minneapolis: University of Minnesota Press, 2006.

Ettinger, Bracha Lichenberg. *The Matrixial Gaze*. Leeds: Feminist Arts and Histories Network, 1995.

Ettinger, Bracha Lichenberg. 'Weaving a Woman Artist with-in the Matrixial Encounter-Event'. *Theory Culture Society 2004* 21:1 (2004): 69–94.

Ettinger, Bracha Lichenberg. 'Wit(h)nessing Trauma and the Matrixial Gaze: From Phantasm to Trauma, from Phallic Structure to Matrixial Sphere'. *Parallax* 7:4 (2001): 89–114.

Evans, E. Estyn. *The Personality of Ireland: Habitat, Heritage and History*. Cambridge: Cambridge University Press, 1973.

Evernden, Neil. 'Beyond Ecology: Self, Place and the Pathetic Fallacy'. In *The Ecocriticism Reader: Landmarks in Literary Ecology*. Edited by Cheryl Glotfelty and Harold Fromm. Athens: University of Georgia Press, 1996, 92–104.

Fakundiny, Lydia. *The Art of the Essay*. New York: Houghton Mifflin, 1991.

Fennell, Desmond. *Beyond Nationalism: The Struggle Against Provinciality in the Modern World*. Dublin: Ward River Press, 1985.

Fennell, Desmond. 'Connacht's View of Itself'. *Iarchonnacht Began* [1969]. Cill Chiaráin: Iarchonnachta 1985, 37.

Fennell, Desmond. 'Cosaint Fhennell ar a Theoric "Iosrael in Iarchonnachta"'. *Inniu* (5 June 1970): 1–2.

Fennell, Desmond. 'Iosrael in Iarchonnachta'. *Iarchonnacht Began* [1969]. Cill Chiaráin: Iarchonnachta 1985, 11–13.

Fennell, Desmond. 'Language Revival: Is it Already a Lost Cause?'. *Irish Times* (29 January 1969): 8.

Féve, Nicolas and Tim Robinson. *Connemara and Elsewhere*. Dublin: Royal Irish Academy, 2014.

Flannery, Eóin. '"The Buried Life": Ecological Criticism and the "Deep" Past'. *Critical Quarterly* 54:4 (2012): 93–109.

Flannery, Eóin. 'Ecology, Memory and Speed in John McGahern's Memoir'. *Irish University Review* 42:2 (2012): 273–97.

Flannery, Eóin. 'Ireland and Ecocriticism: An Introduction'. *Journal of Ecocriticism* 5:2 (July 2013): 1–8.

Flannery, Eóin. 'Ireland of the Welcomes: Colonialism, Tourism and the Irish Landscape'. In *Out of the Earth: Ecocritical Readings of Irish Texts*. Edited by Christine Cusick. Cork: Cork University Press, 2010, 85–107.

Foster, John Wilson. *Between Shadows: Modern Irish Writing and Culture*. Dublin: Irish Academic Press, 2009.

Foster, John Wilson. 'Tim Robinson's Variegated World: *My Time in Space* by Tim Robinson; Tales and Imaginings'. *The Irish Review* 30 (Spring–Summer 2003): 105–13.

Fox, Warwick. 'Deep Ecology: A New Philosophy of Our Time' [1984]. In *Environmental Ethics*. Edited by Andrew Light and Holmes Rolston III. Oxford: Blackwell, 2003, 252–61.

Friedman, Susan Stanford. *Mappings: Feminism and the Cultural Geographies of Encounter*. Princeton: Princeton University Press, 1998.

Gandy, Matthew. 'Landscapes of Deliquescence in Michelangelo Antonioni's *Red Desert*'. *Transactions of the Institute of British Geographers* 28:2 (2003): 218–37.

Geertz, Clifford. 'Thick Description: Toward an Interpretive Theory of Culture'. In *The Interpretation of Cultures: Selected Essays*. Edited by Clifford Gertz. New York: Basic Books, 1973, 3–30.

Gibbons, Luke. 'Peripheral Visions: Revisiting Irish Modernism'. In *The Moderns*. Edited by Enrique Juncosa and Christine Kennedy. Dublin: Irish Museum of Modern Art, 2011, 88–101.

Gifford, Terry. *Reconnecting with John Muir: Essays in Post-Pastoral Practice*. Athens: University of Georgia Press, 2006.

Gillen, Shawn. 'Synge's *The Aran Islands* and Irish Creative Nonfiction'. *New Hibernia Review* 11: 4 (2007): 129–35.

Ging, Debbie, Michael Cronin and Peadar Kirby (eds). *Transforming Ireland: Challenges, Critiques and Resources*. Manchester: Manchester University Press, 2009.

Gladwin, Derek. 'The Literary Cartographic Impulse: Imagined Island Topographies in Ireland and Newfoundland'. Spec. issue 'Text and Beyond Text: New Visual, Material, and Spatial Perspectives in Irish Studies', *Canadian Journal of Irish Studies* 38: 1–2 (2014): 158–83.

Gladwin, Derek. 'Navigating Irish Ecocriticism: Eamonn Wall's *Writing the Irish West: Ecologies and Traditions*'. *Irish Studies Review* 20:3 (2012): 323–8.

Gombrich, Ernst Hans. *Norm and Form: Studies in the Art of the Renaissance*. London: Phaidon Press, 1966.

Gould, Stephen Jay. 'Introduction'. In *Best American Essays 2002*. Edited by Stephen Jay Gould. New York: Houghton Mifflin, 2003, xiii–xx.

Gruchow, Paul. *Grass Roots: The Universe of Home*. Minneapolis, MN: Milkweed Editions, 1995.

Hand, Derek. *The History of the Irish Novel*. Cambridge: Cambridge University Press, 2011.

Hazlitt, William. 'Mr. Wordsworth'. In *The Spirit of the Age* [1825]. Whitefish, MT: Kessinger Publishing LLC, 2004.

Heidegger, Martin. *Being and Time* [1927]. Translated by John Macquarrie and Edward Robinson. Oxford: Blackwell, 2008.

Heidegger, Martin. 'Building Dwelling Thinking' [1954]. In *Poetry, Language, Thought*. Translated by Albert Hofstader. New York: Perennial Library, 1971, 145–61.

Heidegger, Martin. 'Language' [1959]. In *Poetry, Language, Thought*. Translated by Albert Hofstader. New York: Perennial Library, 1971, 187–210.

Heidegger, Martin. 'An Ontological Consideration of Place'. In *The Question of Being*. Translated by William Kluback and Jean T. Wilde. London: Vision, 1956, 18–26.

Heidegger, Martin. 'The Thing' [1951]. In *Poetry, Language, Thought*. Translated by Albert Hofstader. New York: Perennial Library, 1971, 163–82.

Heidegger, Martin. 'What Are Poets For?' [1950a]. In *Poetry, Language, Thought*. Translated by Albert Hofstader. New York: Perennial Library, 1971, 89–142.

Heidegger, Martin. 'What Calls for Thinking?' [1971]. In *Martin Heidegger: Basic Writings*. Edited by David Farrell Krell. London: Routledge & Kegan Paul, 1978, 341–68.

Henry, Paul. *An Irish Portrait*. London: B.T. Batsford, 1951.

Herity, Michael (ed.). *Meath*. Dublin: Four Masters Press, 2001.

Holley, Karri A. *Understanding Interdisciplinary Challenges and Opportunities in Higher Education*. ASHE Higher Education Report 35:2 (2002).

hooks, bell. *Bone Black: Memoirs of a Girlhood*. New York: Holt, 1997.

Huggan, Graham. *Territorial Disputes: Maps and Mapping Strategies in Contemporary Canadian and Australian Fiction*. Toronto: University of Toronto Press, 1994.

Hunt, Alex and Bonnie Roos (eds). *Postcolonial Green: Environmental Politics and World Narratives*. Charlottesville: University of Virginia Press, 2010.

Hyde, Douglas. 'The Necessity for De-Anglicising Ireland'. In *Irish Literature: A Reader*. Edited by Maureen O'Rourke Murphy and James MacKillop. Syracuse: Syracuse University Press, 1987, 137–47.

Ihde, Don. *Listening and Voice: A Phenomenology of Sound*. Albany: State University of New York Press, 2007.

Irish Museum of Modern Art (IMMA). 'Event Horizon Season at the Irish Museum of Modern Art'. Accessed 10 December 2013, www.imma.ie/en/page_19249.htm.

Jameson, Fredric. 'Cognitive Mapping'. In *Marxism and the Interpretation of Culture*. Edited by Cary Nelson and Lawrence Grossberg. London: Macmillan, 1988, 347–57.

Jay, Martin. 'In the Empire of the Gaze: Foucault and the Denigration of Vision in Twentieth-Century French Thought'. In *Foucault: A Critical Reader*. Edited by David Couzens Hoy. Oxford: Blackwell, 1986, 175–204.

Kavanagh, Peter (ed.), *The Complete Poems of Patrick Kavanagh*. New York: Peter Kavanagh Hand Press, 1972.

Kennedy, S. B. *Paul Henry: Paintings, Drawings, and Illustrations*. New Haven, CT: Yale University Press, 2007.

Keogh, Jackie. 'Filmmaker Shows Confidence in the Small'. *The Southern Star* (11 October 2012): n.p. Accessed 8 June 2013, www.southernstar.ie/Home/Filmmaker-shows-confidence-in-the-small-11102012.htm.

Kirby, Peadar. *The Celtic Tiger in Collapse: Explaining the Weaknesses of the Irish Model*. Basingstoke: Palgrave/Macmillan, 2010.

Least Heat-Moon, William. *PrairyErth*. Boston: Houghton Mifflin, 1991.

Leerssen, Joep. 'Monument and Trauma: Varieties of Remembrance'. In *History and Memory in Modern Ireland*. Edited by Ian McBride. Cambridge: Cambridge University Press, 2001, 204–22.

Lloyd, David. *Ireland After History*. Cork: Cork University Press, 1999.

Lodge, David. *Changing Places*. New York: Penguin, 1975.

Lodge, David. *Nice Work*. New York: Penguin, 1990.

Lodge, David. *Small World*. New York: Penguin, 1995.

Lopate, Phillip. *The Art of the Personal Essay*. New York: Anchor, 1995.

Lopate, Phillip. 'The Essay, an Exercise in Doubt'. *The New York Times* (16 February 2013).

Lopez, Barry. *Vintage Lopez*. New York: Vintage Books, 2004.

Love, Glen A. 'Ecocriticism and Science: Toward Consilience?'. *New Literary History* 30:3 (1999): 561–76.

Lysaght, Seán. *Robert Lloyd Praeger: The Life of a Naturalist*. Dublin: Four Courts, 1998.

McCarthy, Caroline. *Greetings*. Dublin: Irish Museum of Modern Art, 1996.

Macfarlane, Robert. *Mountains of the Mind: A History of a Fascination*. London: Vintage, 2004.

Macfarlane, Robert. *The Old Ways: A Journey on Foot*. New York: Penguin, 2012.

Macfarlane, Robert. 'Rock of Ages'. *The Guardian* (14 May 2005). Accessed 9 September 2014, www.theguardian.com/books/2005/may/14/featuresreviews.guardianreview34.

Macfarlane, Robert. *The Wild Places*. New York: Penguin, 2008.

Mac Graith, Uinsíonn and Treasa Ní Ghearraigh. *The Placenames and Heritage of Dún Chaocháin*. Béal an Atha: Comhar Dún Chaocháin Teo, 2004.

McGregor, Robert Kuhn. *A Wider View of the Universe: Henry Thoreau's Study of Nature*. Urbana and Chicago: University of Illinois Press, 1997.

McLucas, Clifford. 'There are ten things that I can say about these deep maps'. *Deep Mapping*. Accessed 1 December 2013, http://cliffordmclucas.info/deep-mapping.html.

Manes, Christopher. 'Nature and Silence'. In *The Ecocriticism Reader: Landmarks in Literary Ecology*. Edited by Cheryll Glotfelty and Harold Fromm. Athens, GA: University of Georgia Press, 1996, 15–29.

Marshall, Ian. *Story Line: Exploring the Literature of the Appalachian Trail*. Charlottesville: University of Virginia Press, 1998.

Maynes, Mary Jo, Jennifer L. Pierce and Barbara Laslett. *Telling Stories: The Use of Personal Narrative in the Social Sciences and History*. Ithaca: Cornell University Press, 2008.

Milosz, Czeslaw. *The Collected Poems, 1931–87*. London: Penguin, 1988.

Minkowski, Eugène. *Vers une cosmologie: Fragments philosophiques*. Paris: Aubier-Montaigne, 1936.

Montague, John. *The Figure in the Cave and Other Essays*. Dublin: Lilliput Press, 1989.

Moore, Jason. 'Ecology, Capital, and the Nature of Our Times: Accumulation and Crisis in the Capitalist World-Ecology'. *Journal of World-Systems Research* 17:1 (2011): 108–47.

Moore, Jason. 'The Modern World-System as Environmental History? Ecology and the Rise of Capitalism'. *Theory & Society* 32:3 (2003): 307–77.

Morrison, J'aime. '"Tapping Secrecies of Stone": Irish Roads as Performances of Movement, Measurement, and Memory'. In *Crossroads: Performance Studies and Irish Culture*. Edited by Sara Brady and Fintan Walsh. Basingstoke, UK: Palgrave Macmillan, 2009, 73–85.

Morton, Timothy. *Ecology Without Nature: Rethinking Environmental Aesthetics*. Cambridge, MA: Harvard University Press, 2007.

Murphy, Robert Cushman. 'The Timeless Arans: the Workaday World Lies Beyond the Horizon of Three Rocky Islets off the Irish Coast'. *National Geographic Magazine* 59:6 (June 1931): 747–76.

Nag Copaleen, Myles. *An Béal Bocht*. Dublin: Mercier, 1999.

Næss, Arne. 'The Deep Ecology Movement: Some Philosophical Aspects' [1998]. In *Environmental Ethics*. Edited by Andrew Light and Holmes Rolston III. Oxford: Blackwell, 2003, 262–74.

Nancy, Jean-Luc. *Listening* [2002]. New York: Fordham University Press, 2007.

Nash, Catherine. 'Irish Placenames: Post-colonial Locations'. *Transactions of the Institute of British Geographers* 24:4 (1999): 457–80.

Nichols, Bill. *Introduction to Documentary*. Bloomington: University of Indiana Press, 2001.

O'Brien, Flann. *The Poor Mouth*. Translated by Patrick Power. London: Picador, 1973.

Ó Catháin, Séamas and Patrick O'Flanagan. *The Living Landscape: Kilgalligan, Erris, Co. Mayo*. Dublin: Comhairle Bhéaloideas Éireann, 1975.

O'Connor, Maureen. *The Female and the Species: The Animal in Irish Women's Writing*. New York: Peter Lang, 2010.

O'Dwyer, Ciara. 'Olwen Fouere's Fire and Ice'. *Independent.ie*. (25 January 2009). Accessed 16 February 2014, www.independent.ie/woman/celeb-news/olwen-foueres-fire-and-ice-26508461.html.

Ó Glaisne, Risteárd. *Raidió na Gaeltachta*. Indreabhán: Cló Chois Fharraige, 1982.

Ó hÉallaithe, Donncha. 'From Language Revival to Language Survival'. In '*Who Needs Irish?': Reflections on the Importance of the Irish Language Today*. Edited by Ciarán Mac Murchaidh. Dublin: Vertias, 2004, 159–92.

Ó hEithir, Breandán and Ruairí Ó hEithir (eds). *An Aran Reader*. Dublin: The Lilliput Press, 1991.

O'Kelly, Aloysius. *Mass in a Connemara Cabin* [1883]. Dublin: National Gallery of Ireland.

O'Leary, Canon Peter. *Papers on Irish Idiom*. Edited by T. F. O'Rahilly. Dublin: Browne and Nolan, 1922.

Ó Muraíle, Nollaig. 'Placenames of Clare Island'. *New Survey of Clare Island Newsletter* 4 (March 1998, RIA): 6.

Ondaatje, Michael. *The English Patient*. London: Bloomsbury, 1992.

Ó Riain, Seán. *Pleanáil Teanga in Éirinn 1919–1985*. Dublin: Bord na Gaeilge, 1994.

Ó Ruairc, Maolmhaodhóg. *Ar Thóir Gramadach Nua*. Dublin: Cois Life, 2006.

O'Toole, Fintan. *Ship of Fools: How Stupidity and Corruption Sank the Celtic Tiger*. London: Faber and Faber, 2009.

Ó Tuathaigh, Gearóid. 'Literature, Language and Culture in Ireland Since the War'. In *Ireland 1945–1970: The Thomas Davis Lectures*. Edited by J. J. Lee. Dublin: Gill and MacMillan, 1979: 111–23.

Ozick, Cynthia. 'Introduction'. In *Best American Essays 1998*. Edited by Cynthia Ozick. New York: Houghton Mifflin, 1999, xv–xxi.

Pascal, Blaise. *Pensées* [1669]. Project Gutenberg, 2009. Accessed 9 November 2013, www.gutenberg.org/files/18269/18269-h/18269-h.htm.

Pearson, Michael and Michael Shanks. *Theatre/Archaeology*. London: Routledge, 2000.

Phillips, Dana. *The Truth of Ecology: Nature, Culture and Literature in America*. Oxford: Oxford University Press, 2003.
Pollard, Natalie. 'The Fate of Stupidity'. *Essays in Criticism* 62:2 (April 2012): 125–38.
Potts, Donna. *Contemporary Irish Poetry and the Pastoral Tradition*. Minneapolis: University of Minnesota Press, 2011.
Praeger, Robert Lloyd. *The Way that I Went*. Dublin: Allen Figgis, 1969.
Preston, Christopher. *Grounding Knowledge: Environmental Philosophy, Epistemology and Place*. Athens: University of Georgia Press, 2003.
Relph, Ted. *Place and Placelessness*. London: Pion Ltd, 1976.
Rhodie, Sam. *Promised Lands: Cinema, Geography, Modernism*. London: British Film Institute, 2001.
Robinson, Tim. *Arann: A Companion to the Map of the Aran Islands*. Roundstone, Co. Galway: Folding Landscapes: 1996.
Robinson, Tim. *The Burren* [1977]. Roundstone, Co. Galway: Folding Landscapes, 1999.
Robinson, Tim. *Connemara*. DVD. Directed by Pat Collins. Cork: Harvest Films, 2011.
Robinson, Tim. 'A Connemara Fractal'. In *Decoding the Landscape*. Edited by Timothy Collins. Galway: Centre for Landscape Studies, 1994, 12–29.
Robinson, Tim. *Connemara Part 1: Introduction and Gazetteer*. Roundstone, Co. Galway: Folding Landscapes, 1990.
Robinson, Tim. *Connemara: The Last Pool of Darkness*. Dublin: Penguin, 2009.
Robinson, Tim. *Connemara: Listening to the Wind*. Dublin: Penguin, 2006.
Robinson, Tim. *Connemara: A Little Gaelic Kingdom*. Dublin: Penguin, 2011.
Robinson, Tim. *Distressed Map of Aran*, notes from Folding Landscapes. Accessed 1 December 2013, www.foldinglandscapes.com/?page_id=21.
Robinson, Tim. 'The Fineness of Things'. In *Best American Essays 2001*. Edited by Kathleen Norris. New York: Houghton Mifflin, 2002, 235–45.
Robinson, Tim. 'The Irish Echosphere'. *New Hibernia Review / Iris Éireannach Nua* 7:3 (Autumn 2003): 9–22.
Robinson, Tim. 'A Land without Shortcuts'. *The Dublin Review* 46 (Spring 2012): 25–44.
Robinson, Tim. *Mapping South Connemara*. Roundstone, Co. Galway: Folding Landscapes, 1985.
Robinson, Tim. *My Time in Space*. Dublin: Lilliput, 2001.
Robinson, Tim. 'Place/Person/Book: Synge's *The Aran Islands*'. In J. M. Synge, *The Aran Islands* [1907]. London: Penguin, 1992.
Robinson, Tim. *Oileáin Árann: A Companion to the Map of the Aran Islands*. Roundstone, Co. Galway: Folding Landscapes, 1996.
Robinson, Tim. 'Orion the Hunter'. In *Best American Essays 1998*. Edited by Cynthia Ozick. New York: Houghton Mifflin, 1999, 187–94.
Robinson, Tim. 'Orion the Hunter', *Tales and Imaginings*. Dublin: The Lilliput Press, 2002, 145–53.
Robinson, Tim. 'The Seanchaí and the Database'. *Irish Pages* 2:1 (2003): 43–53.
Robinson, Tim. 'Seism: Essay Written for Siobhán McDonald on the Occasion of Eye of the Storm'. *Eye of the Storm*. Galway: Glaway Arts Centre, 25 May 2012.
Robinson, Tim. *Setting Foot on the Shores of Connemara and Other Writings*. Dublin: Lilliput Press, 1996.
Robinson, Tim. *Stones of Aran: Labyrinth*. Dublin: The Lilliput Press, 1995.
Robinson, Tim. *Stones of Aran: Pilgrimage*. London: Penguin, 1986.
Robinson, Tim. *Tales and Imaginings*. Dublin: The Lilliput Press, 2002.
Robinson, Tim. *A Twisty Journey: Mapping South Connemara*. Binn Éadair: Coiscéim, 2006.
Robinson, Tim. 'The View from Errisbeg: Connemara and the Aran Islands'. In *The Book of the Irish Countryside*. Edited by Frank Mitchell. Belfast: Blackstaff Press, 1987, 42–52.

Robinson, Tim. *The View from the Horizon: Constructions by Timothy Drever 1972, Texts and Maps by Tim Robinson, 1975–96.* Clonmel, Tipperary: Coracle Press, 1997.

Robinson, Tim. 'William Fiennes, Tim Robinson: La beauté du monde'. Étonnants Voyageurs: Festival International du Livre et du Film. 7–9 June 2014. Accessed 1 October 2014, http://vimeo.com/97608775.

Ronell, Avital. *Stupidity.* Chicago: University of Illinois Press, 2002.

Roorbach, Bill. *Writing Life Stories: How to Make Memories into Memoirs, Ideas into Essays, and Life into Literature.* Cincinnati, OH: Writer's Digest Books, 2006.

Root, Robert. *The Nonfictionist's Guide: On Reading and Writing Creative Nonfiction.* Lanham, MD: Rowman and Littlefield, 2008.

Ruskin, John. 'Preface'. Stones of Venice, Vol. 1 [1951]. Project Gutenberg, 2009. Accessed 3 July 2014, www.gutenberg.org/ebooks/30754.

Ryden, Kent C. *Mapping the Invisible Landscape: Folklore, Writing, and the Sense of Place.* Iowa City: University of Iowa Press, 1993.

Sack, Daniel. 'Walking in and out of Place: The Pedestrian Performances of Tim Robinson'. In *Ireland, Performance, and the Historical Imagination.* Edited by Mary P. Caulfield and Christopher Collins. New York: Palgrave Macmillan, 2014, 19–35.

Said, Edward W. 'Invention, Memory and Place'. *Critical Inquiry* 26 (Winter, 2000): 175–92.

Scott, Yvonne. *The West as Metaphor.* Dublin: Royal Hibernian Academy, 2005.

Sheerin, Pat. 'The Narrative Creation of Place: The Example of Yeats'. In *Decoding the Landscape*, 2nd edn. Edited by Timothy Collins. Galway: Centre for Landscape Studies, 1997: 149–65.

Shepherd, Nan. *The Living Mountain* [1977]. Edinburgh: Canongate, 2011.

Sidaway, James D. 'Shadows on the Path: Negotiating Geopolitics on an Urban Section of Britain's South West Coast Path'. *Environment and Planning D: Society and Space* 27 (2009): 1091–116.

Slater, Eamonn and Terence McDonough. 'Bulwark of Landlordism and Capitalism: The Dynamics of Feudalism in Nineteenth Century Ireland'. *Research in Political Economy* 14 (1994): 63–118.

Slater, Eamonn and Terence McDonough. 'Marx on Nineteenth-century Ireland: Analyzing Colonialism as a Dynamic Social Process'. *Irish Historical Studies* 36:142 (November 2008): 153–72.

Slovic, Scott. *Going Away to Think: Engagement, Retreat, and Ecocritical Responsibility.* Reno: University of Nevada Press, 2008.

Smith, Jos. 'A Step Towards the Earth: Interview with Tim Robinson'. *Politics of Place: A Journal for Postgraduates*, spec. issue of *Maps and Margins* 1 (2013): 4–11. Accessed 1 December 2013, blogs.exeter.ac.uk/politicsofplace/files/2013/08/POP_Issue01_Smith.pdf.

Smyth, Gerry. *Space and the Irish Cultural Imagination.* London: Palgrave Macmillan, 2001.

Smyth, William J. *Map-Making, Landscapes and Memory: A Geography of Colonial and Early Modern Ireland c. 1530–1750.* South Bend, IN: University of Notre Dame Press, 2006.

Snyder, Gary. *Turtle Island.* New York: New Directions, 1974.

Solnit, Rebecca. *A Book of Migrations: Some Passages in Ireland.* London: Verso, 1997.

Solnit, Rebecca. *Wanderlust: a History of Walking.* London: Verso, 2002.

Stuckey-French, Ned. *The American Essay in the American Century.* Columbia, MO: University of Missouri Press, 2011.

Tallon, Frances (ed.). *The Field Names of County Meath.* Navan, IR: Meath Field Names Project, 2013.

Tally, Robert T. Jr (ed.). *Literary Cartographies: Spatiality, Representation, and Narrative.* New York: Palgrave Macmillan, 2014.

Tally, Robert T. Jr. 'On Literary Cartography: Narrative as a Spatially Symbolic Act'. *NANO: New American Notes Online* 1 (January 2011): n.p.

Taylor, Charles. 'Heidegger, Language, Ecology'. In *Heidegger: A Critical Reader*. Edited by Herbert L. Dreyfus and Harrison Hall. Oxford: Basil Blackwell, 1992, 247–69.

Thomas, Edward. 'Lob'. In Edna Longley (ed.), *Edward Thomas: The Annotated Collected Poems*. Newcastle: Bloodaxe, 2008, 76–9.

Thoreau, Henry David. *Walden and Civil Disobedience* [1854, 1849]. New York: Norton, 1966.

Thoreau, Henry David. 'Walking'. In *Great Short Works of Henry David Thoreau*. Edited by Wendell Glick. New York: Harper and Row, 1982.

Till, Karen E. 'Artistic and Activist Memory-Work: Approaching Place-Based Practice'. *Memory Studies* 1 (2008): 95–109.

Tredinnick, Mark. *The Land's Wild Music*. San Antonio, TX: Trinity University Press, 2005.

Trunger, Fred. *Filmic Mapping: Documentary Film and the Visual Culture of Landscape Architecture*. Zurich: Jovis, 2013.

Tuan, Yi Fu. *Cosmos and Hearth: A Cosmopolite's Viewpoint*. Minneapolis: University of Minnesota Press, 1996.

Tuan, Yi-Fu. *Space and Place: The Perspective of Experience*. Minneapolis: University of Minnesota Press, 1977.

Tuan, Yi Fu. *Topophilia: A Study of Environmental Perception, Attitudes, and Values*. Englewood Cliffs, NJ: Prentice-Hall, 1974.

Villiers-Tuthill, Kathleen. *Alexander Nimmo and the Western District*. Clifden: Connemara Girl, 2006.

Viney, Michael. *A Year's Turning*. London: Penguin, 1998.

Walker, J. Crampton. *Irish Life and Landscape*. Dublin: Talbot Press, 1926.

Wall, Eamonn. 'Deep Maps: Reading Tim Robinson's Maps of Aran'. *Terrain.org: A Journal of the Built & Natural Environments* 29 (Spring/Summer 2012): n.p. Accessed 29 May 2013, www.terrain.org/articles/29/wall.htm.

Wall, Eamonn. 'Digging into the West: Tim Robinson's Deep Landscapes'. In *Reflective Landscapes of the Anglophone Countries*. Edited by Pascale Guibert. Amsterdam and New York: Rodopi, 2011, 133–45.

Wall, Eamonn. 'Walking: Tim Robinson's Stones of Aran'. *New Hibernia Review/Iris Éireannach Nua* 12:3 (autumn / fómhar 2008): 66–79.

Wall, Eamonn. *Writing the Irish West: Ecologies and Traditions*. Notre Dame, IN: University of Notre Dame Press, 2011.

Wampole, Christy. 'The Essayification of Everything'. *The New York Times* (26 May 2013).

Whelan, Kevin. 'Reading the Ruins: The Presence of Absence in the Irish Landscape'. In *Surveying Ireland's Past: Multidisciplinary Essays in Honour of Anngret Simms*. Edited by Howard B. Clarke, Jacinta Prunty and Mark Hennessy. Dublin: Geography Publications, 2004, 297–322.

Wilson-Foster, John. 'Tim Robinson's Variegated World'. *The Irish Review* 30 (2003): 105–13.

Winter, Steven L. *A Clearing in the Forest: Law, Life and Mind*. Chicago: University of Chicago Press, 2005.

Wittgenstein, Ludwig. *Tractatus Logico-Philosophicus* [1922]. Project Gutenberg, 2009. Accessed 2 April 2014, www.gutenberg.org/ebooks/5740.

Woolf, Virginia. *A Room of One's Own* [1929]. New York: A Harvest, 1989.

Woolf, Virginia. *To the Lighthouse* [1927]. Oxford: Oxford World's Classics, 1992.

Wylie, John. 'Dwelling and Displacement: Tim Robinson and the Questions of Landscape'. *Cultural Geographies* 19:3 (2012): 365–83.

Yeats, William Butler. 'Anima Hominis' Part V. *Per Amica Silentia Lunae* [1918]. Project Gutenberg, 2009. Accessed 3 July 2014, www.gutenberg.org/files/33338/33338-h/33338-h.htm.

# Index

Abbey, Edward 108, 115–16, 188 n.54
Ackroyd, Norman xviii
Agamben, Giorgio 105, 117 n.10
Andrew, J.H. 9, 17 n.20, n.39, 37 n.8
anthropocentrism 225
Árainn 5, 7 Fig. 2, 16 n.6, 23 Fig. 4b, 26, 31, 34, 55, 65, 70 n.46, 93, 104, 106 Fig. 16, 107, 110
  Fig. 17, 110–11, 120 Fig. 18, 147, 151, 155, 160 Fig. 22, 220–2, 231
Aran Islands xvii, xxii, 1, 2, 14–16 n.6, 21–2, 24–5, 32–3, 42, 53, 65–6, 77, 79, 94, 103, 105, 119–121 Fig. 19, 124–5, 139, 176, 190, 193, 197, 206, 209, 220, 222–3, 230
Arthur, Chris 13, 117 n.8, 127–43
Atlantic Archipelago Research Project xviii, 3

Babine, Karen 4, 10, 13, 16 n.16, 68 n.14, 104, 117 n.8
Berger, John 89, 101 n.1, 161, 170 n.12
Biggs, Iain 38, 57, 67, 69 n.35, n.36, 70 n.39, 70 n.55
bioregion xix, 150, 155
Breton, André 190
Burren 1, 2, 5, 6 Fig. 1, 15, 17 n.19, 21, 24–5, 29, 34, 37, 42, 53, 68 n.13, 76–7, 79, 103, 116, 119, 124, 133, 147, 176, 185, 190, 193, 198, 206
Butler, Judith 205, 214, 217 n.66

cartography 2, 4–6, 8–9, 11–12, 46, 54–5, 58, 66, 75–6, 80–1, 83, 191, 193, 197, 200, 218–19, 224

Celtic Tiger 132, 219–20, 232, 236 n.49
Chatwin, Bruce 148–50, 157 n.2
Cill Mhuirbhigh 5, 65, 151, 193
Coleridge, Samuel Taylor 148, 154
Collins, Pat xiv, xvii, 3, 11, 12, 68 n.14, 73–86, 103, 131 Fig. 20, 174 Fig. 23, 201 n.22, 219, Fig. 26
Conley, Tom 73, 79, 85 n.2
Connemara xx, 1, 2, 4, 5, 8, 12, 14–15, 17 n.19, 22–5, 25 Fig. 5a, 26–7, 29–37, 38 n.29, 41, 42 Fig. 7, 42–3, 45–51, 55, 62, 65, 67, 68 n.13, 73–4, 74 Fig. 12, 75–83, 84 Fig. 13, 84–5, 90, 92, 95, 97–8, 100, 103, 114, 119, 121–2, 124–5, 133, 139, 147–8, 160, 163–70, 173, 174 Fig. 23, 174, 176, 180–3, 185, 189 n.58, 190, 193–4, 196, 198, 204–6, 207 Fig. 25, 208, 212, 215, 219 Fig. 26, 222–3, 226–7, 230–2
*Connemara and Elsewhere* 3
*Connemara Gazetteer* 30, 31, 37 n.10, 38 n.30, 39 n.45, n.48, 40 n.58
Conroy, Jane 3
Cronin, Michael 4, 16 n.16, 55, 69 n.22, 172 n.31, 236 n.53
Cusick, Christine 4, 7, 12, 16 n.16, 17 n.31, 18 n.41, 18 n.44, 39 n.41, 67, 68 n.14, 82, 155, 234 n.5

Davidson, Peter xvii, 11
Deane, Seamus 3, 16 n.12, 220, 235 n.7
Dee, Tim xvii, xviii, xxi n.2, 11
deep ecology 178–9, 185, 187 n.28, 188 n.37, 189 n.57

deep mapping xviii, 8, 12, 17 n.32, 22, 37 n.5, 53–72
DeSilvey, Caitlin xviii
Dillard, Annie 108
Dillon, Brian 221, 235 n.10
*dinnseanchas* 30, 33, 151, 155
Dún Aonghasa 109, 111
DuPlessis, Rachel Blau 91, 101 n.8
dwelling xviii, xix, xxi n.6, 17 n.16, 53, 55, 116, 134, 176, 178–9, 187 n.31, 220, 227, 231

Echosphere 8, 213
ecocritical 56, 67, 91, 218, 220, 224, 228–30, 232–5, 235 n.25
ecocriticism 186, 218, 224–6, 232, 234 n.3, n.4
ecological diversity 2
ecological ethic 56
ecology 2, 14, 15, 49, 122, 178, 218, 225, 227, 231
Elder, John xiv, 3, 4, 12
Ellis, Jack 75
environmental criticism 7, 10, 91, 97, 99
environmental movement xix
environmental studies 105
environmentalist 1, 14, 21, 155, 229
Ettinger, Bracha L. 14, 202–7, 210–11, 213–15

Facio, Bartolomeo 191
Famine, The 23–4, 31, 33, 48, 98, 120, 182, 196, 232
feminist theory 91
Fennell, Desmond 14, 160, 172 n.32, n.34
Féve, Nicolas 3
Fianna Fáil 162
Fine Gael 162
Flaherty, Robert 2, 76–7, 151, 161
Flannery, Eóin 14, 56, 235 n.11, 236 n.53
Folding Landscapes 2, 16 n.10, 41, 45, 49, 67, 72 n.93, 120, 149, 193
Foster, John Wilson 1, 4
Foster, Roy 149
Fouéré, Yann 215
Freud, Sigmund 207, 213
Friedman, Susan Standford 99
Friel, Brian 17 n.20, 32, 47

Gaelic League 163, 165
Gaeltacht 13, 14, 68, 158–71
Gaeltacht Civil Rights Movement 14, 158
Geertz, Clifford 57
geophany xix
Gifford, Terry 4, 16 n.16, 91, 101 n.13
Gladwin, Derek 4, 12, 16, 17 n.37, 34, 39 n.41, 67, 117 n.7
Glotfelty, Cheryll 16 n.16, 68 n.14, 86 n.36, 101 n.10, 157 n.14, 187 n.30
Gombrich, Ernest 193–4
Gunn, Kirsty xvii, 11

Hazlitt, William 42, 52 n.2, 153
Heat-Moon, William Least 8, 55, 57, 155
Heidegger, Martin 14, 173, 175–6, 178–9, 182, 184, 186, 187 n.13, 188 n.35, 188 n.37
Henry, Paul 14, 194–5
Hogan, Colm xiv, 3, 42 Fig. 7, 74 Fig. 12, 84 Fig. 13, 131 Fig. 20, 174 Fig. 23, 219 Fig. 26
hooks, bell 91, 101 n.8
Huggan, Graham 8, 17 n.22
Hyde, Douglas 165

Inis Meáin 9 Fig. 3, 16 n.6, 26, 125, 210
Inis Oírr 16 n.6, 26, 32, 119–20
Inishmore 16 n.6, 25, 56, 109, 113–14, 133, 149, 151–3, 156, 220
Irish language 14, 68 n.14, 158, 160, 163–4, 167, 188 n.50, 193–4, 197, 224, 228, 235
Irish Museum of Modern Art xii, 14, 54, 68, 190, 197
Irish Research Council 67
Irish Tourist Board 196

Jackson, J. B. 22
James Hardiman Library vii, viii, xiv, 3, 16 n.11
Jameson, Fredric 186
Jamie, Kathleen xvii, 11
Joseph, Peter 60, 193

Kavanagh, Patrick 35, 40 n.74
Keating, Seán 14, 194
Kiberd, Declan 149

Lacan, Jacques 202, 207
Lopate, Phillip 130, 133, 138
Lopez, Barry 57, 65, 92, 93, 108, 125
Lorimer, Hayden xviii
Lynch, Tom 16 n.16, 68 n.14, 86 n.36, 157 n.14

Mabey, Richard xvii
Mac Con Iomaire, Liam 68 n.14, 170 n.2
MacDonald, Fraser xviii
Macfarlane, Robert 4, 11, 108, 125
  *Mountains of the Mind* xi, 4
  *The Old Ways* xii, 4, 11, 18
  *The Wild Places* xii, xvi, 4
*Man of Aran* 2, 33, 77, 151, 161
Mandelbrot, Benoît 54, 67 n.4, 203
map-maker 4, 8, 22, 29, 56, 74–83, 85, 153, 168, 191, 200
map-making 2–7, 9, 12, 14, 22, 54, 73–86, 103, 105, 116, 124, 153, 191, 201 n.22, 202
maternal 202, 207, 213
matrixial 14, 202–11, 213–15
McDonald, Frank 36
McLane, Betsy 75
McLucas, Clifford 58, 66, 69 n.37
McNeillie, Andrew xiv, xvi, xviii, 15, 237 Fig. 27
McPhee, John 125
metramorphosis 203, 206, 211, 213
Milosz, Czeslaw 125
Momaday, N. Scott 156
Montague, John 28, 231, 236 n.46
Montaignian 13, 128–9, 135
*Moonfield* 12, 54, 60, 61 Fig. 9, 61, 63, 65, 66, 70 n.55
Moriarty, John 13, 121, 124
Morrison, J'aime 4, 16 n.16, 81
Mother Ireland 205

Nancy, Jean Luc 14, 177
*Nanook of the North* 77
Nash, Catherine 227–9
National University Ireland-Galway 147 *see* NUI-Galway
nature writing 55, 73, 91, 104
Nimmo House 62, 63 Fig. 10, 119, 148 Fig. 21

NUI-Galway xiv, 16 n.11, 67 n.5, 69 n.38, 70 n.46, 147

Ó Cadhain, Máirtín 3, 170 n.2
O'Donovan, John 22, 36
O'Flaherty, Robert 33
Ó hEithir, Breandán 151, 167
*Oileáin Arann* 16 n.6, 17 n.19, 53, 65–6, 68 n.13, 166, 193
oral history 137, 198
Ordinance Survey xiv, 166
Otherness 177, 185, 220

Pearson, Mike 57–8, 68 n.13
phenomenology 14, 176, 181, 186, 187 n.13
placelore 8, 29, 31, 56, 213
placenames 31, 34, 39 n.47, 65, 151, 159, 166, 182, 199, 220, 227, 229–30, 234
Pollard, Natalie 105, 111
postcolonial 10, 14, 56, 68 n.14, 220, 224–8, 234
postcolonialism 14, 218
Praeger, Robert Lloyd 4, 29, 34–5
Preston, Christopher 93, 99
Price, Liam 29
Proust, Marcel xvi
psychoanalytical theory 207

Quinn, Bob 14, 158, 168–9

Relph, Ted 22, 35
Robinson, Tim
  *Connemara: A Little Gaelic Kingdom* 4, 12, 27, 38 n.21, 76, 100, 164, 168–9, 180
  *Connemara: Last Pool of Darkness* 4, 76, 94, 107, 133, 184, 188 n.57, 198, 208
  *Connemara: Listening to the Wind* 4, 12, 30, 41, 43, 47, 49, 76, 84, 100, 166, 173, 175, 180–1, 209, 231–2
  *My Time in Space* xvi, 4, 6, 16 n.16, 53, 54, 59, 68 n.12, 85 n.9, 236 n.51
  *Setting Foot On The Shores of Connemara* xvi, 4, 54, 65, 188 n.50
  *Stones of Aran: Labyrinth* 4, 10, 13, 18, 42, 93, 97, 103, 104, 111, 113–16, 138, 149, 153, 167

Robinson, Tim (*cont.*)
    *Stones of Aran: Pilgrimage* 2, 3, 4, 8, 13, 16, 30, 42, 48, 89, 93, 103, 104, 108, 111–16, 120, 125, 137–8, 149, 153–4, 161–4, 167, 180, 204, 206, 220–1
    *Tales and Imaginings*, xvii, 4, 16 n.16
Ronell, Avital 105, 115
Roundstone 1, 3, 24, 27, 30–1, 33, 41, 43, 45–6, 62, 63 Fig. 10, 67, 80, 114, 119, 121, 125, 131 Fig. 20, 147, 148 Fig. 21, 149, 165, 181, 184, 202, 226
Ruskin, John xvi, 110, 121
Ryden, Kent 9, 78, 218–19, 221, 234

Sack, Daniel 4, 16 n.16
Said, Edward 223
Schama, Simon 91
*sean-nós* 180
Shanahan, Angie xiv, 3, 98 Fig. 15, 121 Fig. 19
Shanks, Michael 57–8, 68 n.13
Sheerin, Pat 29
Shepherd, Nan xvii, 125
Sheridan, Jim 202, 208
Slovic, Scott 91
Smith, Jos 55
Smyth, Gerry 4, 14, 16 n.16, 99, 104, 217 n.62
Snyder, Gary 43, 149, 155
Solnit, Rebecca 36, 149–50

Stegner, Wallace 57
Stenger, Susan xviii
Surrealism 190
Swift, Jonathan 190
Synge, J. M. 114, 188 n.46, 190

*The Living Mountain* xvii
*The View from the Horizon* 54–5, 62, 64 Fig. 11
Thomas, Edward xvi
Thoreau, Henry David 148, 153–5
Thubron, Colin 149–50
Till, Karen E. 58, 67, 70 n.41
*Tim Robinson: Connemara* 11, 12, 39 n.43, 42 Fig. 7, 73–86, 103, 131 Fig. 20, 174 Fig. 23, 219 Fig. 26
*To the Centre* 12, 54, 62–3, 64 Fig. 11, 65–6, 197
Tuan, Yi-Fu 66, 71 n.91, 78

Wall, Eamonn 4, 8, 13, 16 n.16, 17 n.32, 37 n.5, 37 n.7, 39 n.38, 55, 67, 68 n.14, 80, 133, 234 n.5
Whelan, Kevin 18 n.41, 150, 223
Williams, Terry Tempest 149
Wittgenstein, Ludwig 107, 188 n.57
Woolf, Virginia 90, 173
Wordsworth, William 42, 153–5
Wylie, John xviii, 4, 55, 96, 100

Yeats, W.B. 124, 141 n.6, 188 n.46